Environmental Biotechnology for Waste Treatment

ENVIRONMENTAL SCIENCE RESEARCH

Series Editor:
Herbert S. Rosenkranz
Department of Environmental and Occupational Health
Graduate School of Public Health
University of Pittsburgh
130 DeSoto Street
Pittsburgh, Pennsylvania

Founding Editor:
Alexander Hollaender

Recent Volumes in this Series

Volume 34 — ARCTIC AND ALPINE MYCOLOGY II
  Edited by Gary A. Laursen, Joseph R. Ammirati, and
  Scott A. Redhead

Volume 35 — ENVIRONMENTAL RADON
  Edited by C. Richard Cothern and James E. Smith, Jr.

Volume 36 — SHORT-TERM BIOASSAYS IN THE ANALYSIS OF COMPLEX
  ENVIRONMENTAL MIXTURES V
  Edited by Shahbeg S. Sandhu, David M. DeMarini, Marc J. Mass,
  Martha M. Moore, and Judy L. Mumford

Volume 37 — HAZARDS, DECONTAMINATION, AND REPLACEMENT OF PCB:
  A Comprehensive Guide
  Edited by Jean-Pierre Crine

Volume 38 — *IN SITU* EVALUATION OF BIOLOGICAL HAZARDS OF
  ENVIRONMENTAL POLLUTANTS
  Edited by Shahbeg S. Sandhu, William R. Lower, Frederick J. de Serres,
  William A. Suk, and Raymond R. Tice

Volume 39 — GENETIC TOXICOLOGY OF COMPLEX MIXTURES
  Edited by Michael D. Waters, F. Bernard Daniel, Joellen Lewtas,
  Martha M. Moore, and Stephen Nesnow

Volume 40 — NITROARENES: Occurrence, Metabolism, and Biological Impact
  Edited by Paul C. Howard, Stephen S. Hecht, and Frederick A. Beland

Volume 41 — ENVIRONMENTAL BIOTECHNOLOGY FOR WASTE TREATMENT
  Edited by Gary S. Sayler, Robert Fox, and James W. Blackburn

---

A Continuation Order Plan is available for this series. A continuation order will bring delivery of each new volume immediately upon publication. Volumes are billed only upon actual shipment. For further information please contact the publisher.

Environmental Biotechnology for Waste Treatment

Edited by

Gary S. Sayler
Center for Environmental Biotechnology
Knoxville, Tennessee

Robert Fox
International Technology Corporation
Knoxville, Tennessee

and

James W. Blackburn
Center for Environmental Biotechnology
Knoxville, Tennessee

Plenum Press • New York and London

Library of Congress Cataloging in Publication Data

Symposium on Environmental Biotechnology: Moving from the Flask to the Field (1990: Knoxville, Tenn.)
  Environmental biotechnology for waste treatment / edited by Gary S. Sayler, Robert Fox, and James W. Blackburn.
       p.   cm. — (Environmental science research: v. 41)
  "Proceedings of the Symposium on Environmental Biotechnology: Moving from the Flask to the Field, held October 17-19, 1990, in Knoxville, Tennessee"—T.p. verso.
  Includes bibliographical references and index.
  ISBN 0-306-43943-3
  1. Environmental biotechnology—Congresses. I. Sayler, Gary S., 1949-   . II. Fox, Robert. III. Blackburn, James W. IV. Title. V. Series.
TD192.5.S96   1990   91-10620
628—dc20   CIP

Proceedings of the symposium on Environmental Biotechnology:
Moving from the Flask to the Field, held October 17-19, 1990,
in Knoxville, Tennessee

ISBN 0-306-43943-3

© 1991 Plenum Press, New York
A Division of Plenum Publishing Corporation
233 Spring Street, New York, N.Y. 10013

All rights reserved

No part of this book may be reproduced, stored in a retrieval system, or transmitted in any form or by any means, electronic, mechanical, photocopying, microfilming, recording, or otherwise, without written permission from the Publisher

Printed in the United States of America

Preface

The use of biotechnical processes in control of environmental pollution and in hazardous waste treatment is viewed as an advantageous alternative or adduct to physical chemical treatment technologies. Yet, the development and implementation of both conventional and advanced biotechnologies in predictable and efficacious field applications suffer from numerous technical, regulatory, and societal uncertainties.

With the application of modern molecular biology and genetic engineering, there is clear potential for biotechnical developments that will lead to breakthroughs in controlled and optimized hazardous waste treatment for *in situ* and unit process use. There is, however, great concern that the development of these technologies may be needlessly hindered in their applications and that the fundamental research base may not be able to sustain continued technology development.

Some of these issues have been discussed in a fragmented fashion within the research and development community. A basic research agenda has been established to promote a sustainable cross-disciplinary technology base. This agenda includes developing new and improved strains for biodegradation, improving bioanalytical methods to measure strain and biodegradation performance, and providing an integrated environmental and reactor systems analysis approach for process control and optimization.

There remains an identified need to promote cross-disciplinary communication of technology development and application, and to identify choke points that impinge on the effective commercial application of the technology. For these reasons, industrial, federal, and academic partners joined together to sponsor this current dialogue on moving modern environmental biotechnology from the laboratory to successful field application. Unlike other efforts to communicate the technology, this symposium was planned to not only identify current practices and state-of-the-science, but also to identify perceptional and regulatory issues that affect credible applications and evaluation of the technology. In this regard, we must acknowledge the concerned foresight of the sponsors of the symposium, International Technology Corporation; the American Cyanamid Company; the U.S. Air Force, Office of Scientific Research; the University of Tennessee, Waste Management Research and Education Institute; support from the Oak Ridge Waste Management Association; the planning and steering committee and the symposium participants.

A goal of the symposium was to communicate a broad view of environmental biotechnology ranging from conventional practices in biological waste treatment to genetic engineering perspectives in *in situ* treatment technology. From the beginning it was acknowledged that the biology was intimately linked to the environmental application and the engineering design in implementing the technology. This major scale-up consideration is the critical technical hurdle in moving the technology from the lab to practical field use. In this scale up, there are major limitations in monitoring and controlling biotechnical processes, and these limitations further confound societal and regulatory perception of the credibility of the technology.

The outcome of this symposium contributes to identifying applications of fundamental research in emerging technology and to defining industrial research needs. It is also anticipated that strategies will be forthcoming to overcome concerns of the safety and efficacy of the technology. There appear to be numerous opportunities for environmental biotechnology to contribute to integrated waste management, but care must be taken to demonstrate reliable technology in order to capitalize on these opportunities.

Gary S. Sayler

Contents

CURRENT PERCEPTIONS

Environmental Biotechnology: Perceptions, Reality, and Applications .. 1
   Gary S. Sayler and Robert Fox

Media Images of Environmental Biotechnology: What Does the
   Public See? 15
   Amy S. McCabe and Michael R. Fitzgerald

Perspectives on Bioremediation in the Gas Industry 25
   David G. Linz, Edward F. Neuhauser, and Andrew C. Middleton

Considerations in the Selection of Environmental Biotechnology as
   Viable in Field-Scale Waste Treatment Applications 37
   Patricia Taylor Woodyard

The Technical, Economic, and Regulatory Future for Bioremediation:
   An Industry Perspective 47
   A. Keith Kaufman

Removing Impediments to the Use of Bioremediation and Other
   Innovative Technologies 53
   Walter W. Kovalick, Jr.

Bioremediation Research Issues 61
   John H. Skinner

FIELD-SCALE CASE STUDIES

Evaluation of Bioremediation in a Coal-Coking Waste Lagoon 71
   Maureen E. Leavitt, Duane A. Graves, and Craig A. Lang

Evaluation Process for the Selection of Bioremediation Technologies for *Exxon Valdez* Oil Spill 85
    Edgar Berkey, Jessica M. Cogen, Val J. Kelmeckis, Lawrence T. McGeehan, and A. Thomas Merski

Full-Scale Bioremediation of Contaminated Soil and Water 91
    Geoffrey C. Compeau, William D. Mahaffey, and Lori Patras

TECHNICAL ISSUES AND CONCERNS IN IMPLEMENTATION

Feasibility and Other Considerations for Use of Bioremediation in Subsurface Areas 111
    Karolyn L. Hardaway, Mark S. Katterjohn, Craig A. Lang, and Maureen E. Leavitt

Integration of Biotechnology to Waste Minimization Programs 127
    Godfred E. Tong

Bioremediation of Explosives Contaminated Soils (Scientific Questions/Engineering Realities) 137
    Craig A. Myler and Wayne Sisk

Practices, Potential, and Pitfalls in the Application of Biotechnology to Environmental Problems 147
    Carol D. Litchfield

What is the K_m of Disappearase? 159
    Ronald Unterman

Use of Treatability Studies in Developing Remediation Strategies for Contaminated Soils 163
    Michael J. McFarland, Ronald C. Sims, and James W. Blackburn

Biodegradation of Mixed Solvents by a Strain of *Pseudomonas* 175
    J. C. Spain, C. A. Pettigrew, and B. E. Haigler

NONTECHNICAL ISSUES AND CONCERNS IN IMPLEMENTATION

The Field Implementation of Bioremediation: An EPA Perspective ... 185
    Fran V. Kremer and Walter W. Kovalick, Jr.

Contents

An Historical Perspective: Does Good Science or Good Press
    Generate Demand? 191
  T. G. Zitrides

Ways to Identify and Obtain Rights to Technology from Federal
    Facilities 197
  John C. Corey, Gerald J. Hooker, and Cindy K. Tew

An Overview of Current Attitudes on the Use of Biotreatment for
    Cleanup 203
  William J. Lacy

What Are the Critical Issues Necessary to Win the Confidence of
    State Regulators? Views of a Project Manager on
    Bioremediation Sites 211
  Frank R. Peduto

Federal Regulations: How They Impact Research and
    Commercialization of Biological Treatment 217
  Sue Markland Day

INTERNATIONAL ACTIVITIES

Polluted Heterogeneous Environments: Macro-scale Fluxes,
    Micro-scale Mechanisms, and Molecular Scale Control 233
  Geoffrey Hamer and Armin Heitzer

The Pilot Plant Testing of the Continuous Extraction of
    Radionuclides Using Immobilized Biomass 249
  Marios Tsezos and Ronald G. L. McCready

Research and Development Programs for Biological Hazardous
    Waste Treatment in the Netherlands 261
  Esther Soczo and Klaas Visscher

SUMMARY

Environmental Biotechnology—From Flask to Field: A Review 271
  Robert A. Goldstein, Al W. Bourquin, Thomas W. Federle,
  C. P. Leslie Grady, and William D. Mahaffey

Contributors 279

Index 283

Environmental Biotechnology: Perceptions, Reality, and Applications

Gary S. Sayler and Robert Fox

INTRODUCTION

Environmental biotechnology for hazardous wastes can be defined as the use of microorganisms and their processes for socio-economic benefits in environmental protection and restoration. The application of biological processes for disposal and control of waste from human activities is established technology dating back at least 4000 years (1). However, the understanding that microorganisms, "microphytin," were mechanistically responsible for degradation of organic pollutants in wastes is much more recent, being developed in the late 1800's (2). While specific knowledge of the microbial species, biochemistry and genetics mediating biodegradation has only been developed in the later half of this century, practical civil engineering applications of biological processes for control of domestic and some industrial wastes have been developed as highly efficient technology.

The success of these engineered processes for the biological destruction of organic pollutants is largely due to the lability of the pollutants and the relatively broad distribution of microbial species capable of biochemical conversion (biodegradation) of the components of the waste. However, the case for biodegradation of recalcitrant, synthetic and sometimes toxic components associated with hazardous wastes and environmental contamination problems is not transferable immediately from our knowledge of conventional waste treatment process engineering.

Environmental biotechnology encompasses methods that use naturally occurring microorganisms to remove contaminants from the environment. Because it involves little energy input, it has minor chemical needs in the form of nutrients, and it operates at ambient conditions, the cost factors for these systems are believed to be low in relation to other methods for clean-up of environmental contamination. As a result, extensive research and development has occurred over the last several

---

Gary S. Sayler • Center for Environmental Biotechnology, Knoxville, Tennessee 37932.   Robert Fox • International Technology Corporation, Knoxville, Tennessee 37923.

years in identifying and defining ways to use this tool for removing contaminants from the environment.

INDUSTRIAL APPLICATIONS AND CONCERNS

Initially, environmental biotechnology was seen as a relatively straightforward extension of the time-honored biological processes for treatment of wastewater. These processes were concerned with bringing together the microorganism and the waste organic materials under conditions that enabled the microorganisms to utilize the organics as a food source and, by biological activity, bringing about a reduction in the organic contaminants in the wastewater. Reductions were measured by nonspecific parameters such as BOD or COD and by certain compound-specific analyses.

In making this extension of classical biological treatment to new application and new matrices, and in tracking the fate of specific priority pollutants, the initial efforts began to delineate the complex mechanisms involved. Engineers and scientists began to realize that systems for dealing with environmental contamination were much more complex, and that moving from the flask to the field was fraught with potential unknowns and problems.

The activated sludge process was recognized as a treatment system with three major removal pathways for chemicals: stripping, adsorption on biomass, and biotransformation. Use of air or oxygen for biological oxidation required an aeration of the wastewater and provided a removal pathway due to stripping. Similarly, production of a gaseous product in biological treatment, such as the CO_2 formed by mineralization or the CH_4 formed in anaerobic systems, also represented removal pathway by stripping.

The adsorption of specific chemicals on the biomass also represented a removal pathway for organics and inorganics (e.g. metals) when the biomass was separated from the wastewater for disposal. This biomass adsorption pathway was further confused by the fate of the sorbed contaminants upon dewatering of the sludge. Use of heat or chemicals to facilitate dewatering could change the adsorption characteristics of the contaminants or chemically alter the contaminants.

Finally, the removal of organics by biological processes involved an understanding of the microbial community and the changes induced by the microbiological activity. Examination of the microbial community (3) showed that the actual number of degraders of a target contaminant in a mixed culture might only represent 5-10% of the total microbial community. This observation has created the need for the development of methods to measure specific microorganisms in mixed cultures. The gene probe is one such technique (4).

In the microbiological processes, mineralization to CO_2 is the desired result, but the reality is that other biological pathways exist that may or may not be beneficial. Biotransformation of the environmental contaminant may be desirable because the concentration is reduced. However, if the degradation does not proceed to mineralization, daughter products are generated. In aerobic processes, these daughter products are partially oxidized moieties that are more mobile in the environment than the original contaminant. They are also likely to be less toxic than the original contaminant.

Another reality is that these biologically-induced transformations often result in compounds that do not report as contaminants when analyzed by the "standard" method. A case in point, is the analysis of the Total Petroleum Hydrocarbons (TPH) method for characterizing soil contaminated with petroleum products (gasoline, jet fuel, diesel fuel). The TPH method uses a non-polar solvent to extract these non-polar contaminants from the soil. However, as soon as transformation of these compounds begins due to biological, chemical or sometimes photochemical attack, they begin to take on polar substitution groups. They become virtually non-extractable by the TPH solvent and the "standard" method shows high removal of TPH compounds. In reality, if the transformation process is not allowed to go to completion (e.g. mineralization) the product of the process can be a different set of contaminants that are more polar, more hydrophilic, and hence more mobile in the environment.

Other realities that need to be addressed in the application of environmental biotechnology to environmental contamination are the degree of treatment, or removal efficiency, that can be attained and the mass transfer conditions that must be met. Biological degradation, while offering the prospect of lower cost compared to other remediation technologies such as incineration can not achieve the same degree of destruction and removal efficiency (e.g. 99.99%).

In order to achieve successful biological degradation of contaminants, not only must the conditions, i.e. molecular ecology, for activity be defined, but also the various reactants in this system must be brought into contact with each other so the mass transfers involved can take place. The contaminant, present in the environmental matrix, has to be made available to the microorganisms, as do the nutrients and any other chemical needs of the cell. All conditions, defined in the flask, may be right, but if the mass transfer is not abetted in the field, degradation will fall short of expectations.

Therefore, the practice of environmental biotechnology for dealing with environmental contamination requires the integration, or marriage, into one field, of the advances made in several loosely related scientific fields in the 1980's. From an industrial perspective these fields are inclusive of microbiology, biochemistry, molecular ecology, environmental science, chemical/environmental engineering, physical chemistry and analytical chemistry.

Among the important parts of the science of microbiology are the conventional and new emerging techniques for identifying microorganisms and understanding the physiology associated with degradation, e.g. nutrient types and utilization. Biochemistry identifies biochemical pathways, the extent of degradation, and the enzymology involved. Molecular ecology provides insight into the interrelationship among organisms, specific genes, and their environment.

Environmental sciences contribute in a number of ways, ranging from physical and chemical transport and fate processes to the use of geology, hydrogeology and geochemistry in applications involving *in situ* bioremediation of subsurface contamination.

The important parts of chemical/environmental engineering involved in moving environmental biotechnology from the flask to the field are mass transfer (the exchange of chemicals between matrices), mixing, reaction kinetics, adsorption, absorption, and stripping, while physical chemistry provides important information relative to the properties of chemicals, such as octanol-water partition coefficient, vapor pressure, ionization constants, Henry's law constants, water solubility, and adsorption/desorption on surfaces.

Besides providing state-of-the-art methodology to measure specific organics and inorganics at a variety of concentrations in various matrices, analytical chemistry provides valuable understanding of the analytical methods and interferences. Interferences are also dealt with by the use of clean-up procedures on the sample to eliminate them. Characterization and quantitation of process by-products is another important contribution of analytical chemistry to environmental biotechnology.

Through the integration of these scientific disciplines, many wide ranging applications of environmental biotechnology are being visualized and developed.

Groundwater Treatment

Groundwater contamination is characterized by dilute concentration of chemicals with some degree of water solubility.

Significant progress has been made in the application of environmental biotechnology to the *in situ* treatment of petroleum hydrocarbon contamination (e.g gasoline, jet fuel, diesel). The class of organic chemicals typically described as oxygenated (e.g. alcohols, ketones, organic acids) are also readily treatable by biological processes, either *in situ* or in pump and treat situations. Because these compounds are not readily treated by air stripping or carbon adsorption, above ground bioreactors are used in a treatment train for this type of contamination.

Significant research and development resources are being expanded to define systems for biological treatment of the chlorinated solvent chemicals that frequently contaminate groundwater. Results to date show that the use of environmental biotechnology will not necessarily result in systems that are easy to engineer. Chlorinated solvents such as trichloroethylene (TCE) are biodegradable by aerobic microbes that also require another carbonaceous nutrient. Methane, toluene, and phenol have been found to meet this need for some TCE-degrading organisms.

A similar situation has been discovered in the work to utilize the white rot fungus for the difficult-to-degrade environmental contaminants. Degradation activity can only be brought about by first growing the fungus on a rich food source, and then withdrawing the food source and inducing utilization of the contaminant under starvation conditions.

Leachates

Leachates are another form of contaminated aqueous situation treatable by biological systems. Contaminated leachates can span the range of contaminant concentrations from low to high and involve all types of chemicals: organics, metals, and salts. Biological treatments for leachates include aerobic, anaerobic, and the combination of anaerobic and aerobic. For high concentrations, some dilution may be required, especially for aerobic systems.

Soil

Soil contamination represents a very large field of application, and potential cost savings, for environmental biotechnology. *In situ* treatment of soil in the saturated zone by enhancement of the indigenous organisms has also seen significant progress and success. A subsurface bioreactor is set up in a pump and reinjection mode of operation, using the geology and hydrogeology of the site.

The *in situ* biological treatment of the vadose, or unsaturated, zone has been hampered by the lack of a reliable method of providing essential nutrients to degradative indigenous organisms in order to stimulate them to a greater level of biological activity.

Treatment of excavated soil in biological treatment reactor systems is being actively developed utilizing several different types of bioreactors including: 1) slurry reactors (soil slurried in water), 2) controlled land treatment, with engineered confinement of soil, 3) heap leaching, 4) composting, and 5) "layer cake" (soil pile interlaced with piping network for nutrient addition and aeration.

Sludges

Coinciding with developments in the treatment of soils in above ground bioreactors is the application of these same types of systems to the treatment of sludges. Slurry reactors are ready-made for sludge biotreatment. Use of controlled land treatment, heap leaching, or composting require the co-mingling of the sludge with a solid matrix, usually soil.

Sediments

A fruitful area of research and development for environmental biotechnology is its use for the *in situ* treatment of sediments in rivers, harbors, and lakes. The objective is to avoid disturbing the sediments and the possible further spread of contamination. The preliminary work of General Electric in using indigenous anaerobic microbial processes to degrade PCBs in the Hudson River is a prime example of efforts in this area.

Developing Applications

Several other applications for environmental biotechnology are in various stages of development. Some of these applications are summarized as follows: 1) treatment of organic liquids by co-mingling with aqueous phases, 2) bio-scrubbers for air pol-

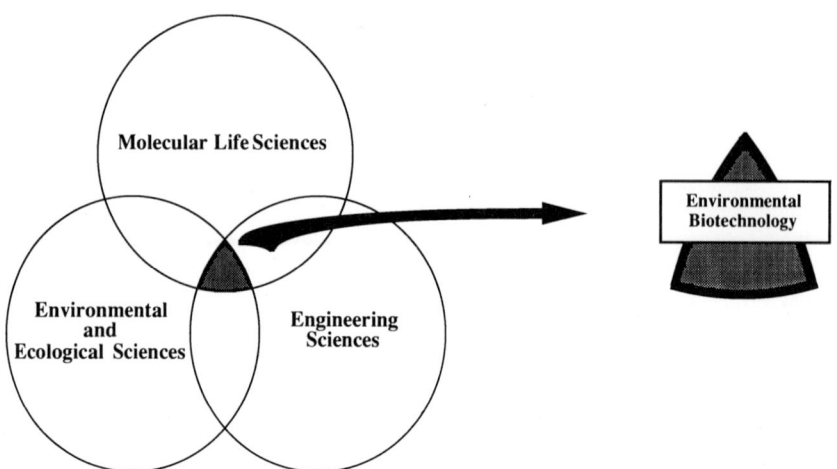

Figure 1. Integration of science and engineering discipline required to achieve practical environmental biotechnology.

lution control, 3) cleaning of coal to remove sulfur, 4) control of acid mine drainage and, 5) separation and recovery of metals.

FUNDAMENTAL RESEARCH DIRECTIONS

Major problems encountered in realizing the application of biodegradation processes relate to the specific environmental scenario and microbiology of the processes involved. There is growing evidence that the ecological physiology of the microbial communities promoting biodegradation is the ultimate determinant of pollutant fate. This is modulated by immediate toxicity of the environment, water activity, aeration, pollutant bioavailability, and availability of appropriate nutrients and electron donors and acceptors.

In 1987, a NSF sponsored research planning workshop identified the need for integrated interdisciplinary research and a basic research agenda for environmental biotechnology of hazardous wastes (5). There was a clear consensus that environmental biotechnology as a research field was evolving at the interface of molecular life sciences, environmental and ecological sciences and engineering (Fig. 1).

This consensus was an affirmation that the problems affecting the development and application of the technology were clearly beyond the bounds of a single discipline and required a coordinated focus among research disciplines. A consensus was also reached on a fundamental research agenda (Table 1). This agenda is comprised of four elements: microbial strain development and improvement, the development of improved bioanalytical methods for measuring biotechnological processes, and the development of environmental and reactor systems analysis techniques leading to better process understanding, control, and optimization. Integra-

Table 1. Fundamental Research Agenda for
Environmental Biotechnology of Hazardous Wastes

Agent (Strain) Development
  Source and Selection
  Characterization
  Modification and Improvement
  Model Systems (mixed and pure)
  Evolutionary Relationships/Diversity
  Stress Response
  Collections and Libraries
Process and System Analytical Tools
  Quantitative Analytical Techniques (Chemical/Physical Measurements)
  Bioanalytical Methods
  Molecular Analysis Methods
  Remote Sensing
  Biomonitors
  Reporter-Signal Analysis/Structure and Function
Environmental System Analysis
  Ecological Interactions
  Environmental Fate and Abiotic Processes
  Population Dynamics (organisms and genes)
  Environmental Stability
  Determination of Kinetic Parameters
  Microhabitats - Niche Invasions
  Organismal or Genetic Mobility
  Controllability and Environmental Modification
  Stress-Induced Effects
Reactor System Analysis
  Reactor Design
  Transient Outcomes and Perturbations
  Dynamic Analysis
  Ecological Interactions
  System Stability and Component Stability
  On-line Analysis and Control
  Kinetic Parameters Analysis

tion across this agenda was considered mandatory in order to make the bench to field scale transition needed for safe and reliable environmental biotechnology.

Microbial Strain Development

Observations made in the laboratory and during environmental fate/treatability studies document the involvement of pure or mixed cultures in the degradation of many of the host of organic chemical in hazardous waste mixture and in contaminated environments. Such chemicals include BTX, PCB, PNA, Dioxin, TNT, DDT, Phthalates, TCE, Chlorobenzenes and benzoates, etc. Yet, the rates and extent of degradation of these compounds may vary by orders of magnitude and individual

Table 2. General Approaches in Microbial Strain Development and Improvement for Degradation of Hazardous Wastes

Approach	Principle Outcome
Batch Enrichment	*In vivo*, pure and mixed cultures capable of growth on specific contaminants.
Chemostat Selection	*In vivo*, selection of high efficiency pure or mixed cultures for specific contaminants.
Plasmid Assisted Molecular Breeding	*In vivo*, pure cultures with new biodegradative pathways.
Plasmid Expansion of Catabolic Pathway	*In vitro*, pure cultures with broadened range of biodegradation.
Genetic Engineering and Pathway Construction	*In vitro*, pure cultures with new and broaden ranges are more efficient and controllable.

species may have no activity on closely related isomers or congeners in the same class.

A variety of approaches are available to develop more efficient microbial agents for degradation of wastes. These approaches are summarized in (Table 2). In general these approaches contribute to the isolation and characterization of new microorganisms with greater activity for degradation, alternative pathways for more efficient degradation, abilities for multiple contaminant degradation, and enhanced environmental fitness and survivability.

In vitro genetic engineering methods can also result in improving the degradation range of new environmental isolates. An excellent example of this case is the degradation range expansion of *Pseudomonas putida* strain B13 (6). This strain recovered from the Rhine river was naturally able to completely degrade 3-chlorobenzoate (and 4-chlorophenol by selection) through a novel modified ortho cleavage pathway for chloro-catechol metabolism. By introduction of the toluene *DL* degradation genes and the *S* regulatory gene, (benzoate metabolism), into this organism, the organism's degradation range was expanded to minerialization of 4-chlorobenzoate (7). Further pathway expansion was achieved by introduction of genes from *Alcaligenes* for 4-methyl lactone isomerization and oxidation to allow B13 to grow also on 4-methyl benzoates and phenols.

While genetically engineered microorganisms are not necessarily the panacea for degradation of hazardous wastes they can make significant contributions to developing controllable degradation processes. A good example in this case falls in the area of TCE degradation by an *E. coli* strain containing cloned genes for a toluene mono oxygenase (TMO) that normally oxidize toluene to p-cresol (8). The TMO is cresol inducible and non-specifically oxidizes TCE as a non-growth substrate. In *E. coli* the TMO was placed under regulatory control of a temperature inducible lambda phage promoter (P_l). The resulting *E. coli* strain can be grown on non-toxic carbohydrates, and the TMO can be temperature induced at 42°C to degrade TCE. Such controllable degradative strains would seem to have immediate applications in confined reactor treatment systems.

These examples of microbial strain development are clearly dependent on substantial basic knowledge of the biochemistry, genetics and molecular biology of degradative pathways. There is a major need for expanded research in this area to exploit more fully applications in environmental biotechnology.

Bioanalytical Measurement Technology

Major rationales for development and application of new and improved measurement techniques for bioprocesses relate to the need for rapid, potentially on-line, analysis of specific activities or cell populations involved in biodegradation/bioremediation practices. Such technology can lead to better reactor and environmental process understanding and, potentially, control and optimization. Both current and future technology in this area can provide critical information demonstrating the credibility of a bioremediation processes.

A major driving force of bioanalytical method development has been the issue of risk associated with engineered organisms in the environment. Numerous methods have been developed and reviewed for their use in measuring specific microbial strains, genes, and processes in the environment (9). Environmental biotechnology for hazardous wastes has directly benefitted from these developments since the issue of measuring the dynamics of degradative organisms, genes, and processes is critical to advanced concepts of bioremediation process control and optimization.

The area of nucleic acid analysis and gene probe technology serves as a good example of the state-of-the-art in measurement of biodegradative microbial communities. DNA probe technology was originally used to detect and quantify bacterial colonies containing degradative genes (4) or metal resistances genes (10) and relate specific gene frequency with degradation and metal transformations in the environment.

More recently, techniques have been developed to rapidly extract DNA and RNA from the environment in order to directly quantify the frequency and activity of degradative genes without cultivation and enumeration of individual microbial populations (11). Such technology currently permits the diagnosis of an environment or engineered waste treatment system as to its capacity for degradation of specific chemicals and instantaneous activity of degradative genes at the time of sampling. This same technology can be applied to the analysis and optimization of waste treatment processes and can provide convincing data that a bioremediation practice is performing as predicted at the biological level.

A parallel development to create on-line measurement of biodegradative activity is bioluminescent reporter technology. This is a specific example of genetic engineering, not for improving degradative strain performance per se, but to develop an *in situ* sensor technology for the presence of specific chemicals, their bioavailability, and their degradation.

The example in this case is the transposon introduction of the bioluminescence (lux) genes from *Vibrio fischeri* into the pathway for naphthalene degradation in a *Pseudomonas* strain (12). A similar genetic construction was also made by cloning

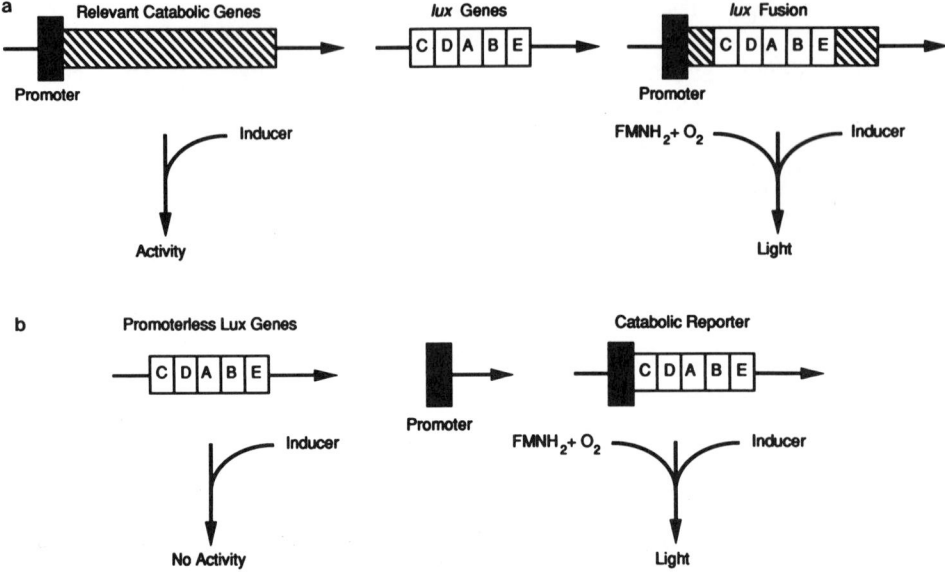

Figure 2. Genetic engineering strategies to develop bioluminescent sensor technology for on-line measurement of naphthalene biodegradation.

the promoter region of the naphthalene dioxygenase directly in front of the *lux* genes in a vector plasmid (13). Both strategies are described in Figure 2.

Naphthalene degradation is induced by the accumulation of the partial degradation product, salicylate. In the presence of naphthalene or salicylate the naphthalene degradative genes are expressed (transcribed at the appropriate promoter) as are the *lux* genes. Expression of the *lux* genes results in enzyme (luciferase) activity and the production of visible light that can be measured with relative ease (Fig. 3). This technology has already been used to measure, with remote sensing, naphthalene biodegradative activity in contaminated Manufactured Gas Plant (MGP) soil (12).

Environmental and Reactor Systems Analysis

The ultimate objective for environmental biotechnology is reliable, effective and predictable application in waste control. To accomplish this objective the biological process will, in most cases, function as a mixed culture system in a complex environmental matrix. There are major needs in identifying and quantifying specific biotic and abiotic processes affecting chemical transformation, the dynamics of biological processes, and dominant variables useful in developing system control strategies. While some of these items can be predicted from highly defined laboratory studies, verification and operation in complex regimes is required.

Figure 3. Bioluminescent response reporting naphthalene degradation in a continuous reactor system (3 minute exposure in total darkness).

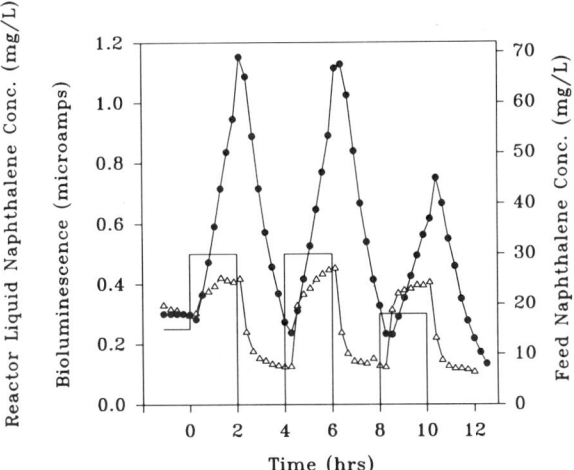

Figure 4. Bioluminescent reporter response to dynamic perturbations in naphthalene concentrations in a continuous slurry biotreatment reactor. • = light response; Δ = naphthalene in effluent; — naphthalene feed.

In part these needs can be met by appropriate bench scale reactor or microcosm simulations in a format more rigorous and informative than a conventional treatability test. One new approach is to apply system analysis methods in bench scale simulations. Perturbation techniques using frequency response analysis have been examined for polyaromatic hydrocarbon contaminated MGP soils. Using a novel continuous slurry reactor system, sinusoidal variations in patterns of naphthalene feed have shown to produce corresponding dynamics in naphthalene biodegradation kinetics. The scale of response does not appear attributable to cell or degradative gene frequency as determined by DNA hybridization methods. Consequently, gene activity may be the main response variable.

This analytical approach and reactor system was chosen for preliminary studies to test the responsivity of naphthalene bioluminescent sensor technology in dynamic analysis of gene expression. A naphthalene degradative *Pseudomonas fluorescens* straining containing a *lux*-naphthalene bioluminescent reporter plasmid produced sufficient light in response to naphthalene exposure for quantitative measurement of dynamic degradative gene activity (Fig. 3). Bioluminescence as a measure of naphthalene biodegradative activity was found to be directly responsive to dynamic fluctuations in naphthalene feed to a continuous bioreactor system (Fig. 4). Extension of this work to naturally contaminated MGP soils demonstrated the potential of this technology to detect *in situ* naphthalene, its bioavailability and biodegradation (13).

These are somewhat limited examples of the potential for developing new biological and engineering insight into biotechnology for waste treatment and control. It should be apparent that the potential for a vastly new and improved form of environmental biotechnology is on the horizon. Yet, fundamental science and engineering contributions will continue to be required to reach this potential.

CONCLUSION

Both industrial needs and the science and engineering research base in environmental biotechnology are in a period of rapid transition. In waste treatment and remediation practice, pollutant removal or disappearance data is no longer sufficient information on which to build a technology. The biotic and abiotic contributors to chemical fate must be identified to insure that pollutants are not being transferred from one environmental compartment to the next or that undesirable environmental transformation products are not being formed or mobilized.

It is also no longer possible to evoke "biodegradation" in an engineered practice without providing concrete data showing that a biological process is actively involved in removal of hazardous wastes from an environmental matrix. Such credibility data will be required by clients as well as informed regulators and the public. This credibility data can be extended to provide quantitative information on critical biological processes that can be used in process control and optimization. It is such information that will ultimately define the regimes in which biotechnology is a reliable and predictable component in an environmental remediation and restoration strategy.

Environmental science and engineering research is on the threshold of environmental biotechnology for hazardous wastes. New analytical tools are available and are being developed that will permit timely and efficient analysis of the specific biological mechanisms and activities involved in waste treatment and bioremediation systems. These tools will permit discrimination of critical biological variables that must be quantified for the analysis and control of an engineered waste treatment process. It is also apparent that cellular or biochemical pathways for biodegradation can be modified to develop more efficient or controllable biodegradation processes.

However, the successful integration of the powerful applications of environmental biotechnology in hazardous waste management is not assured. While the technology need not necessarily be embraced in all cases, there is a practical need for its evaluation in most bioremediation practices. This requires knowledgeable and informed science and engineering practitioners and regulators. Consequently, informational and cross-training efforts are desperately needed, and formal educational efforts in interfacing the science and engineering components of environmental biotechnology should be nurtured for the next generation of practitioners. To expand the applications for environmental biotechnology, there also needs to be a corresponding expansion of the biochemical, genetic, and environmental pathways contributing to the degradation of hazardous waste materials.

REFERENCES

1. Schönborn, W. 1986. Historical development and ecological fundamentals. In W. Schönborn (ed.). *Microbial Degradations*. Biotechnology, Vol. 8, VCH, Weinheim, Germany, p. 23.
2. Dupré, A. 1885. The estimation of dissolved oxygen in water. *Analyst*. 10:156.
3. Sayler, G.S., A. Breen, J. Blackburn and O. Yagi. 1984. Predictive assessment of priority pollutant bio-oxidation kinetics in activated sludge. *Environ. Progress*. 3:153-162.

4. Sayler, G.S., M.S. Shields, E.T. Tedford, A. Breen, S.W. Hooper, and J.W. Davis. 1985. Application of DNA-DNA colony hybridization to the detection of catabolic genotypes in environmental samples. *Appl. Environ. Microbiol.* 49:1295-1303.
5. Sayler, G.S., J.W. Blackburn, and T.L. Donaldson. 1988. Environmental Biotechnology of Hazardous Wastes. NSF Project #CES-8714691.
6. Dorn, E., M. Hellwig, W. Reineke, and H.J. Knackmuss. 1974. Isolation and characterization of a 3-chlorobenzoate degrading pseudomonad. *Arch. Microbiol.* 99:61-70.
7. Timmis, K.N., F. Rojo, and J.L. Ramos. 1988. Prospects for laboratory engineering of bacteria to degrade pollutants. In G.S. Omen (ed.), *Environmental Biotechnology: Reducing Risks from Environmental Chemicals through Biotechnology*. Plenum, New York, N.Y., pp. 72-76.
8. Winter, R.B., K-M Yen, and B.D. Ensley. 1989. Efficient degradation of trichloroethylene by a recombinant *Escherichia coli*. *Bio/Technol.* 7:282-285.
9. Jain, R.K., R. Burlage, and G.S. Sayler. 1988. Methods for detecting recombinant DNA in the environment. In: G.G. Stewart and I. Russell (ed.), CRC *Critical Reviews in Biotechnology*. Vol. 8, pp. 33-84.
10. Barkay, T., D.L. Fouts, B.H. Olson. 1985. The preparation of a DNA gene probe for the detection of mercury resistance genes in Gram-negative communities. *Appl. Environ. Microbiol.* 49:686-692.
11. Sayler, G.S., K. Nikbakht, C. Werner, and A. Ogram. 1989. Microbial community analysis using environmental nucleic acid extracts. In: T. Hattori, Y. Ishida, Y. Maruyama, R. Morita, A. Uchida (eds.) *Recent Advances in Microbial Ecology*. pp. 658-662.
12. King, J.M.H., P.M. DiGrazia, B. Applegate, R. Burlage, J. Sanseverino, P. Dunbar, F. Larimer, and G.S. Sayler. 1990. Rapid and sensitive bioluminescent reporter technology for naphthalene exposure and biodegradation. *Science*. Vol. 249, p. 778-781.
13. Burlage, R.S., G.S. Sayler, F. Larimer. 1990. Bioluminescent monitoring of naphthalene catabolism using *nah-lux* transcriptional fusions. *J. Bacteriol.* Vol. 172, No. 9, pp. 4749-4757.

Media Images of Environmental Biotechnology: What Does the Public See?

Amy S. McCabe and Michael R. Fitzgerald

INTRODUCTION: WHY EXAMINE COVERAGE OF ENVIRONMENTAL BIOTECHNOLOGY?

Environmental biotechnology, as a method for effective and economical treatment of hazardous wastes, is one of the most recent applications of biotechnology to a major societal problem. Research and development in this area continues apace, though fundamental policy questions relating to the successful transfer of technology from laboratories to the private sector remain unanswered. How, and to what extent the government should regulate environmental biotechnology, and educate and inform the public are critical issues that public policy makers must address.

The national television news plays a key role in biotechnology policy direction and redirection in the United States. By communicating biotechnology information directly to a mass audience, network news can educate viewers about topics to which a majority of Americans would not likely otherwise attend. Television news is also influential because of its powerful potential to shape public opinion about biotechnology applications. Finally, by featuring biotechnology stories, news segments often emphasize a critical lack of research support dedicated to conceivably promising technologies. By controlling the flow of information—the context in which it is presented, its position in the broadcast, and the amount of time devoted—the news makes a wide-reaching statement about the relative importance of biotechnology issues.

This chapter examines how environmental biotechnology has been portrayed by major network evening news organizations and in the popular and specialized press. Based on a systematic examination of taped newscasts obtained from the Vanderbilt University Television News Archive and of the written press, the chapter illustrates how biotechnology coverage has evolved, if and how environmental bio-

---

Amy S. McCabe • Energy, Environment, and Resources Center, University of Tennessee - Knoxville, Knoxville, Tennessee 37996, and Vanderbilt University Institute for Public Policy Studies. Michael R. Fitzgerald • Energy, Environment, and Resources Center and Department of Political Science, University of Tennessee - Knoxville, Knoxville, Tennessee 37996.

technology coverage differs from that of other biotechnology applications, and what public perceptions are likely to be fostered by the mass media portrayal of biotechnology issues. Key questions to be considered are, first: why should those concerned with moving "from the flask to the field" be concerned with what the public sees about biotechnology? Second, what does the public "see" about biotechnology in general, and environmental biotechnology in particular? Third, how has the network evening news presented environmental biotechnology stories? Fourth, what are the implications of what the public sees for the future of environmental biotechnology?

THE RELEVANCE OF PUBLIC IMAGES

Those concerned with moving from the flask into the field need to be concerned with the public's images of environmental biotechnology for a variety of reasons. The American political process has evolved into a hyper-pluralistic system in which groups with an inclination to mobilize against biotechnology have effective means of doing so. Nurtured by direct mail contributions, subsidized by non-profit organization tax status, schooled in the techniques of headline-grabbing public demonstrations, and able to call upon the legal expertise of tough, court-wise lawyers, even the smallest interest group can present formidable prospective opposition to any field application of biotechnology. Virtually guaranteed access to legislatures, regulatory agencies, and the courts, small groups will likely be in a position, especially in the near-term, to stop environmental biotechnology in its tracks—if the dominant images of this technology are negative and fear-inspiring.

Ours is now a culture in which ideas and reality for individuals are constantly mediated through a process of communications that is dominated by mass media. The news media trade in the transmission of information that is severely constrained by print space, broadcast time, and the nature of mass audiences. The result is communication devoted to the creation and maintenance of readily transmitted and easily understood images, rather than extended discussion and presentation. In order to remain competitive in mass markets, the media seek to entertain while projecting these images. In America the most entertaining images of science have tended toward extremes in which the fruits of scientific inquiry are portrayed as either the salvation or the damnation of humankind; that is, scientists are about to save us from our most terrible problems (with wonder drugs, radiation therapy, new energy sources, etc.), or scientists are on the verge of unleashing terrible destruction (with altered life-forms, inhuman technologies, and toxic chemicals). Such dramatic images tend to create seriously unrealistic public expectations about science and technology. This can create an atmosphere in which the public expects far more from a given technology than can possible be delivered—leading to disappointment, frustration, and feelings of betrayal. Or, it can create a climate of fear in which the application of technology is delayed and perhaps blocked entirely.

Public images of environmental biotechnology are relevant to those concerned with its field application because these images will significantly affect when and how such applications will be conducted. Most especially, if those who are fearful of biotechnology effectively dominate its image formation and transmission through the mass communications process, and the public lacks a basis for rejecting or modifying that image in a positive way, environmental biotechnology can be brought to a standstill.

For this reason, the Waste Management Research and Education and Research Institute at the University of Tennessee is conducting a long-term study of how the mass media covers biotechnology. In this chapter a preliminary assessment of that coverage is provided.

STUDY DESIGN

A preliminary search was conducted to survey the communication media of mass, elite, and attentive publics to identify biotechnology stories. The Industrial Biotechnology Association's definition of biotechnology—the development of products by a biological process—served as the criterion for selection of stories to be included in the database. Based on this definition, a classification scheme for biotechnology applications was developed. The research team then compared the volume and type of biotechnology stories carried in each medium. At this stage of the project, the search was limited to the major network evening news broadcasts (ABC, CBS, and NBC), a major national newspaper (*The New York Times*), and three specialized science magazine/journals (*Science, Scientific American*, and *Nature*). The data presented in this chapter are preliminary; the final assignment of stories will be subject to a panel review and detailed analysis of every story.

Data Collection from the Network Evening News

Scholars interested in empirical examination of national news are limited by the lack of systematic taping of early news broadcasts. The Television News Archive at Vanderbilt University was founded in 1968 to tape, preserve, and make accessible to the public major network evening news programs. Since 1968, the Archive has published a monthly index that compiles network coverage information by subject area, and an abstract that provides more detailed information about the topics covered in a story, its length, its position within a program, and whether or not a story was covered by more than one network. Based on information gathered from the index and abstract, the actual news program can be viewed. Taping prime time coverage of the Cable Network News has recently been added to the Archive's mission, although the information is not yet available in the index and abstracts. The search, then, for biotechnology stories focused on the primary networks—ABC, CBS, and NBC.

Data Collection from the Print Media

For purposes of comparison a systematic search for biotechnology reports and stories was conducted in selected print media. Using the *Infotrac Database of Current Periodicals and National Newspaper Index*, "biotechnology" and a long series of re-

Table 1. Distribution of
Biotechnology Stories/Articles
by Application, 1987-90
(N=463)

Medical	39.5%
Sociopolitical	35.2
Agriculture	9.3
Environment	7.1
Other	8.9
Total	100.0

lated key words were used to locate relevant stories, reports, and articles in four major print media: *The New York Times, Scientific American, Science,* and *Nature.* These particular sources were selected to provide a preliminary basis for assessing how the popular press and more scientific publications compare with the network news in presenting biotechnology.

FINDINGS

Distribution of Biotechnology Stories

Of the "Big Four" biotechnology areas (Medical, Environmental, Agricultural, and Sociopolitical) environmental biotechnology ranked fourth in relative story incidence during the 1987-90 period; of the 463 total stories and articles found, only 7% specifically related to environmental applications. Medical biotechnology and social/political issues (e.g., law suits; protests; ethical issues; regulatory questions), as can be seen in Table 1, accounted for about three-quarters of all biotechnology articles. As can be seen in Table 2, coverage differs from source-to-source. The Networks since 1987 have focused heavily on medical applications; the Networks account for better than one-half of all the medical biotechnology stories located in the media surveyed. Most of these stories have concerned developments and prospects related to AIDS, where a tragic epidemic offers the most dramatic images of biotechnology. Environmental biotechnology stories and articles, in contrast, were most likely to appear in the *Times* and *Science*—these outlets together contributed about seven-of-ten of the items relating to this application. The tendency, to date, of the Networks to neglect environmental biotechnology is especially apparent in Table 3. Only 3% of the biotechnology stories reported by network news programs concerned environmental biotechnology. As of mid-1990, the outlets providing relatively more balanced coverage of the different areas of biotechnology are *Science* and the *Times*. It appears that medical applications and social/political issues are setting biotechnology's public image.

Table 2. Distribution of Biotechnology Stories/Articles by Application and Media, 1987-90 (N=463)

	Medical	Social Political	Agri	Environ	Other
TV News	55.7%	22.7%	23.3%	15.2%	24.4%
NY Times	14.2	33.7	44.2	39.4	41.5
Sci Amer	4.9	.6	4.7	3.0	2.4
Science	18.0	21.5	14.0	30.3	22.0
Nature	7.1	21.5	14.0	12.1	9.8
Total	100.0	100.0	100.0	100.0	100.0

Table 3. Distribution of Biotechnology Stories/Articles by Media and Application, 1987-90 (N=463)

	TV News	NY Times	Sci Amer	Science	Nature
Medical	62.2%	20.0%	64.3%	35.5%	21.0%
Social/Pol	22.6	42.3	7.1	37.6	56.5
Agri	6.1	14.6	14.3	6.5	9.7
Environ	3.0	10.0	7.1	10.8	6.5
Other	6.1	13.1	7.1	9.7	6.5
Total	100.0	100.0	100.0	100.0	100.0

Nature of Television Coverage

In later papers, every biotechnology story/article appearing in these sources will be systematically analyzed and compared as to its contribution to the public knowledge images of biotechnology. For present purposes we confine detailed review to twelve network news segments. These were analyzed to establish how electronic mass media stories potentially shape public images of biotechnology. Close attention was paid to the language used to explain scientific terms, the titles of people interviewed, the length and depth of the story, and the images associated with biotechnology. Of the twelve chosen for analysis, five were ABC segments, five were CBS, and two were televised on NBC.

Tendency to focus on the extremes. The language used in these the network stories presents a mixed image of biotechnology as both "salvation" and "damnation" science. Anchors, reporters, and those interviewed tend, breathlessly, to predict that biotechnology is "a possible step toward the creation of life itself" or indicate genetic engineering might well "cure inherited diseases, cancer, aging, and [prove an] end [to] birth defects." In those few network stories about environmental biotechnology, viewers are told it can "solve the problem of oil spills, save the salt marshes and estuaries." CBS anchor Dan Rather introduces one medical story by pronouncing biotechnology as "the cutting edge of science." Amid pronouncements that biotechnology is bringing a "new day in agriculture," and segments in which state officials "jubilantly announced the microbes devoured the oil," emerges the public image of biotechnology as the wave of an assured and safe future.

At the same time, language is employed in network stories that creates an unmistakable, and extremely dramatic, sense of dread, uncertainty, and fear about biotechnology—conveying the image of "damnation science." Network biotechnology stories are studded with phrases such as: "it's an assault on the sanctity of life," "how big will this little bug get?," "what happens when it runs out of oil to eat?," "may it be used as an instrument of sabotage?," "scientists working on breeding giant cattle," "how do we know one of these products won't turn out to be another gypsy moth?," "the possibility of damaging life," "playing havoc on the environment," "the government has not put anyone in charge," and "we don't have a science about the effects of living things." That is, the networks tend to use a language of unlimited promise and dreadful risks as dramatic hooks upon which to hang biotechnology stories—stimulating both public hopes and fears, but rarely explaining the technology and associated issues in depth. This may ultimately breed public frustration, as people wonder why is it taking so long to develop the applications, as well as cynicism, when breakthroughs prove few and far-between.

Tendency to focus on "unknown" risks. In many network stories words and visuals project a sense of the mysterious, and even the diabolical about biotechnology. In one early bioremediation story, for example, a laboratory is described in hushed tones as shot at "a former warehouse, in a secret laboratory in Virginia." In another story, the scientist whose patent was the subject of a 1980 Supreme Court decision, describes the bacteria as "exotic chemicals." In covering an early California field-test, a shot of a lone protester confessing that biotechnology scares him is dramatically juxtaposed against the picture of workers in heavy protective clothing and headgear spraying bacteria on strawberries while a project scientist declares "there's no danger here."

Tendency to employ "generic" and poorly identified sources. The networks' choice and identification of experts interviewed may well heavily influence the public's perception of biotechnology; but if so, this may be causing more confusion than clarity. Biotechnology critic Jeremy Rifkin is frequently interviewed for network biotechnology stories, ominously pronouncing doubts and expressing grave fear about the consequences of unintended releases of microorganisms into the environment, but, his credentials and expertise are not identified. Often anchors and reporters do not bother to identify specific sources for their general observations and conclusions, instead employing generic labels such as "proponents," "opponents," "experts," "scientists," "critics," and "government officials." Generic labels such as "government oil spill manager," and "oil pollution expert," in bioremediation stories do not provide the viewing public with sufficient information to make informed judgments about the credibility of the persons being used to frame distinctive perspectives on this technology.

Tendency to overlook bioremediation or focus on oil spills. How, then, does environmental biotechnology stack up in the network news? While the image that emerges from the few stories about environmental biotechnology is not seriously distorted, and seems positive, it is very superficial. As was earlier established, environmental biotechnology does not receive nearly as much network coverage as medical biotechnology, or social/political issues. The only time an environmental application was a "lead" story concerned the 1980 Supreme Court patent decision. Even when significant opportunities other than dramatic oil spills do offer themselves, the networks seem uninterested in covering them. For example, in February 1990, President Bush visited the University of Tennessee where he participated in a bioremediation experiment, but this was not reported by the three major networks.

Tendency toward superficial and incomplete information. Environmental biotechnology, moreover, is not receiving the level or detail of network coverage allocated to medical biotechnology. The public, when it is shown examples of this application, tends only to see aerial views of large oil slicks, oil-covered beaches and wildlife, shots of workers spraying "bacteria" on patches of oil, and later pictures of "remediated" sea or beach after the microbes have enjoyed their oil meal. It is most accurate to say that, insofar as the network news is concerned, environmental biotechnology does not have a well-defined image. Stories tend to focus on oil spills and their cause, rather than remediation techniques and technology. Bioremediation is usually only fleetingly mentioned, if at all, as an application in general biotechnology stories.

CONCLUSIONS

This chapter demonstrates that, based on a very preliminary analysis of biotechnology stories/articles in a variety of media since 1987, where people draw their information (broadcast, print, or specialized print) is likely to seriously affect the type of biotechnology to which they are exposed. As a result, the mass, elite, and attentive publics are likely to differ in their images of biotechnology. Environmental biotechnology lags behind medical and sociopolitical aspects, but this lag may be temporary, especially if environmental biotechnology can reach the field-testing stage. Until that time, it is likely that medical and sociopolitical images are setting environmental biotechnology's public image—for good or ill.

The next five years are crucial to the formulation of the public image of environmental biotechnology as it moves from laboratory research to field testing. Concomitant to the technology's transfer, a far more defined image of the technology is certain to emerge, rooted in a significant growth of stories in the mass and specialized media. The relevant scientific and technical community indubitably will face greater public scrutiny concerning the safety and desirability of applying environmental biotechnology to a variety of problems. Successful transfer depends in large part on how well biotechnologists are prepared to deal with the greater exposure. Are we ready to deal with the American public through the mass communications process? What will it take to be ready?

A Need for Media/Image Awareness

If environmental biotechnology is to develop an accurate, positive public image—the kind of image that will support rigorous research and development as well as effective field work—then all involved in it must become media conscious and media smart. Most especially scientists and technicians used to communicating only with highly sophisticated audiences of like-minded professionals are going to have to assume responsibility for directly interacting with the mass media—especially television, and pressuring the media to "get the story right." Toward that end, every member of the environmental biotechnology community must be aware of the image they project from the outset, and attentive to the techniques of media effectiveness.

A Need for Better within-Community Communication

In support of enhanced media attentiveness and effectiveness, communication *within* the environmental biotechnology community must be improved. This community needs better to share what is going on in the laboratory, in the field, *and in the media*. Centers for environmental biotechnology communication must be developed to track image and reality over time so that the biotechnology community knows what it needs to know and can take steps if the gap between them becomes excessive in the press.

A Need for Better Public Education and Outstanding Science

It must be emphasized that public relations and media awareness are a necessary, but not sufficient condition, for the creation of a supportive public climate for environmental biotechnology. Enhanced public education in basic science and technology is necessary. The biotechnology community must address this with its resources and expertise—provide materials, curriculum revision, adopt-a-school— work at building a community. Ultimately the mass, elite, and attentive publics must have a better foundation in science in order to effectively filter and interpret the biotechnology "reality" that the mass communications process mediates for us all.

While it is true that what the public sees and the images it embraces are important, these are not all that matter. Reality itself is what will ultimately decide the fate of testing environmental biotechnology in the field. Despite a growing familiarity with the technology, it is by no means certain that public understanding is growing at the same rate. This gap is potentially dangerous if the biotechnology community does not recognize the difference between public and scientific perspectives. Because public acceptance will be based on trust that must be earned over time, there is no replacement for successful demonstrations, good science, ethical conduct, and an unflagging commitment to the public interest. Excellent performance, then, remains our best guide to judge the acceptability of applying environmental biotechnology to complex technical problems.

Regulators, Lawyers, and Images

The regulatory framework within which the environmental biotechnology community must operate is defined, set, and implemented by political and career officials who respond to public concerns. The regulatory process, therefore, must be perceived by the public as credible, and the regulations sufficiently stringent to mitigate potential risks associated with field testing. Finally, if the past is any guide at all to the future, the legal delays that permeate the present system are certain to prolong environmental biotechnology's journey from flask to field. While our organisms may or may not flourish on oil and PCBs and then disappear as we wish and expect, we can be sure that the lawyers will be nourished and will not disappear—most especially if opponents of the technology effectively set its public image. Those at the forefront of technology transfer should anticipate, and prepare, to spend a vast amounts of time in court chambers, committee rooms, regulatory agency offices, and in front of news cameras. The extent to which this time is well spent will in large part be determined by the popular images of environmental biotechnology that emerge from network news stories over the next five years.

Environmental Biotechnology in 1995 and Beyond

It is likely that the public image of environmental biotechnology in 1995 will be the product of a series of limited successes and several well-publicized failures. Because some commercial ventures and applications will almost certainly fail to achieve desired results, *how* the environmental biotechnology community manages these failures is likely to significantly influence public perceptions about it. The public must be reminded that promising solutions to long-term environmental problems require continuous trial, error, and modification. If those developing the technology are responsible in their promises of near-term benefits, scrupulously honest about risks and costs, and systematically contribute to public education, environmental biotechnology should be able to weather storms caused by near-misses and outright failures.

Based on its experience with other biotechnology applications, the environmental biotechnology community must pursue its goal of field-testing with *guarded* optimism. It would be wise to ponder what can happen unless supportive popular images of environmental biotechnology, firmly grounded in reality, are developed: the technology can be effectively stymied by a regulatory process and constraints imposed by the official representatives of a cynical, fearful, and hostile public. If this were to happen, America's epitaph might read: "They were a decent, but fearful and contentious people. Their only monument: a despoiled earth, and a million microorganisms incarcerated in a flask."

ACKNOWLEDGEMENTS

We gratefully acknowledge the support of the Waste Management Research and Education Institute (WMREI) of the University of Tennessee-Knoxville, and the Vanderbilt Institute of Public Policy Studies (VIPPS), for this project. The directors of these centers, Dr. E. William Colglazier and Dr. Clifford Russell respectively, merit our special thanks—as do Dr. Scarlett Graham, Director of the Vanderbilt Television News Archive and her outstanding staff. Deserving of special credit are the members of the fledgling Science, Technology, and the Media Project Team: A. Hunter Bacot, Michael Dedmon, Heather Fitzgerald, Patra Rule, and David Spence.

Perspectives on Bioremediation in the Gas Industry

David G. Linz, Edward F. Neuhauser, and Andrew C. Middleton

INTRODUCTION

One of the areas of interest to the gas industry today is the treatment of residues from past gas manufacturing operations. From 1816 into the 1960s gas was manufactured in the United States from coal and oil for baseload distribution to consumers. Initially the gas was used mostly for street lighting. Afterwards it became a fuel for heating, cooking, and industrial use. The manufactured gas industry began to rapidly decline in the 1950s as transcontinental pipelines brought natural gas to more and more of the country. The use of manufactured gas for baseload production decreased to virtually nothing in the 1960s. Some intermittent operation of manufactured gas plants continued after the 1960s to provide peak shaving capacity. During the Era of Manufactured Gas it has been estimated that there were over 1500 operating sites throughout the United States.

Five manufacturing processes were predominant. The coal carbonization and by-product coke processes generated gas by the destructive distillation of coal yielding coke, coal tar, and ammonia as major by-products. The carburetted water gas process generated gas first from coke or coal reacted with steam followed by heat content enrichment with thermally cracked oil. Tar was the major by-product. Finally, the regular and high-BTU oil gas processes generated gas by the thermal cracking of oil. The high-BTU process was used primarily to produce gas which could be mixed with natural gas for peak shaving purposes. Tar and sometimes lampblack were the major by-product of both these processes.

By the 1980s virtually all of these manufactured gas plant (MGP) sites were inactive. As a result of consolidation of gas districts over time, many of the sites are now the responsibility of present gas utilities. As is true of any former site of industrial activity, remnants of the products and process residuals can be found at the site. In the case of MGP sites a major current issue is often the presence of tar or tar constituents in site soils, the tar constituents of interest being polynuclear

---

David G. Linz • Gas Research Institute, Chicago, Illinois 60631. Edward F. Neuhauser • Niagara Mohawk Power Corporation, Syracuse NY 13202. Andrew C. Middleton • Remediation Technologies, Inc., Pittsburgh PA 15219.

aromatic hydrocarbons (PAHs). It is the objective of this paper to present a perspective on bioremediation as a potential technology for remediation of PAHs in MGP site soils.

BACKGROUND

The results of a number of MGP site investigations have shown that PAHs in soils can vary from below detection limits (BDL) to tens of thousands of mg/kg. Typically, the higher levels are the result of free tar still being present as a non-aqueous phase liquid (NAPL). The types of soils present at these sites cover the full spectrum from sands and gravels to silts and clays. Often there is a substantial fraction of fill present.

The ideal remedial process for PAHs in soils would be one that consistently and predictably destroys them to an environmentally sound level at low cost without producing any adverse by-products. Remedial alternatives include incineration and bioremediation. The performance of incineration is generally consistent and predictable. Normally, incineration can achieve at least 99 percent destruction of PAHs in soils in a relatively short period of time. However, incinerators are often difficult to site and permit because of regulatory requirements and community concerns. Incineration is also a high cost alternative costing as much as $1000 per cubic yard or more in many cases.

Bioremediation can destroy PAHs in soils. However, the performance of bioremediation in terms of endpoint analytical concentrations has been variable and difficult to predict without extensive pretesting. Although bioremediation takes relatively longer periods of time to complete than incineration, its cost is low, typically ranging from $50 to $100 per cubic yard for engineered land treatment systems.

Traditionally, the extensive pretesting of bioremediation has been accomplished with pan studies. Pan studies simulate biotreatment of soils in an unsaturated state. Contaminated soils are placed into a pan approximately 8 x 12 x 4 inches deep. Nutrients (nitrogen and phosphorus) and water are added to enhance growth of indigenous bacteria. Soil moisture is maintained at 70 to 80 percent of the soil's field capacity. Periodically, the pan soils are mixed, and the levels of nutrients, pH, and moisture are measured and adjusted as necessary. Soil samples are collected and analyzed initially and typically once per month thereafter for a six month period. The change in contaminant concentration with time forms the basis for estimating the capability of bioremediation to treat the soils. Hence, the time to perform a pan study typically ranges 6 to 9 months allowing for mobilization, time for analysis, and data interpretation along with report preparation. This method takes time and is relatively costly.

As part of the Gas Research Institute's (GRI) program on evaluating remedial alternatives for MGP sites, several pan studies have been performed on MGP site soils contaminated with PAHs. Figures 1 through 4 show the variation of soil PAH concentration with time for four of these pan studies, Soils B, F, J, and D, respectively.

Soil B was taken from an MGP site where a carburetted water gas process operated from around 1840 to 1940. It was collected from a depth of 15 to 20 feet

حدود کلی اصلاح مسائل والی

C 160 mg/kg → 12 weeks

May 28th - SAT

PROJECT DUE 4.4.05 Mon.

ANALYSIS PAPERS
INVENTORY BOARDS

BIOREMEDIATION
$60 - $100 / cubic yd.

engineered land trtmnt systems
PAN STUDIES
NITROGEN + PHOSPHORUS + H_2O added to
enhance growth of indigenous
bacteria
soil moisture maintained 70-80%
of flood c.c.

(page rotated; handwritten notes, largely illegible)

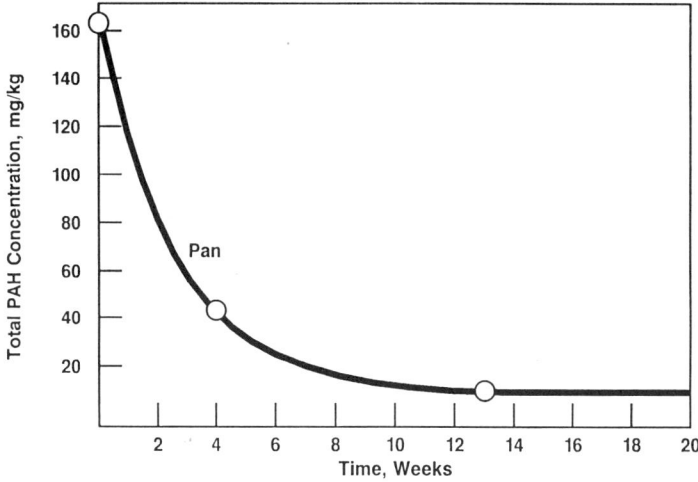

Figure 1. Soil B.

below grade in an area that had been impacted by the former process operations. The soil consisted of coarse sand and gravel and had a tar odor. The finer materials (<0.075 mm) were approximately 3 percent of the particles. The fraction of organic carbon in this soil was 0.6 percent. The initial PAH concentration was approximately 162 mg/kg. After 12 weeks the concentration decreased to approximately 12 mg/kg (Figure 1).

Soil F was taken from a site where coal and oil had been used as feedstocks to manufacture gas from approximately 1890 to 1930. It was taken from 3 or more feet below the surface inside a former gas holder which had been filled in around 1950. The soil appeared to be saturated with tar. The soil consisted of coarse, medium, and fine sands. The finer materials were approximately 7 percent of the particles. The fraction of organic carbon in this soil was 16 percent reflecting the presence of a substantial amount of tar. The initial PAH concentration was approximately 20,000 mg/kg again reflecting the presence of substantial tar. After 22 weeks the concentration decreased to approximately 5,600 mg/kg (Figure 2). This is a substantially higher endpoint than the one for Soil B which was 12 mg/kg.

Soil J was taken from a site where coal had been converted to gas and, later, oil to gas. The samples were taken at a depth of approximately seven to ten feet using a truck-mounted six-inch boring auger. The soil consisted of sands, silts, and clays. The finer materials were approximately 27 percent, by weight, of the samples. The fraction of organic carbon in the soil was nearly 58% which was indicative of the presence of lampblack, a waste that is commonly associated with oil-based manufactured gas plants. The initial PAH concentration was approximately 29,100 mg/kg. After 16 weeks the concentration decreased to approximately 16,900 mg/kg (Figure 3). This is a substantially higher endpoint than either of those for Soils B and F.

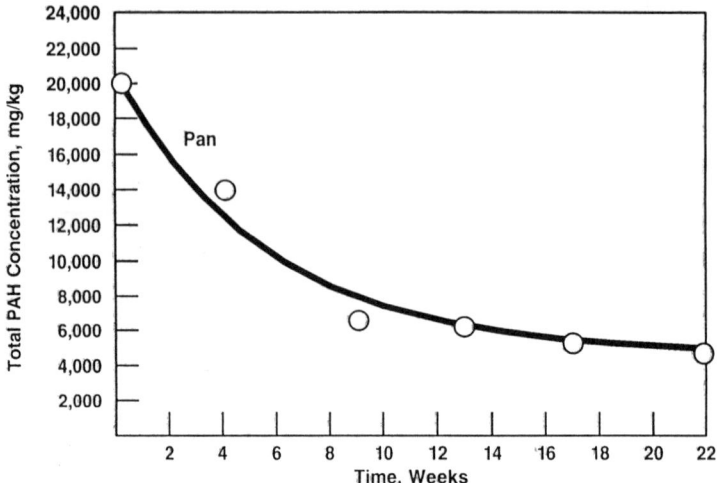

Figure 2. Soil F.

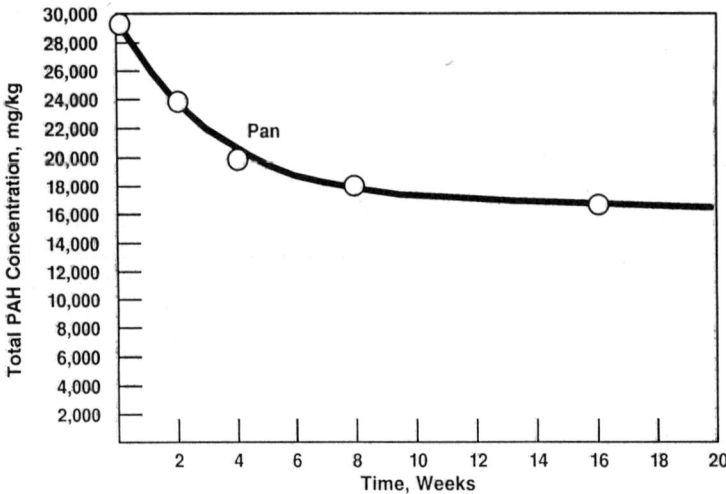

Figure 3. Soil J.

Soil D was taken from a site where a carburetted water gas process operated from around 1881 to 1930. Samples were taken from bore hole cuttings collected during a site investigation. This soil consisted of approximately 48 percent coarse to medium sands and gravels, 26 percent fine sands, and 26 percent finer materials. The fraction of organic carbon in this soil was 6.5 percent. The initial PAH con-

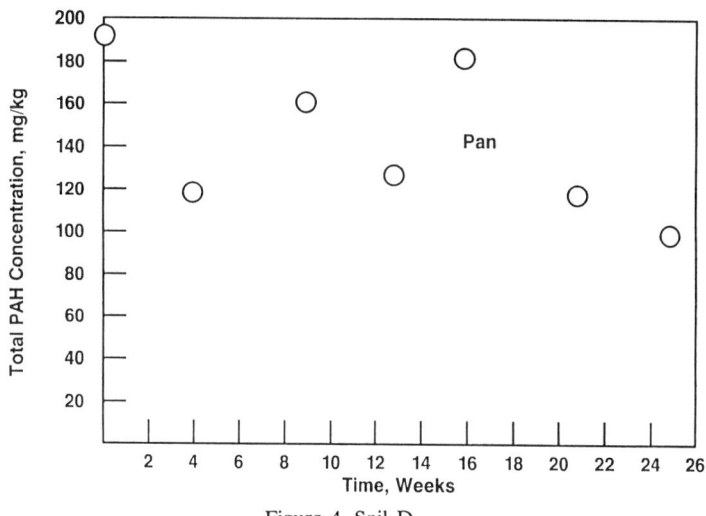

Figure 4. Soil D.

centration was approximately 193 mg/kg. After 25 weeks the concentration decreased to approximately 100 mg/kg (Figure 4). However, the measured concentrations were highly variable during the study. Although this soil had an initial concentration similar to that of Soil B (162 mg/kg), the endpoint was approximately an order of magnitude higher.

The results of these pan studies on MGP site soils show that the treatment endpoint is variable from site-to-site and that typically, at least 4 to 6 months are required for the pan studies to approach a plateau level PAH concentration. In addition, the results for Soil D show that the analytical measurements can be highly variable, and it is not clear after even 6 months that the plateau level was reached. These results clearly illustrate that a key need for the gas industry is a faster, lower-cost, means to evaluate and predict the effectiveness of bioremediation on PAH contaminated soils.

GRI'S ACCELERATED TREATABILITY PROTOCOL

In 1988, GRI initiated a research program to develop a faster, lower-costs means of evaluating potential treatability of soils with bioremediation. In the initiation of this program, it was clear that bioremediation is a function of traditional parameters, such as moisture content, nutrient levels, oxygen levels, microbial population, and microbial ecological factors. However, shortly into the program it became evident that biodegradation rates and endpoints can be greatly affected by the mass transfer of the compounds from the soil-waste matrix to the macropore water phase where microbial action primarily takes place. Specific components of tar may be contained in different soil matrices to different degrees.

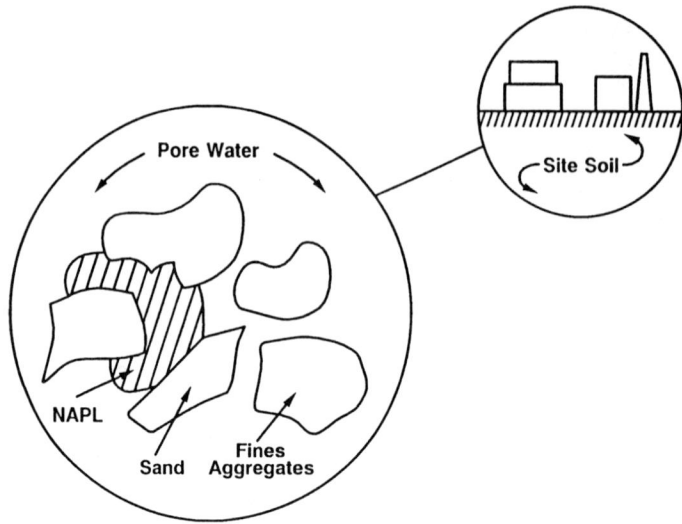

Figure 5. Conceptual model.

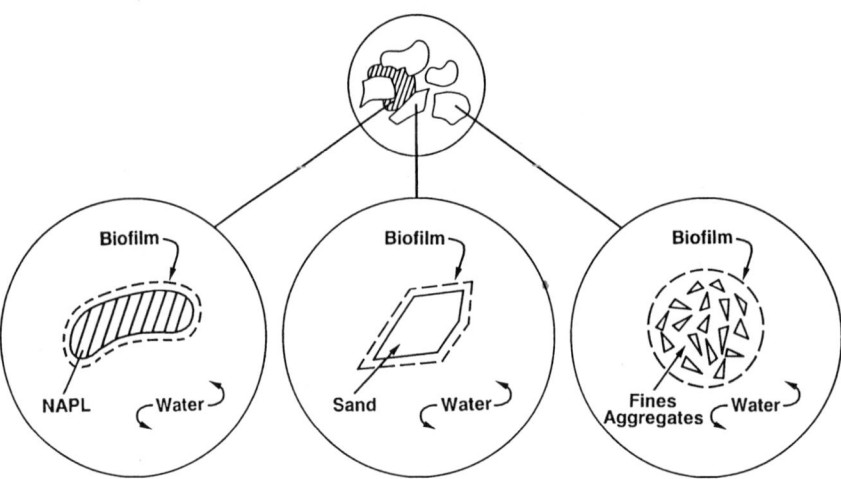

Figure 6. Conceptual model.

A general conceptual model was first developed to provide a framework of analysis for this situation. Figure 5 shows a schematic diagram of a site soil at a microlevel. Site soils are made up of individual sand particles, aggregates of finer particles, and pockets of NAPL, i.e. tar in this case, trapped in the void space. In a saturated situation as shown in Figure 5, the voids are filled with water. Figure 6 shows is a schematic diagram of this situation at a greater magnification. In the first circle NAPL is shown with a film of bacteria (i.e., biofilm) around its surface.

Biodegradation proceeds through bacterial utilization of organics that diffuse to the surface of the NAPL where they can dissolve into the biofilm. Once in the biofilm they are degraded. Initially, the tar NAPL contains a mixture of PAHs. As the more mobile PAHs diffuse to the biofilm, the nature of the tar changes to a mixture of the more immobile PAHs. As these changes take place, the rate and endpoint of biodegradation greatly alters. Hence, the amount of tar NAPL present in the soil is a variable influencing the bioremediation process in so far as the amount of fines relate to aggregate/agglomerate structures which can impede mass transfer of PAHs to the macropore (soil) water.

In the second circle in Figure 6, an individual grain of sand is shown surrounded by a biofilm. The surface of the sand particle provides an area for PAHs to adsorb. In this situation the PAHs must desorb and diffuse into the biofilm in order for biodegradation to proceed. In contrast to the situation with the tar NAPL, there is not a similar change over time with this process.

In the third circle in Figure 6 an aggregate of finer particles (i.e., silts and clays) is shown also surrounded by a biofilm. Since silt and clay particles are substantially smaller than bacterial cells, bacteria are unable to migrate into the interior of these aggregates. In this situation PAHs adsorbed to particle surfaces at the periphery of the aggregate have a much shorter pathway to the biofilm than ones inside it. The ones inside must desorb from the surface and then diffuse through the micropore structure of the aggregate to the bulk aqueous phase where they are available to the biofilm. In this pathway there is continuous opportunity for adsorption/desorption throughout the aggregate to slow the rate of movement. Hence, the amount of fines present in the soil is another variable influencing the bioremediation process.

Figure 7 is a box-model of the bioremediation process of PAHs in soils. At the far left PAHs are shown in the NAPL, on the macrosurfaces, i.e. those of sands and outer aggregates of fines, and on the microsurfaces within the aggregates of fines. The second column is the micropore water. The PAHs must desorb from the

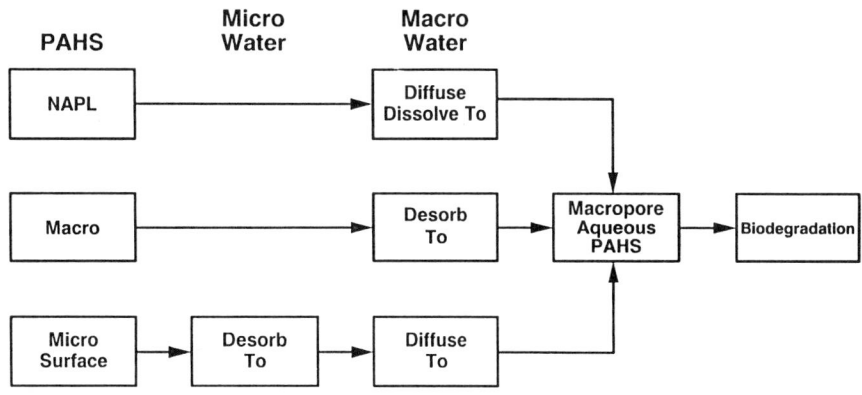

Figure 7. Bioremediation process.

Table 1. GRI Accelerated Treatability Protocol

Characterization
  soil particle distribution
  TOC, PAHs, O&G
  microbial PAH degraders
Desorption
  PAH desorption isotherms
  K_p determination
Bench-Scale Bioslurry Reactor Testing
  maximize transport of PAHs to biodegradation

---

microsurfaces and diffuse through this to reach the macropore water. The third column is the macropore water. PAHs reach this by diffusing to the surface of the tar NAPL and dissolving into it, by desorbing from macrosurfaces, and by diffusing from the micropore water. Once in the macropore water, the PAHs are available to the biofilm for degradation.

In analyzing the conceptual model in Figure 7, it becomes clear that in addition to the biological processes that PAH degradation in soils is most likely a function of the concentration of tar NAPL and of the content of soil fines (i.e., mass transfer processes). It would seem reasonable to expect slower degradation and higher endpoints as the mass transfer effects increase.

In seeking the goal of accelerating bioremediation testing, it also became apparent that the degree of water saturation would also be a major factor in the rate of bioremediation. This is the case because if mass transport via the various steps shown in Figure 7 is a limiting function, then operating in a saturated environment should cause faster rates than in an unsaturated one. All of the processes are water-based transport phenomena. At saturated conditions all of the pore spaces are filled with water and transport should be at the maximum possible. In addition, if the aggregates of fines could be dispersed exposing the microsurface areas to macropore water, then transport rates should be further enhanced.

All of these considerations led to a protocol consisting of soils characterization, desorption testing, and biodegradation in a soil/water slurry. Table 1 lists the elements of this GRI Accelerated Treatability Protocol (GRI ATP). Soil particle size distribution is determined by sieve analysis and hydrometer testing. Chemical analysis is performed for total organic carbon (TOC), PAHs, oil and grease (O&G), and other chemicals as necessary. Microbial PAH degraders are enumerated using agar plate counts with media containing phenanthrene as the carbon source. Soil desorption testing consists of sequential batch washing of a soil sample. At the end of each batch wash the PAH concentration in the wash water is determined. The wash water is replaced and a subsequent wash is carried out. Typically, washing is repeated five times on a sample. PAH concentrations are determined on the soil sample itself initially and at the end of the washes. The wash data are analyzed as isotherms and the soil water partitioning coefficients (K_p) for the individual PAHs are calculated.

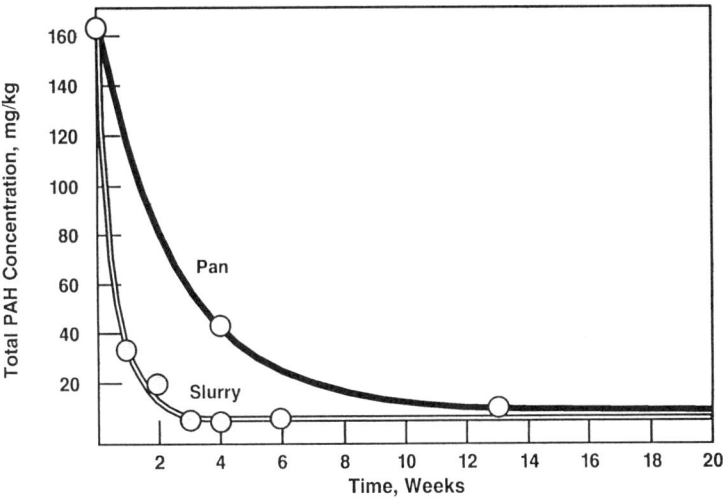

Figure 8. Soil B.

The final step in the protocol is the bench-scale biodegradation testing of the soils in a soil/water slurry reactor. Soils are mixed with water in ratios of 1:9 to 1:4 and added to a completely mixed batch reactor. Nutrients are added and the pH is controlled within a range of 6.5 to 7.5. The reactor is aerated to maintain oxygen levels in the neighborhood of 6 to 7 mg/L. This reactor is designed to achieve the maximum transport possible of PAHs, oxygen, and nutrients to the biofilm. PAH concentrations are measured as a function of time to track the bioremediation process. Typically, weekly analyses are performed for 3 to 5 weeks.

As part of the program the soils that were tested with the pan studies were also tested with the soil/water slurry reactors to evaluate bioremediation. Figures 8 through 11 show the variation in PAH concentration with time of Soils B, F, J, and D for these reactor tests, respectively. The results of the pan studies are also shown on each graph for comparison. In each case the biodegradation curves are similar in shape. The PAH concentration decreases in an exponential manner from the initial concentration to some apparent plateau concentration. The decrease for the slurry reactors is always much faster than for the pan reactors. This is consistent with the expectation that transport processes are maximized in the slurry reactor, while the pan reactor is operating in an unsaturated mode. The plateau concentrations are similar between the individual slurry reactor and pan tests for all of the studies except that for Soil D. The similarities support the hypothesis that in many cases slurry reactors can be used to estimate the approximate endpoints of bioremediation regardless of the mode of operation. However, the results of Soil D indicate that the hypothesis may have its limitations. The limitation may be related to the amount of fines present. Soil D contained about 26 percent fines,

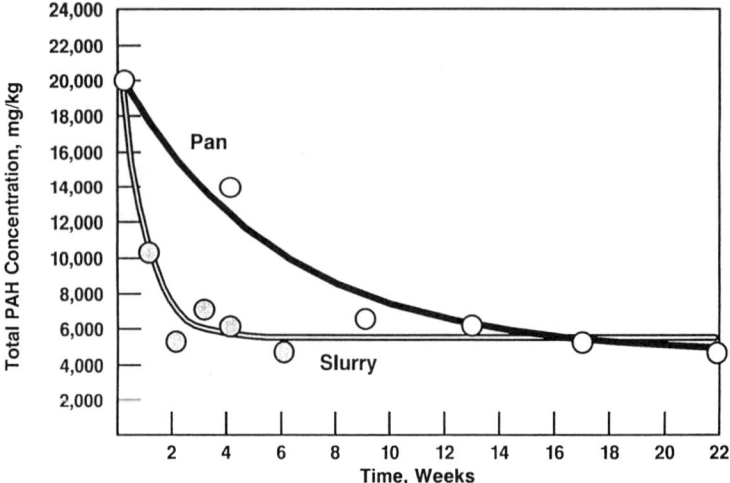

Figure 9. Soil F.

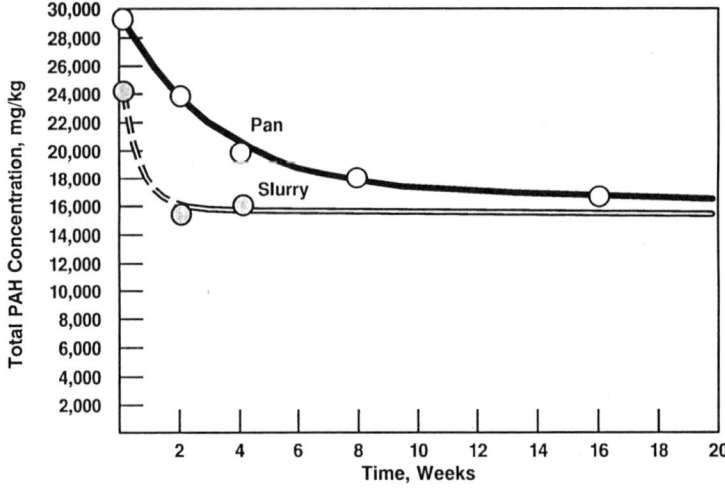

Figure 10. Soil J.

an amount substantially greater than the other soils. In the slurry reactor the fines were dispersed exposing virtually all the microsurfaces within the aggregates. In the pan study, unsaturated mixing could not expose anywhere near the same degree of microsurfaces. Hence, it is likely that the pan study could not reach a similar plateau in a reasonable amount of time, if ever.

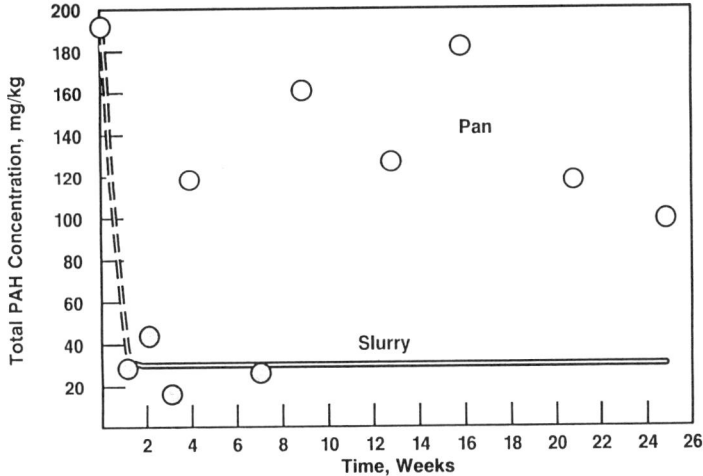

Figure 11. Soil D.

CONCLUSION

In reviewing the work performed to date in the GRI program on bioremediation, it is clear that bioremediation can be a viable process to treat PAH contaminated soils. Different soil and tar situations will result in different bioremediation rates and endpoints. For soils tested thus far, the GRI Accelerated Treatability Protocol predicts the approximate bioremediation endpoint in a shorter period of time than the traditional pan studies. However, the exception to this conclusion is situations where the content of fines exceeds 5 to 10 percent. It is clear from the work to date that wide variability of performance exists, and that both mass transfer and biological processes contribute to this variability.

In the future more work is needed in several areas. An innovative approach is required in situations of high fines content and/or high free NAPL phase concentrations. Mass transfer limitations in the presence of high amounts of fines appear to be substantial. There is virtually no work to date to predict actual field performance based on the results of protocol testing. Hence, field scale up is an area requiring attention. The protocols must be further refined to be able to screen for treatability for all soil types, for both mass transfer and biological activity, key parameters, and for various alternative process applications (i.e., land-based, tank-based, or in situ).

Consideration needs to be given to slurry reactor treatment itself as a bioremediation alternative. Clearly, at bench scale it is capable of substantially reducing PAH concentration. Finally, a key issue is whether the bioremediation endpoint is a sound one in itself regardless of what the PAH concentration is. There is support for this premise because a bioremediated soil has virtually destroyed the readily water available PAHs. The remaining PAHs are tightly bound to the soil matrix or within the remaining tar NAPL as it may have been modified. Hence, the remaining PAHs are highly unlikely to leach into water and become mobile only at de minimis rates. Testing in all these areas is planned as part of the future program. It is certain that acceptable treatment endpoints must be demonstrated and full-scale performance, both technical and economic, reliably predicted if bioremediation is to be a widely used technology for MGP sites.

Considerations in the Selection of Environmental Biotechnology as Viable in Field-Scale Waste Treatment Applications

Patricia Taylor Woodyard

Environmental biotechnology is a technology with increasing potential for use in degrading toxic contaminants in water and soil media at hazardous waste sites. The term biotechnology encompasses a number of approaches in utilizing microorganisms to degrade contaminants both above ground and in situ. Some of the technology's applications are well understood with extensive field data such as biological land treatment of some petroleum wastes. On the other hand, others such as microbial degradation of PCBs in soil are in the initial stages of field experimentation.

Biological treatment technology's use on contaminants in water has an extensive history of application resulting in a technology that is very predictable. The mechanics of treating soils with microorganisms differ greatly from those of treating water. Fewer soil treatment technologies have been field demonstrated resulting in a limit to the predictive capability in assessing the technology's feasibility.

When biotechnology applications involve treatment of soils, additional variables aside from the contaminants themselves add complexity to the technology's performance ability. Such variables include soil matrix complexity (differing soil structures, moisture contents), presence of elements toxic to the bacteria, presence of metals, rate of parent compound disappearance, supply of nutrients, temperature, pH, and formation of intermediates. The successful application of biotechnology at hazardous waste cleanup sites, whether through microbial treatment in water or soil is directly tied to the technology's ability to degrade contaminants to cleanup levels. Additionally, when multiple contaminants are present in onsite soils, biotechnology may be applicable to only a fraction of the cleanup area. These variables from site to site should not restrict the consideration of biotechnology. Even with these complexities, biotechnology shows promise at selected sites as a lower cost option to other more proven alternative technologies.

Patricia Taylor Woodyard • CH2M HILL, Inc., Emeryville, California 94506.

PERCEPTIONS OF ENVIRONMENTAL BIOTECHNOLOGY

The use of environmental biotechnology in cleanup actions has applicability in a number of regulatory programs such as the Comprehensive Environmental Reclamation and Liability Act (CERCLA), Resource Conservation and Recovery Act (RCRA), Clean Water Act (CWA), Toxic Substance Control Act (TSCA), and voluntary actions. The formal technology evaluation and selection process can vary depending on the jurisdiction and regulatory involvement. To date, the successful application of biotechnology can be readily found in voluntary cleanups and underground storage tank remediation. Such applications may be a result of reduced roadblocks and the types, diversity, and distribution of contaminants at these sites. Greater consideration of this technology is being made at CERCLA, RCRA, CWA, and TSCA sites than ever before, although such considerations do not automatically result in its selection.

The perceptions of the viability of environmental biotechnology have been both an asset and detriment to the selection and implementation of biotechnology for field-scale applications. These perceptions exist due to varying levels of understanding over the technologies applicability and effectiveness. Since this lack of understanding can result in overly optimistic appraisals and premature endorsements of the technology's effectiveness, a lack of *frankness* on the realistic effectiveness of the technology under site-specific conditions can result. Alternatively, a "black box" explanation of the technology leads to a lack of credibility of the scientific basis for its effectiveness. With limited explanation of the technology's application an overly conservative evaluation of its effectiveness can result.

Perceptions as to the reasons why environmental biotechnology is not considered and selected on more cleanup sites vary. Perceptions include:

- Cleanup levels are not achievable
- Developers of the microbiology technology are not considering the impact of site-specific technical and nontechnical considerations
- Some agency representatives are advocating the technology without a sufficient information base to be sure of its applicability and effectiveness
- Other agency representatives set up institutional roadblocks to its application
- Entities with cleanup needs expect only controllable uncertainties and costs with limited risk potential
- Consultants are evaluating the technology from an overly conservative viewpoint
- Vendors and contractors are selling the technology's application without consideration of pre- and post-processing technical and non-technical implications
- Public has mixed biases toward biotechnology's ability to destroy contaminants at the same effectiveness as incineration and against biotechnology because of the uncertainty of leaving residuals that may have environmental impact.

All of these represent perceptions that set up roadblocks to effective communication on the **realistic** applicability of environmental biotechnology.

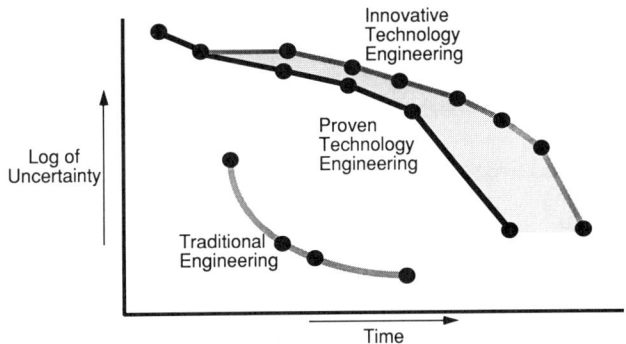

Figure 1. Uncertainties impact on remediation engineering and implementation.

In reality, effectiveness of the technology will need to be evaluated on a site by site basis. The effectiveness and applicability will vary depending on 1) site conditions, 2) type of biotechnology used, 3) biodegradability, 4) treatment duration, 5) logistical constraints, 6) cost-effectiveness, 7) technical uncertainties present, and 8) the ability to meet treatment and cleanup goals. The selection of the technology will be impacted by both technical and nontechnical considerations.

As more scientifically based data is gathered on the varying technology's effectiveness under field-scale conditions, it will be possible to present a realistic viewpoint on the "state of the technology." In turn this will reduce perceptions that are negatively affecting the technology's selection.

EVALUATION OF ENVIRONMENTAL BIOTECHNOLOGY

In the identification, evaluation, and selection of remedial and corrective action alternatives, an unbiased assessment is undertaken of each technology's applicability from both a technical and nontechnical perspective. Technical and nontechnical issues are considered so roadblocks can be identified, perceptions reduced, and methods for resolving technical uncertainties can be determined. The technology evaluation often compares more traditional technologies (i.e., excavation and removal) and proven technologies (i.e, incineration) with innovative technologies (soils bioremediation of PCBs in situ).

Biotechnology applications themselves vary from using more traditional engineering/construction methods to innovative applications. This variability can confuse perceptions because of the wide range of knowledge base existing *depending* on the biotechnology application being considered, contaminants present for remediation, and cleanup levels to be reached. Figure 1 shows the variability that can exist between traditional engineering, proven technology engineering, and that of innovative technology engineering where the greatest uncertainty is usually present.

TECHNICAL CONSIDERATIONS

In general, it is recognized in evaluation of environmental biotechnology that a reduction in contaminants below cleanup levels represents a permanent remedy. The ability of the technology's application to meet cleanup goals is a major hurdle in its acceptability. With institutional constraints such as Land Disposal Restrictions under the Resource Conservation and Recovery Act, biotechnology often must compete with technologies such as incineration for acceptance. The ability of incineration to destroy contaminants under proven applications has provided advantages to its selection in the past. Bioremediation may be a comparable technology but in many applications there is not a proven track record of performance. Without a track record of performance sufficient to resolve uncertainties identified in the technology's 1) biodegradability, 2) reliability, 3) byproduct generation, 4) duration of performance, 5) cost impact, 6) ability to treat multiple contaminants under differing soils conditions, and 7) ability to meet cleanup goals, bioremediation selection will remain limited.

The resolution of these technical uncertainties is found in the performance of scientifically-based treatability studies where laboratory and field-scale experimentation can resolve many of these technical uncertainties. More technology evaluations currently are recommending such treatability studies so that enough information is gathered to determine under what site-specific conditions is biotechnology a viable solution to remediation needs.

With the implementation of laboratory and field-scale experimentation for innovative applications, the need grows for an **iterative** process of collecting data to resolve technical uncertainties and an allowance of time necessary to accomplish this task. This data-gathering approach, utilizing **incremental steps** to reach the ultimate implementation goal (effectively remediate soils to established cleanup levels), can allow for identification and resolution of uncertainties that could ultimately affect performance success. This approach, which has been used in the geotechnical engineering field for years is called the "Observational Approach." It uses a **learn-as-you-go** process which needs *flexibility* to allow for necessary adjustments as more data are collected.

This **Learn-As-You-Go** process is a way of reducing the endless study approach being undertaken currently. In order for this process to be utilized at biotechnology sites, all parties (agencies, public, client, vendor, consultant) need to accept this approach and allow for the *flexibility* needed to gain the knowledge of the technology in field operations. Flexibility can mean the need for more incremental steps, extended schedule before field implementation, and additional lab or bench-scale testing. Upon initiating this process, it is necessary to develop the *most probable* approach to implementation (not necessarily the exact approach) and identify reasonable deviations from this approach. Part of the agreement needed from all parties is the acceptance of possible deviations in the approach and allowance for schedule and testing changes through field- and full-scale implementation. In turn, changes in schedule and approach need to be allowed for along the way.

Unless a **Learn-As-You-Go** process is followed, there is a risk that conservative technology selection will prevail due to too many technical uncertainties in comparison to proven technologies with less uncertainties and risks. The greater the uncertainties, the greater must be the potential benefits for the technology to be given serious consideration and ultimate selection.

NONTECHNICAL CONSIDERATIONS

This technology's *technical* viability is often weighed against *nontechnical* issues and the risks associated with its implementation. This technology is often then compared to issues and risks arising from implementing field- and full-scale alternative technologies.

The nontechnical issues under consideration often include the level of uncertainty associated with the technology as stated previously in combination with legal, regulatory, sociopolitical, and business implications and uncertainties. Any nontechnical issue or combination of issues can at times outweigh the potential technical feasibility of environmental biotechnology. Identifying and evaluating the considerations and implications associated with these technical and nontechnical issues is a valuable step in determining the site-specific viability of environmental biotechnology.

Guidance documents and standards of practice have been established for assessing technical feasibility of technologies, but there is little written on the implications of nontechnical issues. Because nontechnical considerations can play a large role in the selection of biotechnology under varying regulatory jurisdictions, it is important to note some of these considerations. Table 1 identifies some of these nontechnical considerations along with the implications for selection of this technology. It should be recognized that every site has its differences and similarities so that what could be a negative nontechnical consideration against biotechnology at one site could be a positive consideration toward biotechnology at another site. In turn, every regulated entity (i.e., Department of Defense, Department of Energy, private sector companies, utilities) with clean up needs places different priorities on the importance and implication of one over another.

The viability of this technology's application at sites may depend on the ability to reduce or eliminate the negative perceptions and implications presented in Table 1. It will be important for scientifically based, realistic, and factual information to be provided by and *to all* entities in this process to reduce conflicting perceptions. To accomplish these tasks, all parties involved in the evaluation, selection, and implementation of the technology need to do their part in addressing the issues and reaching practical cost-effective resolutions.

RECOMMENDATIONS

The application of environmental biotechnology at cleanup sites nationwide has increased as the issues hindering biotechnologies utilization are being resolved. The current perceptions that are hindering the implementation of biotechnology need to be eliminated. By gathering scientific data from laboratory and field-scale

Table 1. Selective Nontechnical Considerations in Evaluation of Biotechnology's Viability

Potential Considerations	Perceptions/Implications
Legal Considerations	
Scheduling	Potential for agency pressure and fines due to delays
	Need for flexible schedules
Liability Risk	Potential to negatively impact operational compliance
	Potential for greater environmental damage if technology application failure
Property Transfer	Uncertainty that cleaned property may have future acquisition impacts
Flexibility Potential Written Into Permits/Orders/Consent Decrees	Potential ability to negotiate flexibility and "Learn-As-You-Go" approach into legal documents
	Uncertainty as to reasonableness of agency to allow flexibility into legal documents
Regulatory Considerations	
Achievement of Cleanup Goals	Potential for cleanup goals not to be achieved
	Potential for cleanup goals to be changed during process
Implementability under Compliance Schedule	Tight schedules can impede flexibility needed to allow technology to work
Hazardous waste designations	Need for treatment of listed waste to still require delisting prior to land disposal
Chemical/nutrient incorporation	Agency(s) uncomfortable with in situ addition
Design and operation	RCRA places hurdles and constraints to operation of treatment systems
Land Ban Impact	Potential that treated soil will not meet treatment standards resulting in disposal constraints
Multiple Contaminants	Regulations expect multiple contaminants to meet cleanup standards where biotechnology may only apply to a few
	Potential for multiple treatment train
Regulatory Acceptability of Results	Agency's potential uncertainty with data results
Permit Acquisition Potential	Permit development, acceptance, maintenance costs, and impact for consideration
Cost-Effectiveness	Consideration of cost-effectiveness varies with regulation (CERCLA vs. RCRA)
Onsite Treatment Bias	Bias toward onsite treatment technologies varies with regulation (CERCLA vs. RCRA/TSCA)
ARAR's impact on cleanup goals (similar issue with risk assessment)	Potential for lowest ARARs to be selected potentially inhibiting viability of biotechnology
Innovative Technology Encouragement	Variability between regulations in providing incentives and flexibility needed to use innovative technologies (CERCLA vs. RCRA/TSCA)
Flexibility Potential within Confines of Regulations	Variability between regulations and regulators in allowance for reasonable flexibility
State and Local Regulation Variability	Agencies may have additional or different requirements than federal regulations resulting in additional hurdles
Treatment Residues	Handling and disposal of treatment residues (i.e., wastewater) may be hindered by NPDES or POTW discharge restrictions

(continued)

Table 1. *(continued)*

Potential Considerations	Perceptions/Implications
Sociopolitical Considerations	
Political uncertainty toward technology	Politicians can impact pressure against or toward technology application
Public's faith in process and parties	Basic mistrust of actions taken at cleanup sites
Public's focus on cleanup goals	Impedious for most conservative goals resulting in reduction of technology viability
Public's concern over chemical/nutrient addition	Variable perceptions on acceptability toward additions in situ
Business Considerations	
Acceptable uncertainty	Regulated party(s) acceptability that a "Learn-As-You-Go" process has inherent risks
Communication Roadblocks	Regulated party(s) ability to address the decisionmaking to management in recognition of uncertainties
Cost Accuracy	Need for good grasp of costs to translate into financial impact to regulated party(s) in advance of actions
Cost Control	Need for technologies where differentiation between scope creep versus unknown change in conditions can be made Vendor's emphasis on "black box" technology increases concern over ability to control costs
Treatment Train Implications	Technology conversion of contaminants from soil to aqueous phase can translate to additional risks/uncertainties and costs not always identified in initial discussion of soils remediation
Multiple Party Concurrence (CERCLA)	Uncertainties with technology leads to varying opinions on viability such that multiple party sites may have difficulty reaching agreement on technology
Liability Risk	Concern that failed technology application may not reduce risk but can confuse or increase regulated party's risk
Environmental Risk	Concern that failed technology application may alter or increase environmental risk
Flexibility Realities	Concern that agency's intent for flexibility is lost at the agency decisionmaking levels
Negotiating Compromises	Roadblock potential during agency negotiations of remedy implementation (variable by regulation and regulator)
Permitting	Need for altered or additional permits may be perceived overly burdensome on regulated party or a hindrance to implementation progress
Remediation Time	Uncertainties on time allotted for remediation affects costs and potentially routine facility operations
Insurance Coverage	Concern over whether insurance company will consider innovative technology's uncertainties acceptable cost-effective solutions

applications and verifying the effectiveness of biotechnology under a variety of applications and conditions, such perceptions should be reduced. All parties involved in biotechnology selection need to contribute to resolving issues, perceptions, and technical uncertainties inherent in the implementation of some of this technology's application under field-scale conditions.

Those who have been instrumental in developing the technology need to understand and/or accept that the microbiology aspects of this technology are partners with the technical and nontechnical issues that arise in site-specific applications. The vendors selling biotechnology applications need to address the nontechnical and site-specific technical issues from a factual and experienced viewpoint providing reasonable options for solutions to the roadblocks that arise.

Local and state agencies and federal EPA regional offices need to do their part to reduce regulatory roadblocks and implementation hurdles so that the applications of this technology can be cost-effectively tested. The public needs to recognize that using innovative technologies *means* there are unknowns and uncertainties that will need to be worked out during field applications.

The governmental and/or private sector entity with cleanup needs should recognize the technical uncertainties associated with applications of this technology because of limited field-scale experience with site-specific contaminants and diverse subsurface conditions. Legal counsel needs to assess how the risk of uncertainty can be minimized during negotiation of orders, consent decrees, and permits. Insurance companies that evaluate (after remediation is complete) whether biotechnology was the most viable and cost-effective technology choice need to accept that uncertainties existed at time of selection. The client should not be risking insurance coverage when opting to apply this technology onsite. The engineering consultant who is typically a problem solver should identify 1) **reasonably possible** considerations, issues, and **most probable** conditions, 2) methods of reducing negative perceptions, and 3) potential uncertainties focusing on how to reach resolution.

CONCLUSION

As more data are gathered during field-scale implementations, there will be a point where the uncertainties will be reduced sufficiently to have the more innovative applications of this technology move in the direction of proven remediation technology design and construction engineering (see Figure 1). At such a point, predictive capabilities on the potential for successful implementation of the technology to meet cleanup goals should be available. With this ability, the negative perceptions of this technology will ultimately be reduced or eliminated.

The consultant as an unbiased technology evaluator will be watching and participating in the progress of this technology as more field experiences are gained. In the meantime, the consultant will weigh all the considerations (both technical and nontechnical) and provide recommendations including how to reduce the uncertainties and roadblocks associated with this technology. Ultimately, it will be the regulated entity with cleanup needs that will have to make the decision on how much uncertainty and roadblocks represent an unacceptable risk.

BIBLIOGRAPHY

Brown, Stuart M., Lincoln, David R., and Wallace, William A. CH2M HILL, INC. April 1989. "Application of the Observational Method to Remediation of Hazardous Waste Sites."

Brown, Stuart, M., Lincoln, David R., and Wallace, William, A. CH2M HILL, INC. November 1988. "Application of Engineering under Uncertainty to Remediation of Hazardous Waste Sites." Presented at "Superfund 88," A Conference Sponsored by the Hazardous Materials Control Research Institute, Washington, D.C.

Kahn, K.A., Krishman, R., O'Gara, T.F., et al., June 24-29, 1990. "Soil Bioremediation Treatability Studies." Presented at the Annual Meeting of the Air and Waste Management Association, Pittsburg, Pennsylvania.

RCRA Corrective Action Plan. June 1988. Interim Final (OSWER Dir. 9902.3).

Safferman, Steven I., Glaser, John A., and Shelton Daniel, June 24-29, 1990. "Evaluation and Testing of a Protocol to Determine the Aerobic Degradation Potential of Hazardous Waste Constituents in Soil." Presented at the Annual Meeting of the Waste Management Association, Pittsburg, Pennsylvania.

Wallace, William, A. and Lincoln, David, R. CH2M HILL, INC. April 20-21, 1989. "How Scientists Make Decisions about Groundwater and Soil Remediation," in Remediating Groundwater and Soil Contamination. Are Science, Policy, and Public Perception Comparable, National Research Council, Water Science and Technology Board, Washington, D.C.

The Technical, Economic, and Regulatory Future for Bioremediation: An Industry Perspective

A. Keith Kaufman

As recently as five years ago, the word "bioremediation" was rarely, if ever, seen in the literature, let alone understood or appreciated. Back then, biological processes as they relate to waste management fell under the almost exclusive domain of sewage and wastemater treatment. Only a handful of companies existed nationwide which openly advertised bioremediation as an exclusive service offered for soil and/or groundwater cleanup. Finding it difficult to successfully compete in a market dominated by the more traditional and still widely accepted forms of remedial technologies, some of these companies left the market, while others persevered by, temporarily or permanently, aligning themselves with more diversified (albeit traditional) remediation service-oriented companies. Those that chose to stay in the bioremediation field were to face significant technical, regulatory, and public relations challenges. Though many of these challenges remain today, tenacity and perseverance, coupled with a firm conviction of the value and efficacy of the technology, have brought about the evolution of biotreatment to a relatively high level of understanding and respect among environmental and regulatory professionals worldwide.

Today, there are an estimated 200 companies within the United States that offer bioremediation as an exclusive or adjunctive remedial service, and this number is expected to grow substantially over the next five years. To further underscore the level of understanding and acceptance of bioremediation within the environmental industry, requests for proposals are being generated with increasing frequency which specifically designate bioremediation as the cleanup method of choice. Estimates as to the national market size for bioremediation vary from 200 million to one billion dollars. In all probability, this discrepancy could be related to the source of the numbers. Clearly, only a small percentage of sites which could be bioremediated are actually being considered as bioremedial candidates. The lowest estimate, therefore, likely reflects those projects nationwide which are now undergoing bioremediation as well as those which have been approved for bioremedial

---

A. Keith Kaufman • Applied BioTreatment Association, Anaheim, California 92807.

implementation. The highest estimate is more likely reflective of the "perfect world" scenario, wherein all technical, practical, and regulatory obstacles are removed such that all sites which *could* be bioremediated in the United States are, in fact, approved for bioremediation. While the biotreatment industry has come a long way in a relatively short amount of time, pursuit and attainment of this "perfect world" acceptance (if indeed such is possible) will be challenged by significant obstacles yet remaining, some of which could actually threaten the continued existence of the biotreatment industry.

There is little question that, technically, biotreatment technologies are still in their early stages of development. Although, the efficacy of bioremediation for cleanup of petroleum hydrocarbon-impacted soil and water is rarely disputed, its efficiency on some of the more refractory of environmental pollutants such as PCB's and other halogenated compounds remains in debate. Much of the technical concern today relates not so much to whether or not a compound is biodegradable (under appropriate conditions, most are), but rather to the degree to which a compound can actually be mineralized. For example, organisms have been found which can biologically degrade a variety of halogenated compounds. While this fact is not insignificant, it is only a part of the picture. The questions which must be asked for environmental cleanup applicability relate to whether these organisms can degrade such compounds under site-specific conditions to achieve preestablished cleanup levels and without the generation of undesirable end-products. Furthermore, can such a feat be accomplished in a reasonably timely, cost-effective, and safe manner? Such are issues which are receiving widespread attention from both academic and industrial sectors. In fact, each year brings with it increased emphasis on Federal, State, and private funding of research and applied programs which are designed to address these concerns. Recent news headlines associated with biologically-mediated cleanups of Prince William Sound, Alaska and Galveston Bay (Texas) have further heightened technical awareness and the need for more finite answers to questions relating to actual bioremedial processes. The biotreatment industry remains highly supportive of these endeavors, for it is only with answers that the industry can hope to retain and build upon public and regulatory confidence.

Perhaps more than any other parameter, confidence is the key to bioremediation's survival. Since confidence is directly proportional to the technical performance of bioremediation, it stands to reason that the technical competency of the biotreatment service provider will largely influence the survival and prospective growth of the industry. Biotreatment has historically been viewed as a high growth, potentially high profit industry. As such, it has gained the attention of a variety of corporations, investment groups, venture capitalists, and the like who, quite understandably, want to "jump on the bandwagon" in order to acquire a market advantage over companies electing to provide only traditional forms of remedial technologies. Since there have never been overt barriers to entry into the biotreatment industry, even small undercapitalized companies have found out that by just saying "they do" bioremediation, their economic prosperity and outlook are vastly improved. What is all too frequently lacking in these scenarios however, is the requirement for "competency". Although snake-oil salesman are not nearly as frequently encountered as they were in the past (in part due to increased technical sophistication within the client base), they still exert an influence that the biotreatment industry can little afford to tolerate. Perpetuation of myths, promulgation of technical inaccuracies, and making economic and/or performance promises which cannot be kept are examples of end-products which continue to damage the biotreatment industry. Like most technology-based services, competency is assumed. Unlike most technology based services in today's marketplace however, proof of competency in relationship to bioremedial treatment has not heretofore been required. The result, in many cases, has been the undermining of confidence and acceptance of the technology, severely restricting growth potential and, more importantly, the implementation of a technology which could, under the direction of competent personnel, achieve the goals originally desired. As mentioned earlier, this problem is not as prevalent as it once was due to an increasing ability of the client base to discern fact from fiction. Although, it is not the intent of the biotreatment industry to unduly thwart the attempts of those desiring to get into the business of bioremediation, it is industry's belief from both ethical and economic standpoints that there be emplaced standards by which at least a modicum of competency and expertise be required of professionals wishing to practice bioremediation. Such standards will ensure public, private, and regulatory sectors that, at the very least, such personnel and/or companies have the qualifications of providing a service they advertise. By extension, the biotreatment industry could better retain at least a level of professionalism reflected in other remedial industries. In addressing this very serious issue, the Applied BioTreatment Association (Washington, D.C.), is presently drafting guidelines for the development of a national certification program for biotreatment specialists.

One of the most serious threats to the continued viability of the biotreatment industry relates to regulation, particularly at the Federal level. Two years ago, industry representatives were incensed to learn of a pending draft rule under the Toxic Substances Control Act (TSCA) which would have required comprehensive reporting criteria for the prospective use of naturally-occurring microorganisms for environmental bioremediation. Clearly, this set of regulations would place intolerable burdens on a yet economically unhealthy profession, substantially lengthening the time which would be required to gain implementation approval (if in fact, such approval would be ultimately obtained), and costing thousands of more dollars per individual project. This was perceived as a potential death sentence to the industry, since most, if not all of the practical and economic advantages associated with biotreatment would have been negated by its implementation and enforcement. The biotreatment industry was quick to respond to this threat. In fact, the Applied BioTreatment Association was formed as a direct result of this development. Lobbying efforts were mounted throughout Washington, D.C. and, with the support of other professional, trade, and scholarly groups, the inclusion of naturally-occurring microorganisms within the regulated universe appears unlikely.

However traumatic this period of time was for those trying to establish and/or maintain a biotreatment business, it did bring about the unity of a heretofore unconsolidated industry. Furthermore, these events focused the attentions of the public and regulatory sectors on many of the obstacles facing the continued development and utilization of bioremedial technology for pollution control. After all, it appeared on the surface that everyone (e.g., Congress, EPA, the public) wanted to encourage and enhance the development of alternative methods for hazardous waste management. How then, could this be reconciled with the development of prospective regulations which would, in effect, undermine this objective? This contradiction fortunately did not go unnoticed by the EPA. In February of 1990, the first ever symposium on the future of biotechnology for pollution control was hosted by the EPA in Washington, D.C. Leaders from industry, academia, and government were invited for the purpose of identifying barriers to biotechnical development and process utilization and how best to remove such barriers in the short and long term. Although few answers could be expected from a single eight hour meeting, such an effort provided the mechanism for much needed dialogue among regulators, as well as those affected by prospective regulations. It was indeed significant that EPA Administrator William Reilly attended and actively participated in this meeting, underscoring the importance of such dialogue. In the months which have followed this meeting, numerous committees and sub-committees have been formed to address specific concerns previously identified. It is anticipated that another general meeting will be held in the near future where at least some resolutions will be presented. At the very least, the biotreatment industry applauds this effort as a very big step in the right direction and is optimistic that the forward momentum established will continued to grow and yield positive results.

It is all too obvious however, that the road up ahead will not be easily traveled. Whereas a draft TSCA regulation can be modified prior to it becoming law, regulations which are already law are not so easily changed. The biotreatment industry's most pressing challenge presently relates to the Resource Conservation and Recovery Act (RCRA) under which looms the Land Disposal Restrictions (LDR's) and the Best Demonstrated Available Technology (BDAT) criteria. At first glance, the LDR's might appear to promote the utilization of bioremedial technology for site cleanup. After all, a major advantage afforded by such technology relates to its ability to provide remediation for contaminated soils and water without the need for off-site disposal. Upon closer scrutiny however, the regulations would appear to once again negate the practical and economic benefits of biotreatment. In particular, *in situ* bioremedial treatment systems of soil and water which rely on the injection and recirculation of groundwater through soil may, in fact, violate the LDR's, since contaminated groundwater is applied to soil (usually through infiltration galleries or injection wells). Although this method has been effectively utilized for many projects in the past, such a process could be precluded from further utilization by the LDR's. While treatment of groundwater prior to its application to the soil may circumvent the perceived restriction, such action would substantially minimize the benefits of *in situ* bioremediation and would necessitate substantially higher project costs due to the air stripping and/or carbon treatment which would be required for the water prior to its reapplication to soil. The most logical question here is, why bother to utilize biotreatment for water at all if air stripping or other methods are going to be required?

As discussed earlier, the ability of bioremedial technology to reduce concentrations of contaminants to environmentally acceptable levels is an issue of serious concern. If, for example, it can be shown that bioremediation can effectively reduce

1,000 ppm of a contaminant under site-specific conditions to 1.0 ppm in a timely and cost-effective manner, would this be considered "environmentally acceptable", even if a previously performed risk analysis showed that 5.0 ppm was a safe level from the public health standpoint? Under BDAT guidelines, the answer likely would be no, since such criteria are generally guided by minimum achievable concentrations by a technology presently available. In most cases, these concentrations are not health/risk-based, but rather are related to a specific technology capable of reducing concentrations to a greater degree than any other, irrespective of cost, time, and public sentiment (parameters which are not legislatively a part of the RCRA decision tree). Thus, in the above example, a technology such as incineration would be selected since it could reduce the contamination to below 1.0 ppm, even though such technology is substantially more expensive and less generally accepted by the public than bioremediation. For RCRA wastes then, utilization of biotreatment may be severely restricted unless and until new legislation specifically addressing bioremediation is promulgated and adopted.

CONCLUSION

The biotreatment industry can look back on its relatively short history with a true sense of accomplishment. No longer is bioremediation looked upon with such skepticism or trepidation. Rather, it is now viewed as an acceptable and highly desired form of remedial technology, affording users with an economic and environmentally responsible form of site mitigation.

Although the industry has risen above many of the challenges it faced in its infancy, obstacles remain which could seriously jeopardize biotreatment's continued growth and utilization. However significant, these obstacles are not insurmountable and industry retains great optimism that through unified strength, continued education, and inter-discipline cooperation, biotreatment's future will remain bright.

Removing Impediments to the Use of Bioremediation and Other Innovative Technologies

Walter W. Kovalick, Jr.

Increasing the diversity of technologies used to remediate contaminated soils and groundwater is one of the goals of EPA's Office of Solid Waste and Emergency Response (OSWER). While conventional methods of waste remediation, such as stabilization, containment, and incineration, are certainly valid approaches to resolving waste problems, statutory and economic considerations are now, more than ever, encouraging the entire remediation community to consider change in thinking and practice.

OSWER is working on several initiatives to promote the application of innovative technologies at Superfund clean-up sites. Impediments that must be overcome include inhibiting regulations; conservative attitudes in the remediation community that foster fear of risk; and a lack of information on technology performance and costs that reinforces negative attitudes.

Efforts are underway to streamline or revise, where necessary, those regulatory programs that impede the use of innovative technologies. Efforts are underway that will offer incentives to innovative technology users that will dispel many of the conservative attitudes that impede the use of innovative technologies. And, efforts are underway to improve the availability of information on technology performance and costs that will likewise dispel negative attitudes that impede the use of innovative technologies.

THE REMEDIATION MARKETPLACE

One of the forces that dictates an expanded use of innovative technologies at hazardous waste clean-up sites is the large number of sites requiring remediation. These sites represent a large marketplace for new technologies. In many instances, responsible parties, consulting engineers, and others are looking closely at

---

Walter W. Kovalick, Jr. • Technology Innovation Office, Office of Solid Waste and Emergency Response, U.S. Environmental Protection Agency, Washington, D.C. 20460.

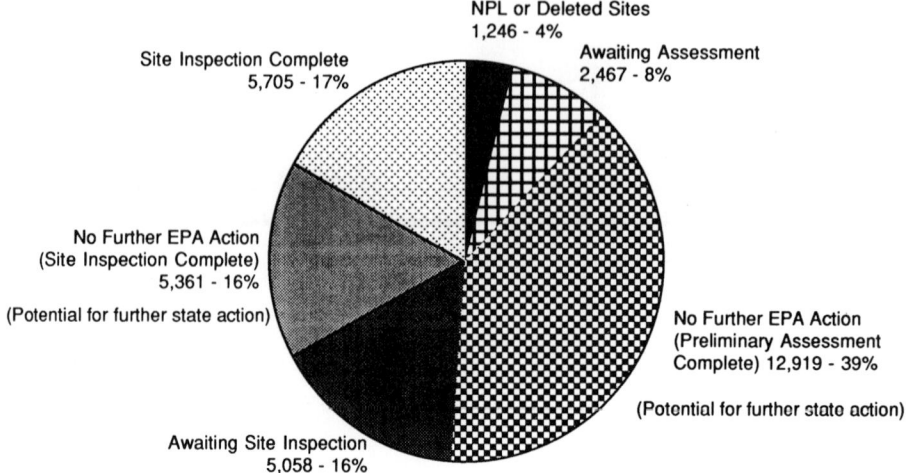

Figure 1. Most sites in the Superfund inventory have been assessed. CERCLIS Inventory: 32,756.

bioremediation as a remedy because of its potential for cost-effectiveness and permanence compared to historical technologies. From a federal standpoint, the job ahead of us is an enormous one.

The Superfund program National Priority List (NPL) includes more than 1,200 sites—those sites judged to be a priority for Federal action. In addition, a number of states have their own lists of sites that may not rank with the nation's worst, but are nonetheless scheduled for remediation because they pose a threat to human health or the environment. (Figure 1 depicts the universe of sites EPA is evaluating for NPL listing. Several significant populations of sites may be acted upon by states in the future). Since many of these sites do not actually qualify as federal Superfund sites, clean-up is the responsibility of state programs and will be financed by state Superfund programs.

In both federal and state Superfund programs, the identified Superfund sites contain a broad mix of chemical substances, contaminated soils, debris, and contaminated groundwater. Our early analysis (Table 1) of NPL sites reveals the types of industry and waste types that contributed to contaminated sites. Sites may fall into more than one category. Further analysis (Fig. 2) reveals the types of treatment remedies selected for the specific problems found at sites for which Records of Decision (RODS) have been signed. Table 2 demonstrates how the use of biotreatment has increased over the last few years—some improvement, but not enough. Figure 3 illustrates the types of contamination addressed by bioremediation.

Additional markets for innovative technologies are the Underground Storage Tanks program and the Resource Conservation and Recovery Act (RCRA) Corrective Action sites program. To illustrate:

- It is estimated that 15 to 20 percent of the five to seven million underground storage tanks in the United States are leaking. Conservative estimates give an average cost figure of $180,000 per site. (Since most of these problem

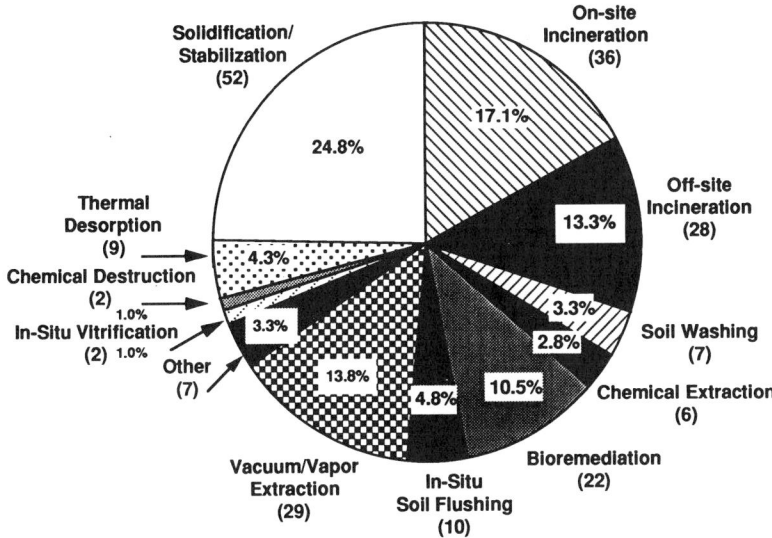

Figure 2. Source control treatment. Fiscal years 1987-1989. Sources include solids, soils, sludges, and liquid wastes; waste sources do not include ground water or waste water. On-site incineration includes sites where the location of incineration is yet to be determined. Figures in parentheses are the number of times a technology was selected; the numbers include actual implementation decisions subsequent to ROD when available.

Table 1. Number of NPL Sites in Each Site Category (Total number of sites with RODs = 465; total number of NPL sites = 1218)

Category	Number of sites with RODs	Total number of sites
Wood preserving	25	60
Battery/lead	8	25
Plating	10	48
PCB	63	156
Petroleum	16	43
Mining waste	18	37
Municipal landfill	42	145
Industrial landfill	124	361
Dioxin	20	30
Volatile organics	237	702
Mixed waste	7	39
Asbestos	8	16
Pesticides	39	114
Others	92	178

Note: Analysis based on information from RODs and NPL site summaries. Each site may fall under more than one category.

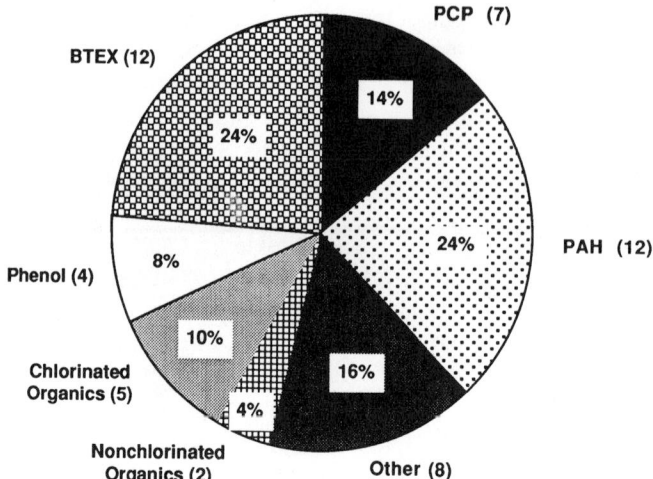

Figure 3. Source control bioremediation sites. Fiscal years 1982-1989 by major contaminant. "Other" includes pesticides, dioxins, and other organics not specified. Number of RODs: 21. Some RODs have more than one major contaminant.

sites involve petroleum-saturated soils, biotreatment is a logical option for remediation.)
- More than 4,700 facilities in the United States treat, store, or dispose of hazardous wastes. Of these, about 3,700 facilities that house approximately 64,000 solid waste management units may need corrective action.
- The Departments of Energy and Defense are responsible for remediation at their own facilities. According to Government Accounting Office testimony, the DOE, alone, has more than 3,500 inactive sites requiring environmental clean-up—the universe of federal facilities represents a large and complex clean-up problem.

Table 2. Selection of Bioremediation for Source Control

FY	Source control RODs	Bioremediation RODs
82-86	48	3 (6%)
87-89	165	18 (11%)

THE RCRA IMPLEMENTATION STUDY

One major impediment to the application of innovative technologies, such as bioremediation, is that posed by regulations. Unlike many other technology oriented programs (e.g. transportation, health), the development of new waste treatment technologies is regulated under the Resource Conservation and Recovery Act (RCRA). RCRA is the statute that controls how solid and hazardous wastes are treated, stored, and disposed of. Regulations, policies, and guidance developed pursuant to RCRA largely determine the types of technologies that will be used to treat hazardous waste.

This year OSWER conducted a broad study of how we implement RCRA. This study, called the RCRA Implementation Study (RIS), recommended many areas in the program that could use streamlining or revision for greater clarity and a more effective regulatory program. It also identified those elements that impede the use of innovative treatment technologies such as permitting priorities and standards for demonstrated and available technologies.

The RIS recognizes that we need to steer the RCRA program in a direction that will encourage innovation in hazardous waste clean-up. The report points out the need to balance prevention and clean-up. Both of these elements are objectives that require considerable resources for implementation. Our current regulatory program is almost entirely dedicated to prevention. The new clean-up—or corrective action—program will address thousands of operating sites that have leaked in the past. Corrective action provides an opportunity to introduce innovative technologies, including biotreatment, to a broad spectrum of contamination problems. Many of the recommendations found in the RIS will help to remove regulatory barriers to utilization of innovative technologies, and several provide policy background for the upcoming reauthorization of RCRA.

Some of the more important recommendations in the RIS that relate to technology development and application are:

- An 8-year action plan to evaluate the universe of corrective action sites should be considered. The plan would take into account site stabilization, development of performance standards for clean-up actions, and national criteria for establishing priorities among facilities.
- Corrective action and permitting should be "decoupled" so that EPA can address the worst contamination first, and not be driven by an automatic schedule for permit review.
- The 1992 Congressionally-mandated deadline for issuing permits should be changed to enable EPA to devote resources to corrective action.
- Correction action plans should clearly specify what is expected of facilities in corrective action, so that owners/operators understand what are their clean-up targets and what technologies are required to reach those targets.
- Experiences from the Superfund program should apply to the RCRA Corrective Action program. (e.g. The issue of whether the RCRA program needs a demonstration program that is separate from the Superfund Innovative Technology Evaluation ((SITE)) Program could be raised.)

Some of the recommendations relate to the permitting process that would foster increased use of innovative technologies include:

- Address the problem that the current Research and Development (R&D) permitting process does not effectively promote innovation by exploring the potential for Testing and Evaluation facilities to reduce the permit workload for regions and vendors;

- Consider centralizing the process of evaluating and issuing R&D permits to streamline the permitting process;
- Elevate the importance of R&D permits by providing credit in EPA management systems for R&D permit issuance;
- Promote additional use of mobile treatment units by using "permit-by-rule";
- Provide additional guidance to permit writers and vendors on how to issue a "Subpart X" (miscellaneous) permit for new technologies; and
- Recognize the potential for innovative technologies in the proposed Best Demonstrated Available Technology (BDAT) rule for Soils and Debris.

Finally, the RIS suggests that the states should play a major role in promoting acceptance of innovative technologies. "Innovative relief" mechanisms that can be adopted by the states include:

- Requesting authority to issue R&D permits;
- Promoting the 1000 kg exemption for treatability studies among states; and
- Encouraging state economic development offices to promote innovative technology firms and their associated service industries as financially beneficial to the state.

Although involving a great amount of "retooling" on the part of EPA as it approaches RCRA reauthorization, the goals outlined above are attainable, and it is a direction that is certainly being pursued.

THE TECHNOLOGY INNOVATION OFFICE

The Technology Innovation Office (TIO) was formed this year in the Office of Solid Waste and Emergency Response for the purpose of stimulating development and application of innovative treatment technologies. The parties that are integral to the acceptance of such technologies—consulting engineers, federal and state project managers, potentially responsible parties, and technology vendors—are closely linked in the day-to-day decision-making related to site remediation (Figure 4). TIO's expressed mission is to remove those impediments that inhibit increased use of innovative technologies.

To accomplish this task, TIO is pursuing several major projects to dissolve those barriers:

- A Market Assessment Project is now underway. This is an effort to profile the remediation market retrospectively and over the next several years. Our objective is to provide developers and investors with information on the worst site problems so that development dollars can be channelled more productively. Site profiles may also help vendors market their technologies to site managers. I expect an initial monograph this winter on Superfund and other federal sites.
- A Technology Incubators project has also begun. This effort will identify the most successful models for facilities that conduct testing and evaluation of new technologies. An attempt is being made to determine if incubators

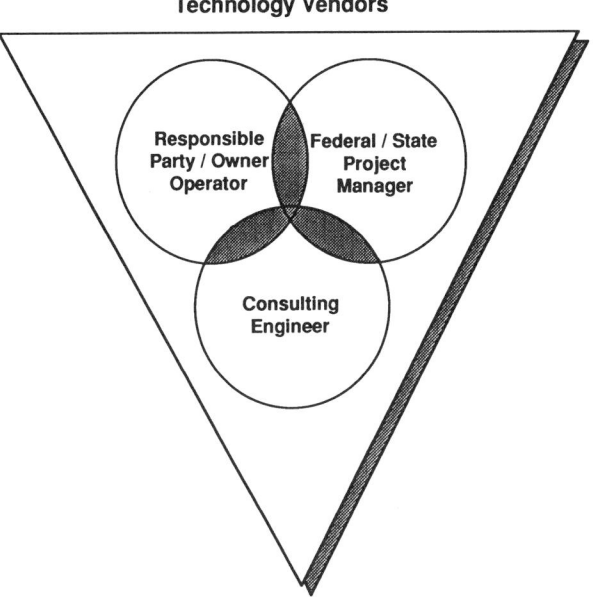

Figure 4. The Technology Innovation Office: Making innovative remediation technologies happen.

provide a valuable service as a third party evaluator of performance — if they streamline the process to develop data and if their use can speed the introduction of new technologies.
- A Vendor Identification Database project has been initiated. This is an effort to compile information from vendors and their clients concerning the most useful performance and cost data. These data will provide an added element of confidence in vendor data and, when provided in a centralized fashion, should reduce the uncertainty consulting engineers, reponsible parties, and project managers have about new technologies.
- An Action Plan for Reducing Barriers to Innovative Technology Use (especially by the engineering community) is being instituted. Information is being sought from consulting engineers concerning impediments unique to their industry. These impediments may include liability and risk, lack of information on performance and cost, and financial barriers. Defining this information will help to formulate new projects and an action agenda with consulting engineers, professional societies, and others.
- Of special interest to those concerned with biotreatment is the two year Bioremediation Field Initiative recently initiated. This is a joint effort between OSWER and the Office of Research and Development (ORD). The program is designed to 1) more fully document performance of full-scale field applications of bioremediation, 2) provide technical assistance for sites in a feasibility or design stage to facilitate the conduct of treatability studies and field pilot studies, and 3) regularly provide the EPA regions with in-

formation on treatability studies, design, and full-scale operations of bioremediation projects in other regions. This program is intended to improve the decision-making and subsequent selection and operation of biological systems and provide current cost and performance data.

This effort will produce a regular bulletin on Bioremediation Field Initiative projects already completed or in progress. The bulletin will provide useful information to the remediation community and can speed the potential application of bioremediation remedies at planned sites.

Because of TIO's special interest in bioremediation, I invite developers of biotreatment technologies to provide us with information on performance and cost. In addition, TIO is interested in hearing your views on impediments for this important technology. Information that we receive will be shared with vendors, federal and contractor project managers, universities, and trade and professional organizations.

INCREASING DIVERSITY/LOWERING COSTS

Conventional methods of waste remediation have been shown as effective means of hazardous waste site clean-up. However, soaring costs of these clean-up actions compel us to expand our thinking and certainly our practices. Innovative technologies appear to be the answer to our problems because they promise permanent solutions at reduced cost.

EPA's Office of Solid Waste and Emergency Response is working to overcome the barriers that impede the use of innovative technologies with programs underway and on the planning table. However, success will not be reached unless all members of the remediation community work together. By exchanging ideas and information, federal and state governments and the private sector can identify problems and find solutions. I am here today to reinforce that message, and let you know that EPA is actively pursuing your support in this challenging arena.

Bioremediation Research Issues

John H. Skinner

BACKGROUND

Bioremediation is one of the most promising technologies on the horizon for more cost-effective cleanup of hazardous wastes. The basic concepts are not new. Bioremedial processes have been recognized and used for many decades. Most notable are the processes involved in wastewater and sewage treatments. Microbiologists have been familiar with biodegradation (organic breakdown) and bioconversion (action on inorganics) for nearly a century. The microbial actions involved in these processes have also been used extensively in the food processing industry—in the production of items such as beer, wine, yogurt, cheese, and sauerkraut—as well as in chemical science.

In spite of this awareness, the transfer of biological treatment technology to the environmental sector for use in soil and groundwater cleanup has been tediously slow. In the past, there was a limited use of innovative remedial techniques. Conventional cleanup methods, such as landfill disposal and incineration, were generally used to manage hazardous wastes. However, with increasing costs associated with traditional technologies, closure of landfills, and more stringent restrictions pertaining to air pollution, the impetus had arrived for developing and utilizing new forms of remedial methods. It was at this point that bioremediation began to receive the serious attention of the environmental community.

Over the past several years, EPA has tested new applications of bioremediation, using it successfully in emergency responses to oil spills and cleaning up Superfund sites. Natural biodegradation of wastes occurs in most environments. However, sometimes the natural rate is so slow that it is not considered an effective cleanup mechanism. Fortunately, there is much we can do to enhance the process. For example, we can enhance the efficiency of the process by adding nutrients to encourage proliferation of microorganisms, or by adding cultured or acclimated microorganisms to the waste materials.

John H. Skinner • Office of Research and Development, U.S. Environmental Protection Agency, Washington, DC 20460.

While bioremediation has many potential advantages in treating hazardous wastes, our experience with these technologies to date has been limited. Questions remain among environmental professionals as to where and how bioremediation fits into the remedial action decision tree. What factors need to be considered in evaluating the appropriateness of bioremediation for a given site? How much will it cost, and how long will it take? What are the required steps in developing and implementing an effective bioremediation cleanup program? The lack of reliable information on bioremediation applications and performance is a major impediment to its use. Research is needed to improve our understanding of bioremediation and how it works. Field demonstrations are also needed to establish reliable performance data and a track record of bioremediation successes. Finally, this information must be widely disseminated to motivate site managers to consider bioremediation along with other, more conventional technologies in planning cleanups.

POTENTIAL ADVANTAGES OF BIOREMEDIATION

The potential advantages of applying biodegradation principles to the cleanup of contaminated sites have been recognized for some time. Benefits include:

- *Reduced cost.* Bioremediation can be much cheaper than other technologies (perhaps 1/3 to 1/2 the cost of incineration).
- *Reduced risk of exposure.* When used *in situ*, bioremediation reduces the risk of exposure during cleanups by avoiding the need for excavation.
- *Reduced residual contamination.* Bioremediation shows promise for further reducing the low levels of contaminants left after excavation of high level contaminants.
- *Minimal environmental impact and liability.* Bioremediation is a natural process that has the potential of degrading toxics to harmless products—carbon dioxide, water, and fatty acids—when the process is completed.

If *in situ* application can be further developed, significant cost savings could be possible, and the risks associated with the cleanup of contaminated sites should be reduced. In addition, *in situ* bioremediation holds promise as an alternative to the sometimes costly and time-consuming practice of pumping and treating contaminated groundwater. Another potential benefit of bioremediation is the long-term prospect of developing improved microorganisms that can be tailored to degrade specific wastes or groups of wastes.

Bioremediation is not, however, a panacea for cleaning up hazardous wastes. It is not suited to all situations. For example, it does not destroy toxic metals. Also, based on our current experience, bioremediation technologies may not be capable of achieving the very high destruction rates obtainable through thermal and chemical treatment. Finally, bioremediation is a slow process, taking from days to months, depending on the wastes, the microorganism(s), and the method of application. Competing treatment processes can usually be completed in less time.

THE ROLE OF RESEARCH

Given the potential benefits of bioremediation, why is it not being used more widely to clean up hazardous wastes? In addition to the limitations noted above, one major factor is a lack of information. Our limited experience with bioremediation results in a lack of information on the actual performance and cost of bioremediation technologies. This information is essential to those responsible for selecting cleanup technologies—to enable them to evaluate bioremediation along with other methods, to assess its applicability to specific wastes, to compare the effectiveness of various methods, and to compare their costs. Information on performance is also needed to instill confidence in bioremediation technologies. The potential liabilities associated with unproven technologies are substantial. For site managers and contractors to select bioremediation instead of more proven methods, they must be confident that bioremediation will be as good as or better than other available approaches.

Research can play a major role in providing this important information. Research programs can develop treatability protocols to assist in the decisionmaking process. Field demonstrations can provide information on the applicability and effectiveness of specific bioremediation technologies, and can begin to build a track record of successes for these technologies. Research on the basic science of bioremediation can contribute to better understanding of biodegradation processes, methods of enhancing these processes, and the byproducts of biodegradation.

EPA is actively pursuing research in all of these areas. To realize the full potential of bioremediation, we must:

- Learn to address complex mixtures at sites—most applications thus far have been on relatively simple waste and siting combinations.
- Learn to handle difficult physical forms of waste such as contaminated debris and cobbles.
- Understand the impact of varying environmental conditions and nutrient availability on effectiveness of bioremediation of various wastes by differing microbes.
- Develop the capability to monitor and control *in situ* applications.

The goal of EPA's research program is to develop new scientific knowledge to take bioremediation from its current stage to routine application. Some examples of our current bioremediation research activities are described in the next section along with the process that EPA has established to identify bioremediation research needs and develop an agenda for action for the 1990s. Our plans for future research initiatives are presented in the final section.

CURRENT RESEARCH PROJECTS

Demonstrations at Superfund Sites

EPA has selected bioremediation as a primary cleanup technology at 25 Superfund sites to date, but few of these have progressed to the point where results can be reported. The results of two demonstrations are discussed below.

In one example, at the Traverse City Coast Guard Station in Michigan, EPA conducted a successful *in situ* bioremediation demonstration. A groundwater plume was contaminated with benzene, toluene, and xylenes as a result of a 95,000 liter spill of aviation fuel. Hydrogen peroxide was added as an oxygen source to stimulate growth of indigenous microorganisms. As a result, the groundwater in the plume was brought within U.S. drinking water standards within six months (Wilson, et al., 1990).

In another example, at the French Limited site in Texas, petrochemical wastes had been disposed of in a seven acre pit for a number of years. The sludges contained 10 different metals and over two dozen organics, many of them volatile. The remedy chosen for the contaminated lagoon was *in situ* bioremediation. One end of the lagoon was diked off to test the concept. First, the liquid was treated with air spargers to encourage aerobic biodegradation. Nutrients were added and centrifugal pumps were used to emulsify the sludge. Then the subsoil was mixed in with a hydraulic dredge.

At the French Limited site, the demonstration process took 120 days to treat a small portion of the contaminated lagoon. In the sludge, total volatiles were reduced from 3,400 ppm to 150 ppm, benzene concentrations from 300 ppm to 12 ppm, vinyl chloride from 600 ppm to 17 ppm. The sludge volume was reduced by 85%. The remaining sludge will be chemically stabilized and left in place. This demonstration illustrated not only the effectiveness of the technology, but also demonstrated the attractiveness of bioremediation from a cost standpoint—costs for bioremediation were projected to be $47 million compared to $63 to 167 million for the other options evaluated. However, current estimates that include groundwater treatment are approximately $75 million (Clark, 1987).

The Alaska Bioremediation Research Project

A widely publicized application of bioremediation to oil spill cleanup is EPA's research project in Prince William Sound, Alaska. In March of 1989, the supertanker *Exxon Valdez* ran aground on Bligh Reef in Prince William Sound, flooding one of the nation's most pristine and sensitive environments in less than five hours with approximately 11 million gallons of crude oil. The spilled oil affected an estimated 900 miles of shoreline in the Sound. The Alaska Bioremediation Research Project was initiated in the aftermath of this oil spill to evaluate the feasibility of using bioremediation to assist in cleanup operations.

The Alaska Bioremediation Research Project has been a cooperative research effort involving scientists from EPA, Exxon, the Alaska Department of Environmental Conservation, and the University of Alaska in Fairbanks. In 1989, the project objective was to demonstrate the feasibility of augmenting, in an environmentally safe manner, shoreline cleanup by accelerating natural microbial oil degradation processes through the application of fertilizers. Based on the success of the 1989 and 1990 research, more than 100 miles of Alaskan coastline have been successfully treated with bioremediation by the Exxon Company, USA.

Results of field activities in 1989 (Pritchard, et al., 1990) were:

- Visual inspection of beaches treated with the fertilizer showed that within three weeks following fertilizer application, considerably less oil was observed on the rock surfaces than in the untreated control beaches. This condition became more pronounced with time and remained visually apparent through the end of the summer season (five weeks).
- No oil slicks were observed in the near-shore seawater following application of the inorganic fertilizer, indicating oil was not released.
- Samples of the oil taken from the beach surfaces when the oil was visually beginning to disappear showed changes in composition indicating extensive biodegradation.
- Visual disappearance of the oil as a result of inorganic nutrient application can only be attributed to enhanced biodegradation.
- Addition of fertilizer to oiled shorelines did not cause any increases in planktonic algae or bacteria or any measurable nutrient accumulation in adjacent areas.
- No oil was detected in tissues taken from mussels that had been placed (in flow-through plastic containers) just offshore of the fertilizer-treated beaches; the bioremediation treatment did not disperse the oil, but kept it available for biodegradation.
- Samples taken from the bioremediated beaches showed that as the oil was biodegraded, the slight mutagenicity of fresh Prudhoe Bay crude was eliminated. Thus, no mutagenic byproducts result from enhanced biodegradation.

In 1990, the project is addressing two additional needs. First, the project is seeking to advance our understanding of the science of bioremediation as it applies to oil spill cleanup, particularly to subsurface cleanup. Second, a range of commercial products are being assessed for their applicability to shoreline cleanup.

To date, the 1990 results are very encouraging. A single application of fertilizer has been shown to increase the rate of oil biodegradation by two to three times over the rate of an untreated shoreline. This accelerated rate has been sustained for several weeks, even after nutrient concentrations return to background levels. We can now estimate the rate at which nutrient-enhanced bioremediation occurs. Data from our test plots show enhanced oil degradation proceeding at daily rates of approximately 10 mg of oil per kg of beach sediment. This advancement represents an important step forward in oil spill bioremediation since it will permit us to compare more precisely the results from different fertilizer treatments and different environments. Such comparisons will allow a greater degree of process "fine-tuning", leading to more effective use of this treatment tool. The most recent field research on risks examined the survival of mysids (shrimp-like crustacea) which were maintained for 96 hours in seawater taken from treated and untreated segments of shoreline. These tests indicated that there were no significant differences in survival.

EPA also plans to continue research on the application of bioremediation to oil spills, concentrating on beach bioremediation as an outgrowth of the Alaska project. EPA is evaluating the ability of several commercial products to enhance bioremediation in Alaska. To identify candidate products, EPA enlisted the assistance of the National Environmental Technology Applications Center (NETAC), a cooperative venture of EPA and the University of Pittsburgh. NETAC assembled a Bioremediation Products Evaluation Panel to evaluate proposals that were submitted by product developers. The panel recommended 11 products—two nutrients, one dispersant, eight microbial cultures—for testing. Ten vendors supplied products for testing in EPA's laboratories. The Panel then evaluated the results of these tests and recommended products for further demonstration. Of the 10 products, three were forms of fertilizer while the rest include specialized naturally occurring microorganisms. From this work EPA selected the two most promising for field studies on weathered crude from the Valdez accident in Alaska. Both products contain naturally-occurring microbes. Data on oil chemistry degradation, microbial population counts, and microbial activity will be available late in October 1990. As a followup over time, EPA intends to perfect its screening protocols for routine use in evaluating the effectiveness of bioremediation products. Later, similar screening protocols will be developed for oil spill cleanup in marshes. Also, EPA will investigate the potential for application of bioremediation for oil on open waters.

EPA'S BIOREMEDIATION ACTION COMMITTEE

The success of the Alaska Bioremediation Research Project sparked interest within EPA, state and local governments, and others involved in cleaning up hazardous waste sites and oil spills. In response to this interest, EPA conducted a meeting on the Environmental Applications of Biotechnology on February 22, 1990. The purpose of the meeting was to recommend a "biotechnology agenda for action for the 1990s" to guide EPA, other government agencies, industry, and academia in collaborative efforts to expand the application of biotechnology.

As a follow up to this initial meeting, EPA established a process that will continue to provide guidance on biotechnology research and implementation issues. A joint federal agency, academic, and industry committee, called the Bioremediation Action Committee, was created to implement the actions from the February meeting and to guide future research. On June 20, 1990, the first meeting of this committee was held to discuss the recommendations from the February meeting. The group established four Subcommittees to address different issues. Each of Subcommittees include representatives from industry, government, and academia. These Subcommittees are:

- *Research and Education Subcommittee.* This Subcommittee will review federal, state, and university research on bioremediation to determine the current status of research in this area, and to recommend appropriate actions to expand and coordinate this research. This Subcommittee will also identify existing or needed education efforts.
- *Data Identification and Collection Subcommittee.* This Subcommittee will work with states and industry to collect biotechnology field demonstration data and make it available in a centralized data base source, the EPA supported Alternative Treatment Technology Information Center (ATTIC). The

National Governors Association and Applied BioTreatment Association will coordinate collection of the data.
- *Treatability Protocol Subcommittee.* This Subcommittee will provide technical input into the development of treatability protocols for testing the applicability and effectiveness of bioremediation as a cleanup technology at specific sites. The protocols currently under consideration address both aerobic and anerobic treatment of soils and liquids containing hazardous contaminants.
- *National Bioremediation Spill Response Subcommittee.* This Subcommittee will review the concept of establishing a National Bioremediation Spill Response Plan to address oil spills. The concept—advanced by the Applied BioTreatment Association—proposes to characterize a wide range of potential oil spill scenarios and the effectiveness of various microbial products for addressing them. This would provide the basis for rapid decision-making by oil spill response authorities and rapid development of a bioremediation response.

EPA RESEARCH INITIATIVES

Routine application of bioremediation will require improved knowledge of the effectiveness of different techniques in varied environmental settings. This knowledge will allow us to optimize field applications and be able to predict effectiveness with some confidence. For instance, much remains to be learned about how to effectively apply microorganisms and nutrients in the subsurface environment, how to ensure good contact with contaminants, and how to monitor bioremediation progress *in situ*. EPA intends to significantly expand its bioremediation research program to develop the necessary information and technology in these areas.

EPA established a formal biotechnology research program in Fiscal Year 1985. The overall purpose and scope of these early research efforts in biotechnology were to provide a basis for estimating the risk of biotechnology products, such as microbial pesticides, to the environment and public health.

Since FY 1985, EPA has broadened the scope of its biotechnology research efforts to include pollution control methods and bioremediation. The Agency's current research program is being conducted cooperatively by five EPA laboratories and it addresses four research areas including: engineering science, microbiology/biology, measurement/analytical methods, and toxicity of organisms. The research program also has a demonstration component and a technology transfer/information dissemination component.

Engineering Science

The ultimate success of bioremediation depends on microorganisms staying in close physical contact with the substance to be degraded. Some methods do not provide such contact and this has reduced the efficiency of treatment. The polluted surface may have different abilities to hold microorganisms. Pretreating the waste with some additional natural element, such as oxygen, may provide a more hospitable environment for the microorganism, assuring maximum viability and effectiveness.

An ongoing EPA research project is seeking to determine the ability of granular activated carbon to hold organic contaminants under a variety of environmental conditions. This project stems from a discovery that, in the presence of dissolved molecular oxygen, the activated carbon adsorbs greater quantities of organic compounds then in the absence of molecular oxygen. Complete knowledge of this interaction will contribute to better design of waste treatment facilities (Vidic, et al., 1990).

Microbiology/Biology

In the area of microbiology, research is examining, at a cellular level, the process of breaking down complex substances. This will require study of the metabolism, genetics, and biochemistry of different species of microorganisms. For example, branched-chain aromatic hydrocarbons are generally more difficult to degrade than straight-chains. Also, degradation is often incomplete. In such situations, the end products are not solely carbon dioxide and water, but a hydrocarbon that is less complex than the starting material. Are these hydrocarbons benign or are they creating additional environmental problems? How do species vary in their completeness of degradation? Information concerning the ability of selected species to break down specific waste will be used to guide species selection for a particular site.

Recent EPA studies have resulted in the isolation of bacterial cultures capable of degrading chlorinated hydrocarbons, providing both carbon and energy for organism growth. With a more thorough understanding of these unique abilities, it is reasonable to expect the development of biological systems capable of destroying particularly problematic pollutants, such as complex hydrocarbons, that have not previously responded to bioremediation. For bioremediation to be recognized as a legitimate tool for hazardous waste treatment, progress must be made in degrading the more complex compounds.

Measurement/Analytical Methods

Measurement research is an important component of bioremediation investigation. To determine the best site and microorganism combination, the ecology of the polluted site must be characterized. Description of the subsurface, especially with respect to its ability to sustain microbial activity and the biodegradation reaction, is receiving growing attention. Especially important will be the development of improved sampling procedures that are capable of taking subsurface samples, and accurately evaluating the biodegradation progress in these samples.

Fundamental to the assessment of biodegradability and biotreatability is the ability to distinguish the biological from the physical processes. To bring order to this area, EPA is planning a program to evaluate the accuracy, sensitivity, and reproducibility of commercially-available instrumentation. Standard systems, whose reaction rates can be independently established, will be used to validate our ability to rapidly and accurately monitor bioremediation efforts.

Toxicity of Organisms

To assure the public regarding the safety of bioremediation, it will be necessary to test the microorganisms and breakdown products for toxicity to humans and the ecosystem. A number of high priority compounds for disposal by bioremediation are known to be carcinogenic and many are also mutagenic. Some of these compounds are procarcinogens; that is, they must first be metabolized to the carcinogenic form before they will have their genotoxic effect. Determination of these toxic and carcinogenic properties requires the development of dependable testing methods.

Previous research has resulted in the development of short-term bioassays that can monitor for organic pollutants and their genotoxic byproducts within actual complex field samples. EPA intends to take advantage of these assays to measure the degradation products of chemicals, and the decrease in mutagenicity that occurs during biodegradation.

Demonstration

To build confidence in bioremediation among decision makers and the general public, more credible field experience is needed. There are two facets to this field demonstration program. First, EPA will identify cleanup projects that appear amenable to bioremediation and will then attempt to design and implement bioremediation approaches. Some of these projects may involve modifications of already planned pump-and-treat groundwater cleanup remedies.

The second facet of the field demonstration program will involve refocusing of the ongoing Superfund Innovative Technology Evaluation (SITE) program to include more innovative bioremediation products or systems developed by the private sector. Currently, the program includes five reactor based techniques and an *in situ* process using acclimated microorganisms and nutrients. Also, the SITE program may demonstrate several bioremediation systems developed in EPA laboratories. One promising organism is white rot fungus (*Phenerocheate chrysosporium*) which has been used successfully at pilot scale in a rotating biological contractor. White rot fungus generates 17 enzymes that are capable of degrading a wide range of organic compounds. In a pilot-scale field test, the fungus was added through inoculated wood chips to test plots contaminated with a pentachlorophenol based wood preservative. After 146 days the pentachlorophenol concentration was reduced 82-86% (Glaser, 1990).

The SITE program's Emerging Technologies Program has developed an artificial wetlands approach to removing metals from acid mine drainage. The mine drainage filters through the root zone of marsh plants where bacteria cause the metals to precipitate in a highly insoluble sulfide form. Zinc, cadmium, copper, and iron are readily removed; manganese is not (Haz TECH News, 1990). The process will be used at a Superfund site in Colorado.

Technology Transfer/Information Dissemination

Results of this research must be made available to potential users of bioremediation technologies. Therefore, a key component of this program is information dissemination. Information from tests, demonstrations, and actual cleanups needs to be compiled, reviewed, and distributed. A clearinghouse is needed to provide a centralized resource for all potential users of bioremediation technologies. EPA is planning to expand our Alternative Treatment Technology Information Center (ATTIC) to incorporate this information. ATTIC is an on-line, keyword searchable database that includes information on innovative hazardous waste treatment technologies. It currently includes performance and cost data on technologies that have been demonstrated in the Superfund Innovative Technology Evaluation (SITE) program or addressed by Superfund treatability studies. ATTIC could easily be expanded to include a file on bioremediation technologies as well.

CONCLUSION

EPA is optimistic about the promise of bioremediation in helping us to address some of our most difficult environmental problems. We believe that sound science is the foundation upon which the future of bioremediation resides. We are committed to conducting the necessary scientific and engineering research that is needed to enable bioremediation to fulfill this promise.

REFERENCES

Clark, H., 1987, In Situ Biodegradation Demonstration Report, Volume 1-Executive Summary, French Limited Site, Document No. 275-21.

Glaser, J. A., 1990, White Rot Fungus Applications to Hazardous Waste Treatment, USEPA Risk Reduction Engineering Laboratory, Cincinnati, Ohio.

Haz TECH News, Colorado Artificial Wetlands Project Receives Engineering Excellence Award *Haz TECH News*, p. 74, May 17, 1990.

Pritchard, P.H., Araujo, R., Clark, J.R., Claxton, L.D., Coffin, R.B., Costa, C.F., Glaser, J.A., Haines, J.R., Heggem, D.T., Kremer, F.V., McCutcheon, S.C., Rogers, J.E., and Venosa, A.D., 1990, Interim Report: Oil Spill Bioremediation Project (February 28, 1990).

Vidic, R.D., Suidan, M.T., Sorial, G.A., and Brenner, R.C., 1990, Effect of Oxygen on Adsorptive Capacity and Extraction Efficiency of GAC for Three Ortho-Substituted Phenols, prepared for presentation at the 1990 Annual American Institute of Chemical Engineering Meeting and for submission to the *Journal of Hazardous Materials* (October 5, 1990).

Wilson, J., Leach, L., Michalowski, J., Vandergrift, S., and Callaway, R., 1990, In Situ Reclamation of Spills from Underground Storage Tanks: New Approaches for Site Characterization, Project Design, and Evaluation of Performance, Proceedings: Environmental Research Conference on Groundwater Quality and Waste Disposal.

Evaluation of Bioremediation in a Coal-Coking Waste Lagoon

Maureen E. Leavitt, Duane A. Graves, and Craig A. Lang

INTRODUCTION

Bioremediation is a new and relatively unproven alternative for the destruction of complex organic wastes. While bioremediation of relatively simple hydrocarbons, such as the constituents of gasoline and diesel fuel has been documented for soil and ground water systems, complex molecules such as polyaromatic hydrocarbons (PAHs) represent a considerable challenge for biodegradation. Furthermore, the chemistry relative to the sorption/desorption of these compounds from coal and weathered solids is not understood. Bioremediation of this class of compounds, particularly in atypical matrices, is currently the focus of exploration within the bioremediation industry.

The present study describes initial efforts to demonstrate bioremediation of coal-coking byproducts. Remediation of the subject former coal-coking site is currently regulated under Superfund. During the Remedial Investigation/Feasibility Study, the coal-coking waste lagoons, consisting of coke byproducts and slag concentrated in PAHs, were considered to be candidates for bioremediation.

Bench-scale and pilot-scale demonstrations were designed and conducted to provide site-specific data to determine if bioremediation could effectively reduce the concentrations of carcinogenic PAHs. The results of this phase and the current status of the project are reported herein.

BENCH-SCALE BIOTREATABILITY STUDY

Approach

The purpose of the biotreatability study was to document biodegradation of PAHs in lagoon solids and water using indigenous microorganisms. A computerized respirometer was used to assess the metabolic activity of microorganisms by measuring the consumption of oxygen. Since the primary carbon sources in the solids and ground water were the contaminants, oxygen consumption was an indication of microbial transformation of these compounds.

Table 1 summarizes the treatment scheme employed in the respirometer study. Treatment 1 evaluated biodegradation of organics in nutrient-supplemented ground water alone. Treatment 2 was a biologically-inhibited control for Treatment 1. Treatment 3 contained a 50:1 (water:solids) mixture. This dilute treatment was used to determine if the relatively high concentration of PAHs in the lagoon solids adversely affected microbial growth and metabolism. Treatment 4 examined a 10:1 (water:solids) slurry. Treatment 5 was a biologically-inhibited control for treatment 4. Treatment 6 was an untreated control consisting of a 10:1 mixture of water and solids, respectively. This treatment was placed in a respirometer vessel, sealed, and incubated at 25°C. No nutrients were added and the contents were not stirred. This sample represented any changes caused by experimental manipulation.

All treatments were vigorously stirred using a magnetic stirrer, except Treatment 6, the untreated control. All treatments except the biologically-inhibited controls were conducted in duplicate. Nutrient concentrations, pH, microbial enumerations, and contaminants were analyzed at the beginning and end of the study. Contaminant analyses included total petroleum hydrocarbons, total dissolved organic carbon, volatile organic compounds, and base-neutrals and acid extractable compounds. The study was conducted for 500 hours.

Results

Analysis of untreated samples at the beginning of the study indicated that: a) the pH ranged between 7.1 and 7.6, b) nutrients were already present in significant concentrations, and c) a viable bacterial population existed (Table 2 and 3). The target contaminants, PAHs were present in hundreds of parts per million for individual species, totaling more than 5,000 ppm (Table 4). Total dissolved carbon in filtered water before treatment was 99 ppm (Table 5). Initial total petroleum hydrocarbon concentrations were less than 2 ppm in water and 1900 ppm in the solids (Table 6).

Oxygen consumption was monitored at regular intervals throughout the study. Figures 1 a and b illustrate the cumulative oxygen consumption versus time. Treatment 1 (ground water) exhibited immediate oxygen consumption that ceased after 40 hours. Treatments 3 and 4 (50:1 and 10:1, respectively) experienced a minor lag phase and consumed oxygen at a steady rate throughout the study. Both treatments exhibited short plateaus that were subsequently terminated by nutrient addition (see arrows, Figure 1).

Oxygen consumption in the inhibited controls is shown in Figure 1b. The water-only control (treatment 2) exhibited no oxygen consumption throughout the study. However, the 10:1 slurry (treatment 5) consumed oxygen at nearly ten times the rate in the live samples after an 80 hour lag phase. This sample also reached a plateau that was terminated after nutrient addition. The lack of inhibition could

Table 1. Experimental Design for Coal-Coking Waste Biotreatability Study

	Contents			Initial nutrients			Nutrient addition		
Treatment[a]	Water (mL)	Soil (g)	Inhibitor[b]	Conc[c] (ppm)	Time (h)	Conc (ppm)	Time (h)	Conc (ppm)	Time (h)
1-Water	500	0	–	200	0	500	212	500	336
2-Water, abiotic	500	0	$HgCl_2$	200	0	500	212	500	336
3-50:1[d]	500	10	–	200	0	500	189	500	336
4-10:1	500	50	–	200	0	500	212	500	336
5-10:1, abiotic	500	50	$HgCl_2$	200	0	500	212	500	336
6-10:1, untreated	500	50	–	0	0	–	–	–	–

[a]Data represent the mean of duplicate samples in Treatments 1, 3, 4, and 6.
[b]Saturated $HgCl_2$ was added to treatments 2 and 5 to inhibit biological oxygen consumption. The final $HgCl_2$ concentration was 100 gmL^{-1} (ppm).
[c]A sterile solution of Restore brand nutrient was used to provide nutrient amendments.
[d]Water:soil ratio.

Table 2. Physical and Chemical Solids and Water Parameters before and after Treatment

Treatment[a]	pH	Ammonia µg · mL^{-1} (ppm)	Phosphate µg · mL^{-1} (ppm)
Initial soil[b]	7.1	46	71
Initial water	7.6	106	< DL[c]
1-Water	9.3	231	151
2-Water, abiotic	8.7	229	144
3-50:1	9.0	216	144
4-10:1	8.3 (6.3)[d]	174 (55)	124 (931)
5-10:1, abiotic	7.3	154 (30)	195 (302)
6-10:1, untreated	6.7 (6.8)	145 (63)	< DL (39)

[a]Data represent the mean of duplicate samples in Treatments 1, 3, 4, and 6.
[b]Initial solids and water measurements were conducted after equilibration as a 10:1 slurry.
[c]DL, detection limit was 0.5 ppm for phosphate.
[d]Values in parenthesis are for solids recovered from the treatments at the end of the study. For ammonia and phosphate the data are given us gg^{-1} dry wt. (ppm). Insufficient solids were available for analysis in those treatments where solids data are absent.

Table 3. Microbial Population Size in Soil and Water from the Laboratory Composite and following Treatment

Treatment[a]	Carbon utilization mode	CFU[b] · mL^{-1} (x 10^6)
Initial	Heterotrophs	1.3
Composite	BTX degraders	1.2
1 (water)[d]	Heterotrophs	0.73
Water	BTX degraders	0.24
2 (water)	Heterotrophs	0
Water, abiotic	BTX degraders	0
3 (water)	Heterotrophs	1.5
50:1	BTX degraders	3.9
4 (water)	Heterotrophs	8.7
10:1	BTX degraders	2.0
4 (soil)	Heterotrophs	16
10:1	BTX degraders	3.3
5 (water)	Heterotrophs	69
10:1, abiotic	BTX degraders	17
5 (soil)	Heterotrophs	41
10:1, abiotic	BTX degraders	18
6 (water)	Heterotrophs	3.4
10:1, untreated	BTX degrader	0.31
6 (soil)	Heterotrophs	47
10:1, untreated	BTX degraders	4.4

[a]Data represent the mean of duplicate samples in Treatments 1, 3, 4, and 6.
[b]CFU, colony-forming units.
[c]NA, not applicable.
[d]Matrix evaluated (soil or water).
[e]Water:soil ratio.

Table 4. Quantitative Analysis of Priority Pollutants from Water and Solids Derived from the Laboratory Sample Composite

Method of analysis	Compound	Initial concentration Water $\mu g \cdot L^{-1}$	Soil $\mu g \cdot kg^{-1}$
VOA[a]	Benzene	< DL[b]	98
VOA	Ethylbenzene	< DL	46
VOA	Toluene	< DL	81
BNA[c]	Acenaphthene	22	92000
BNA	Acenaphthylene	13	180000
BNA	Anthracene	14	280000
BNA	Benzo(a)anthracene	14	350000
BNA	Benzo(b)fluoranthene	12	240000
BNA	Benzo(k)fluoranthene	17	340000
BNA	Benzo(a)pyrene	14	300000
BNA	Benzo(g,h,i)perylene	10	66000
BNA	Chrysene	12	290000
BNA	Fluoranthene	57	990000
BNA	Fluorene	33	270000
BNA	Indeno(1,2,3-cd)pyrene	DL	170000
BNA	Naphthalene	39	600000
BNA	Phenanthrene	72	1200000
BNA	Pyrene	42	77000

[a] VOA, volatile organic analysis.
[b] < DL, detection limit for VOA, 5 ppb, for BNA, 20 ppb.
[c] BNA, base neutrals and acid extractables. All other compounds were below the detection limit.

Table 5. Total Dissolved Organic Carbon in Filtered Water from the Laboratory Composite and Treated Samples

Treatment[a]	Total dissolved carbon ($\mu g \cdot mL^{-1}$)	Total dissolved inorganic carbon ($\mu g \cdot mL^{-1}$)	Total dissolved organic carbon ($\mu g \cdot mL^{-1}$)
Initial composite	99	0	99
1-Water	27	21	5
2-Water, abiotic	13	5	8
3-50:1	20	10	10
4-10:1	56	37	19
5-10:1, abiotic	109	81	28
6-10:1, untreated	88	75	13

[a] Data represent the mean of duplicate samples in Treatments 1, 3, 4, and 6.

Table 6. Total Petroleum Hydrocarbon Analysis
(TPH) of the Untreated Solids, Water, and
Water/Solids Slurries

Treatment[a]	TPH[b] (ppm)[c]	Calculated initial hydrocarbon concentration[d] (ppm)	Calculated hydrocarbon remaining[d] (ppm)
Initial water	< 2.0	NA[e]	NA
Initial soil	1920	NA	NA
1-Water	< 3.5	4.3	< 3.5
2-Water, abiotic	4.3	4.3	4.3
3-50:1	< 3.5	~11.5	<7
4-10:1	7.9	~58	~47
5-10:1, abiotic	9.8	~58	~29
6-10:1, untreated	6.1	~58	~37

[a]Data represent the mean of duplicate samples in Treatments 1, 3, 4, and 6.
[b]Total recoverable petroleum hydrocarbon analysis.
[c]ppm, μg/g for dry soils and mg/L for water samples. Water content of the soil was 41%.
[d]Calculations of the initial and final concentration of petroleum hydrocarbons were performed based on a water content of 41% and the efficiency of solids removal during preparation of the samples for analysis.
[e]NA, not applicable.

Table 7. Initial and Final Concentration of Priority Pollutants in 10:1 Water:Solids Treatments

Compound	Initial conc. mass (ng)	Final conc. (Tmt 4) mass (ng)	Final conc. (Tmt 5) mass (ng)	Final conc. (Tmt 6) mass (ng)
Benzene	2891	37	8	74
Ethylbenzene	1357	7	8	74
Toluene	2389	37	8	74
Acenaphthene	2725000	1239	144	2950
Acenaphthylene	5316500	9440	1164	9588
Anthracene	8267000	5531	781	7375
Benzo(a)anthracene	10332000	16225	1196	11948
Benzo(b)fluoranthene	7086000	12206	1100	16078
Benzo(k)fluoranthene	10038500	9366	1260	13423
Benzo(a)pyrene	8857000	12132	1435	11726
Benzo(g,h,i)perylene	1952000	8739	1068	8555
Chrysene	8561000	11616	1084	9956
Fluoranthene	29233500	33188	3508	36875
Fluorene	7981500	1881	287	6933
Indeno(1,2,3-cd)pyrene	5015000	6111	909	7375
Naphthalene	17719500	7633	1021	29500
Phenanthrene	34536000	9698	1021	28763
Pyrene	2292500	22863	2870	23600

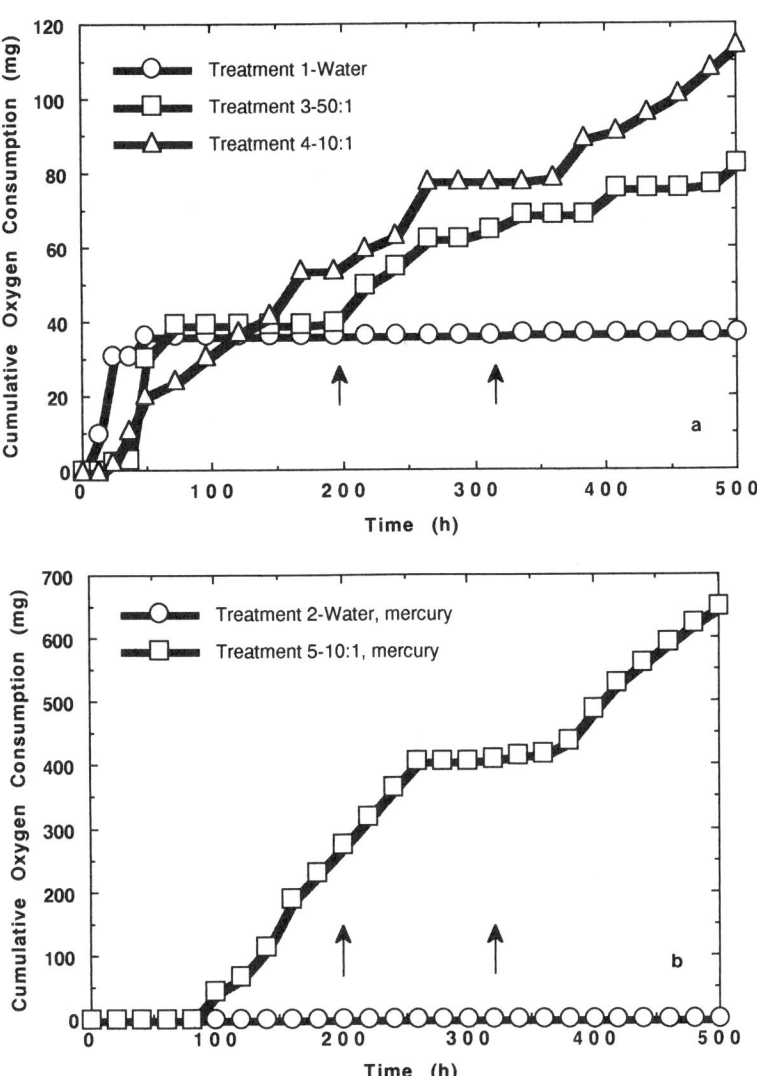

Figure 1. Cumulative oxygen consumption during biodegradation of coal-coking waste. Arrows indicate points of addition of 500 ppm Restore brand microbial nutrient to amend the nutrient content of the treatments. a) Biologically-active treatments, b) Biologically-inhibited treatments.

have been caused by mercury binding or other mercury inactivation due to the solids present. Alternatively, a mercury-resistant population may have developed during the lag phase and proliferated as the sole active population.

Contaminant analyses by gas chromatography/mass spectrometry of base neutral and acid extractable compounds are listed in Table 7. A reduction in PAHs occurred in all treatments, including the slurry abiotic control and the untreated control. Generally, there was a greater reduction in the treated samples than the untreated sample for compounds that are readily biodegraded.

Conclusions

The conclusions drawn from this elementary study were that a viable hydrocarbon-degrading bacterial population existed, and that nutrients and oxygen would be required to accelerate the rate of biodegradation.

PILOT-SCALE DEMONSTRATION

Approach

The purpose of the pilot-scale demonstration was to simulate bioremediation of a full-scale system, and provide data on the reduction of contamination within a six month period. The resulting data will be used to determine if bioremediation should be included in the Record of Decision for this site.

Two tank systems were designed and implemented. One system addressed the saturated zone, simulating a conventional aquifer remediation system. The second system simulated a potential application for remediation of the unsaturated (vadose) zone. Each tank was approximately ten feet in diameter, six feet high, holding four to five feet of solids from the lagoon. Schematic diagrams of each system (Figure 2) illustrate the nutrient and oxygen delivery mechanisms.

The saturated system, labelled the "X" tank, consisted of three injection wells in the center of the tank to deliver nutrients and oxygen as hydrogen peroxide. Water injected through these wells flowed outward toward the rim of the tank and was collected by two recovery wells. A geotextile fabric provided a barrier between the soil and the tank wall, forming a reservoir so that water could be pulled evenly from around the tank. Water was collected, augmented with necessary nutrients or hydrogen peroxide, and recirculated.

The unsaturated system, labelled the "Y" tank, consisted of a surface sprinkler, delivering nutrient-augmented water to the surface solids. Oxygen enrichment was provided by the aerosol sprinkling mechanism. Water then percolated from the surface through approximately three feet of unsaturated solids, to a two foot thick saturated zone. Water was collected at the bottom of the tank, augmented with nutrients and recirculated. Hydrogen peroxide was not used in this system.

Physical parameters (pH, redox, nutrients, chloride, hydrogen peroxide, dissolved oxygen) were measured in the recovered water weekly. Monthly soil and water samples were collected and analyzed for nutrients, and microbial density. Also at monthly intervals, solids and water samples were submitted for TPH, and total phenolics analyses. These analyses were used as indicators of contaminant concentrations. Base neutrals and acid extractable compounds were analyzed at the beginning, middle, and end of the six month period.

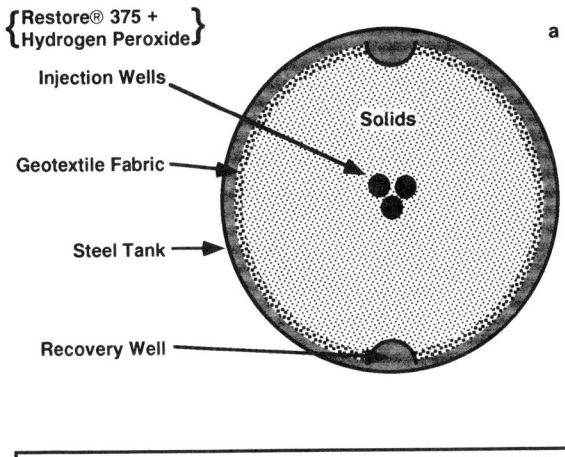

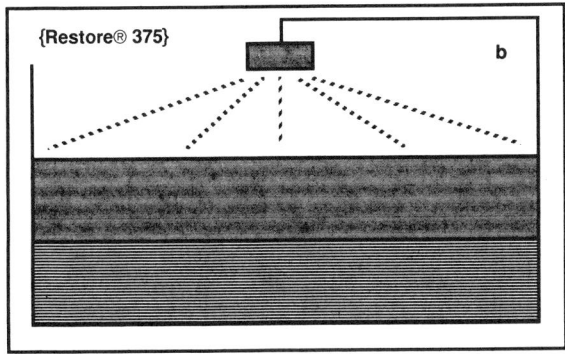

Figure 2. Schematic diagrams of bioremediation demonstration tank systems. a) X tank view of surface configuration, b) Y tank lateral cross-section indicating unsaturated and saturated zones, as well as sprinkler system.

To determine the efficiency of oxygen transport through the vadose zone in the Y tank, soil gas and water samples were collected from varying depths within the system and analyzed on site for oxygen content. The results were compared to a standard curve generated from oxygen-saturated and oxygen-free water, as well as differing quantities of atmospheric air.

Results

The pH and nutrient contents of solids and water are summarized in Table 8. The pH levels remained within the ideal range for bioremediation, with a low of 5.8 and a high of 7.0. Ammonia levels ranged in water ranged from 100 to 1,000 mg/l, and 70 to 400 mg/kg in solids. Phosphate levels in water ranged from 2 to 9 mg/l and 500 to 1,000 mg/kg in solids. Overall, it was unlikely that nutrients were limiting biological activity in either of the tank systems.

Table 8. X and Y Tank Monitoring Indicators during Five Months of Treatment

	pH	Ammonia (ppm)	Phosphate (ppm)
X Tank Water	6.1-7.0	400-1000	2-8
X Tank Soil		150-400	500-1000
Y Tank Water	5.8-6.7	100-400	3-9
Y Tank Unsat.		70-120	800-1000
Y Tank Sat.		70-100	650-850

Table 9. Microbial Population Densities in the X and Y Tanks at Time Zero and after 5 Months of Treatment

	Heterotophs (log cfu/g or mL)		Hydrocarbon degraders (log cfu/g or mL)	
	T_0	T_5	T_0	T_5
X Tank Water	5.0	5.0	5.7	4.7
X Tank Soil	6.9	7.5	6.2	6.4
Y Tank Water	4.0	5.3	5.8	5.0
Y Tank Unsat.	7.3	8.3	6.8	7.2
Y Tank Sat.	7.3	7.4	6.9	6.4

Bacterial enumerations in water and solids at time zero and after five months are presented in Table 9. Total heterotrophs and hydrocarbon degraders were measured by the spread plate technique. Overall, there was very little change in the density of either population in the treatment tanks. This may be an indication that the organisms are not thriving in the system, but are gaining enough energy through their metabolism to maintain their current population density. This non-growth mode is typical for organisms utilizing recalcitrant molecules as their carbon source.

Oxygen concentrations in the unsaturated zone of the Y tank are shown in Figure 3. Oxygen concentrations within 10 inches of the surface remains at virtual full-atmospheric levels. Oxygen was reduced by approximately 50 percent within the first two feet of vadose zone. As much as 70 percent of atmospheric oxygen was consumed as the saturated zone approached, and only 1-4 ppm oxygen was observed in any water sample. While this concentration appears to be much less than in the vadose zone, it is present in sufficient levels to avoid oxygen limitation in the saturated zone.

Oxidation-reduction measurements from throughout the X tank and at the recovery wells showed that the oxidation-reduction potential could be manipulated by altering the hydrogen peroxide amendment concentrations. These data, and dissolved oxygen measurements at the recovery well indicate hydrogen peroxide was effective in maintaining elevated dissolved oxygen in the X tank.

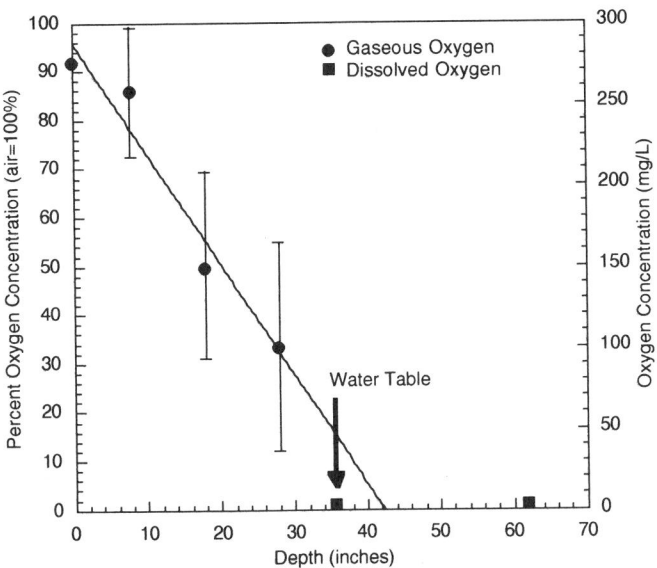

Figure 3. Oxygen concentration in the vadose and saturated zones of the Y tank. The mean and standard deviation of three measurements at each depth is shown. The line shows the least squares best fit line for vadose zone measurements.

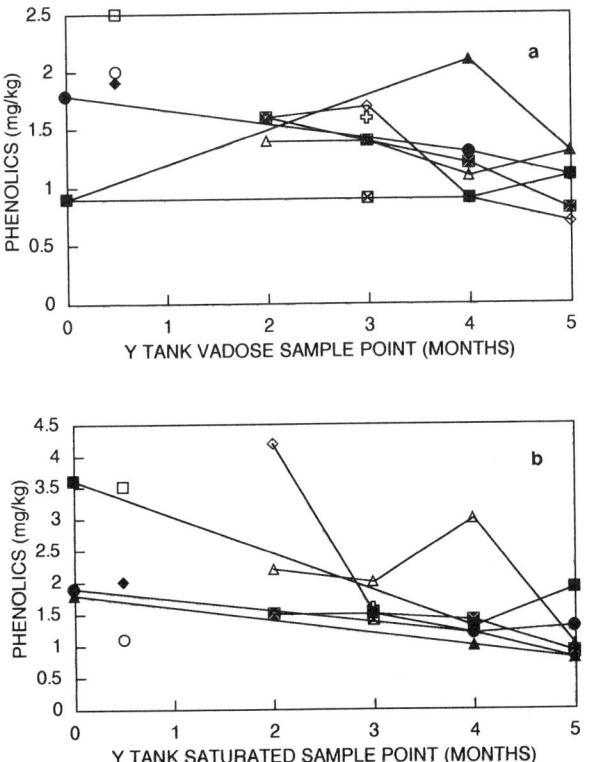

Figure 4. Phenolics analysis of Y tank solids. Samples which were taken from the same area at multiple time points are plotted with the points connected. a) vadose zone samples, b) saturated zone samples.

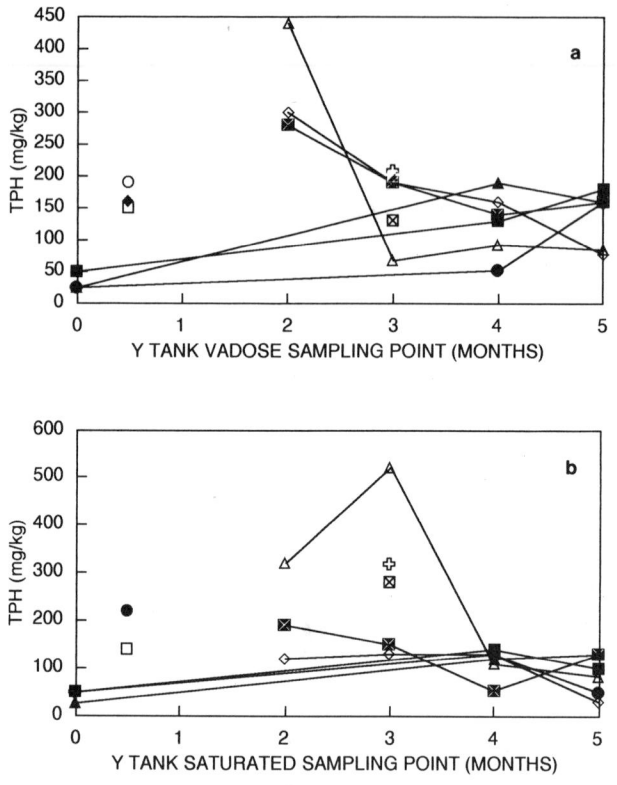

Figure 5. Total petroleum hydrocarbon (TPH) concentrations in the Y tank solids. Samples which were taken from the same area at multiple time points are plotted with the points connected. a) vadose zone samples, b) saturated zone samples.

Contaminant analyses collected from water samples indicated that all analytes were near or below the detection limits for the duration of the study (data not shown). This suggests that contaminants mobilized into the ground water are rapidly degraded.

Representative contaminant indicator analyses from solids are listed in Figures 4 through 6. In each figure, samples which were taken from the same area at multiple time points are plotted with the points connected. Figure 4a and b exhibits total phenolics concentrations collected from the Y tank vadose and saturated zones, respectively. Concentrations initially ranged between 0.5 and 5 ppm, and after five months ranged between 0.5 and 4 ppm. Samples collected from the same areas at different time points show a generally declining trend, particularly in the saturated zone.

Total petroleum hydrocarbon content (TPH) for the Y tank samples is illustrated in Figure 5a and b. Initial concentrations were found to be within the range of 25 to 50 ppm. The observations during the first five months of operation ranged between 25 and 500 ppm. The samples collected after five months of treatment

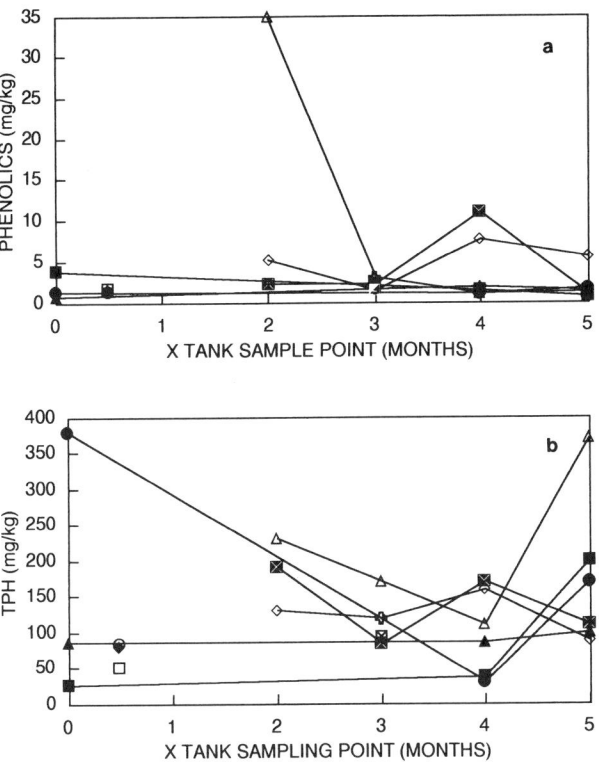

Figure 6. Phenolics and total petroleum hydrocarbon (TPH) analyses in X tank solids. Samples which were taken from the same area at multiple time points are plotted with the points connected. a) Phenolics, b) TPH.

were between 10 and 175 ppm. As stated above, some sample series collected from the same area indicate a declining trend.

Data collected from the X tank solids are illustrated in Figure 6a and b. Total phenolics initially ranged between 1 and 4 ppm. After five months of treatment, sample values ranged between 1 and 6 ppm. TPH values in the x tank varied widely, initially ranging between 25 and 375 ppm. This range represents the range observed over the course of the study. After 5 months of treatment, values ranged between 100 and 375 ppm. Again, some series of samples collected from the same area exhibited a declining trend.

Conclusions

The data collected from the first five months of operation of the pilot test systems at the coking waste lagoons demonstrate several important aspects of the process which will impact the ultimate application of bioremediation at this site. First, nutrients and oxygen can be effectively transported through both the unsaturated and saturated solids. Secondly, the subsurface environment can be manipulated to provide conditions conducive to microbial activity. Thirdly, contaminant analysis of the ground water suggests that organic compounds are quickly metabolized once they enter the dissolved phase. Contaminant analysis also indicates that substantial heterogeneity exists within the material. The pilot study is scheduled to continue for an additional 12 months during which time close attention will be given to the changes in contaminant concentration in the solid phase.

ACKNOWLEDGEMENTS

The authors wish to acknowledge Satish Gupta, David Shott and Brian Conroy of IT Corporation for project management and engineering design of the described pilot systems.

Evaluation Process for the Selection of Bioremediation Technologies for *Exxon Valdez* Oil Spill

Edgar Berkey, Jessica M. Cogen, Val J. Kelmeckis,
Lawrence T. McGeehan, and A. Thomas Merski

BACKGROUND

On March 24, 1989, approximately 11 million gallons of Prudhoe Bay crude oil were spilled into Prince William Sound, Alaska, as a result of the grounding of the *Exxon Valdez* tanker. The Prince William Sound area is a harsh and diverse environment with poor accessibility. According to the Alaska Department of Environmental Conservation, the shoreline is geologically young, composed largely of metamorphic rock, and ranges from vertical cliffs to boulder and pebble beaches. High-energy beaches are common and tides in these areas vary from +4 to -1 meters. In some areas, glacial and snow melt introduce large amounts of fresh water to the nearshore water of the Sound. Prince William Sound has a large population of seals and sea otters, extensive herring spawning areas, and significant numbers of sea and shorebirds. In addition, the area supports a substantial population of migratory birds that feed at beaches and inter-tidal areas.

The spilled oil from the *Exxon Valdez* ultimately spread over an estimated 350 miles of shoreline in Prince William Sound. Major contaminated shoreline areas included Knight Island, Eleanor Island, Smith Island, Green Island, and Naked Island. Knight Island, the largest and one of the most heavily polluted of these islands, has restricted tidal flushing action in some bays and coves. The oil settled into the beach gravel and on rock surfaces and the faces of vertical cliffs. In some areas, oil penetrated into the subsurface beach material. Contamination occurred primarily in the inter-tidal zone.

Initial weathering of the spilled oil resulted in a loss of approximately 15 percent to 20 percent of the oil by volatilization. Volatilized components included normal aliphatic hydrocarbons of less than 12 carbon atoms; aromatic hydrocarbons of less

---

Edgar Berkey, Jessica M. Cogen, Val J. Kelmeckis, Lawrence T. McGeehan, and A. Thomas Merski • National Environmental Technology Applications Corporation, University of Pittsburgh Applied Research Center, Pittsburgh, Pennsylvania 15238.

than 12 carbon atoms; and aromatic hydrocarbons—benzene, toluene, xylene, and methyl-substituted naphthalenes. The resulting residue consisted of approximately 40 percent to 50 percent high molecular-weight waxes and asphaltenes. On most beaches in Prince William Sound, the weathered oil was black and viscid rather than brown and mousse-like.

During the Summer of 1989, many of the beaches were physically cleaned by a combination of periodic flooding and the application of heated water under high pressure. Vacuum extraction and physical skimming were then used to remove the released oil from the water surface. The cleaning process partially removed oil from the surface of rocks and beaches, particularly the pools of oil, but did not effectively remove the oil trapped in and below the matrix of gravel and cobble, nor did it remove the thin layers of oil that visibly coated rock and beach surfaces. The extent of physical treatment was dependent upon the degree of contamination.

In the early cleanup stages of the Valdez spill, the U.S. Environmental Protection Agency (EPA) and Exxon recommended that two fertilizers, Inipol and CustomBlen, capable of enhancing the natural biodegradation of the oil, should be applied to supplement these more traditional physical cleaning methods. For the *Valdez* spill, bioremediation was intended as a supplemental technology, to support chemical and physical cleaning techniques. Nonetheless, the cleanup would eventually become the largest field bioremediation effort ever undertaken in response to a marine oil spill.

Several weeks after Inipol and CustomBlen were applied to beaches in Prince William Sound, the beaches where they were used appeared distinctly cleaner than control beaches nearby. However, the results were difficult to verify scientifically in the field, even though tests conducted in the laboratory also showed positive results.

Soon after the spill, private sector bioremediation companies began generating pressure to use their products in Alaska. There was an active bioremediation industry in the U.S. asking, "Why aren't our products being given a chance?" The EPA and Coast Guard had received many proposals from these companies, but there was no mechanism to evaluate the effectiveness of the products.

One of the priorities of the EPA under Administrator William K. Reilly has been the encouragement of innovative environmental technologies. Reilly believed bioremediation might hold significant promise in Alaska, and EPA's Office of Research and Development proposed using a new approach to pave the way for the private sector to demonstrate their bioremediation products in Alaska. To accomplish this, the Agency came to NETAC.

NETAC ESTABLISHES EVALUATION CRITERIA

The National Environmental Technology Applications Corporation (NETAC) was created in 1988 under a cooperative agreement between EPA and the University of Pittsburgh Trust to help commercialize innovative environmental technologies.

Due to this unique relationship, the U.S. EPA, in November 1989, requested that NETAC establish criteria by which bioremediation products for cleaning up oil spills could be evaluated. NETAC's goal was to develop a standardized procedure to evaluate bioremediation products on a consistent basis and to provide a mechanism to evaluate the numerous proposals submitted for the Alaskan beach remediation. NETAC's work during the Winter of 1989 offered the hope that some other effective bioremediation technologies might be identified which could then be tested when cleanup resumed in the Spring and Summer of 1990.

The evaluation criteria were to be practical and applicable to oil spills in general as well as the *Exxon Valdez* spill. "The bioremediation protocol development is urgently needed to enable EPA to compare competing technologies," Dr. John H. Skinner, EPA's Deputy Assistant Administrator for Research and Development, stated in requesting NETAC's assistance.

NETAC CONVENES EXPERT PANEL

In December 1989, NETAC assembled an independent panel of expert scientists to meet in Pittsburgh under the chairmanship of Dr. Edgar Berkey, NETAC's then Executive Vice President. The goal of the panel was to define criteria and to develop a protocol which could be used to evaluate products which potentially could be used on the beaches of Alaska. Members of the panel represented academia, the bioremediation products industry, and independent research institutes involved with bioremediation of environmental problems. They were selected by NETAC to reflect a broad spectrum of disciplines including microbiology, ecotoxicology, chemistry, as well as experience in remediation methodologies.

The panel was divided into subgroups, according to professional background and expertise, to develop criteria in three distinct areas: Microbiology, Remediation Methods, and Ecotoxicology. Each of these subgroups looked at the question of how to evaluate an oil bioremediation product for its efficacy and environmental safety, and which criteria were needed to adequately provide this evaluation. After two days of discussion, the panel decided that two types of baseline criteria need to be considered when evaluating products: (1) criteria which would automatically eliminate a product, and; (2) criteria required to adequately evaluate the product's capabilities to clean up oil spills. To accomplish these two tasks, the panel developed the criteria identified in Tables 1 and 2.

NETAC SOLICITS PRODUCT PROPOSALS

With criteria established by the expert panel, EPA next took steps to open the door to the private sector, by requesting proposals on promising bioremediation products that could be used to clean the contaminated beaches. The proposals were to be submitted to NETAC according to criteria set forth in a *Commerce Business Daily* announcement on February 12, 1990. The companies and individuals who had previously submitted proposals were contacted and asked to resubmit according to the established criteria. Thirty-nine proposals were ultimately submitted for review as a result of the solicitation.

Table 1. Criteria for Non-Feasibility

Product must not contain any genetically engineered organisms
Product cannot contain any known or suspected human or animal carcinogen
Product cannot contain any known pathogen to humans or indigenous flora or fauna
Product cannot contain any chemical listed in 40 CFR §268 Land Disposal Restrictions

Table 2. Product Capability Evaluation Criteria

1. Those data and other information used to support the application for product listing in the National Contingency Plan's (NCP) Product Schedule, as required under 40 CFR ?300, Subpart H (any confidential business information should be clearly identified to prevent its unintentional release).
2. A description and results of any laboratory or field tests performed in a laboratory or the field on crude or weathered oil that indicate that the proposed method enhances biodegradation.
3. Results of any acute or chronic toxicity tests performed on agents used in the method.
4. A description of how the proposed biological method is to be used on a large scale, including a justification of its practicality.
5. From current knowledge, is it likely that the proposed use of the method can comply with all applicable federal, state, or local laws and regulations?
6. A statement of corporate or organizational qualifications, including previous experience with hydrocarbon degradation, observed results, personnel resources, and capabilities.
7. Test results using the following bioassays:
    a. *Mysidopsis bahia* acute toxicity tests, 4 days; and
    b. *Bivalve/echinoderm larval* test, 2-5 days.

The criteria shown in Table 2 were intended to be the minimum information needed to determine the feasibility of the product or process for the environmental conditions at hand. EPA accepted the panel's recommendations.

NETAC RECONVENES EXPERT PANEL

NETAC reconvened the expert panel in March 1990 to review the proposals for demonstrated effectiveness of the products in degrading oil, toxicity, feasibility of applying the product on a widespread basis, organizational experience, and capability. Eleven promising bioremediation technologies were identified and recommended to EPA for laboratory testing. The purpose of this laboratory evaluation was to further qualify the technologies for field evaluation and possible use on the remnants of the *Exxon Valdez* oil spill in Alaska during the Summer of 1990. Table 3 identifies the products selected by the NETAC panel.

EPA accepted NETAC's recommendations, and Dr. Skinner directed EPA's Risk Reduction Engineering Laboratory in Cincinnati to conduct the laboratory tests. In June, the NETAC panel reviewed the results of the tests on the ten technologies (one of the eleven dropped out) and recommended two to EPA for field testing on the weathered crude oil in Prince William Sound. Test results indicating alkane degradation, for these two products were clearly better than the test results for the other products and for the organic fertilizer.

The two bioremediation treatments recommended by NETAC were Bi-Chem ABR Petroleum Blend, manufactured by Sybron Chemicals, Inc. (Salem, Virginia) and Micro Pro Marine "D", manufactured by Environmental Remediation, Inc.,

Table 3. Bioremediation Products Recommended by the NETAC Panel for Further Testing by U.S. EPA

Product name	Corporate proposer
ALPHA BIOSEA PROCESS	Alpha Environmental, Inc.
BI-CHEM ABR PETROLEUM BLEND	Sybron Chemicals, Inc.
BIOVERSAL	BioVersal USA, Inc.
DBC R5	Imbach Corporation
INIPOL EAP 22	Elf Aquitaine - M&T Chemicals Inc.
MICROPRO:D,G,SUPERGEE,NOW BAC	Environmental Remediation, Inc.
MUNOX^{TM} 101,201,501	Microlife Technics
PETROBAC, HYDROBAC	Polybac Corporation
PETRODEG 100 & 200	Insatech
TOXIGON 2000/CUSTOMBLEN	Woodward-Clyde Consultants
WMI CULTURE	Waste Microbes, Inc.

(Baton Rouge, Louisiana). Both products contain naturally occurring microbial cultures with the addition of fertilizer. The vendors of these products had experience using their cultures for the remediation of petroleum wastes in lagoons, bioreactors, and land farming.

EPA accepted NETAC's recommendation to field test these two products and started bioassay testing to evaluate potential toxicity to marine organisms. A field test protocol developed by the EPA was submitted to the agencies responsible for cleanup of the Alaskan spill. EPA, Coast Guard, and exxon all agreed to assist with a field test of these two technologies in Prince William Sound. Field tests of the two products was initiated in July 1990 and completed in September 1990. Test results are expected to be competed by the end of October 1990 and a final report issued by the end of 1990.

PROJECT SUMMARY

Conditions on the oil-stained beaches have changed dramatically since 1989. The oil has penetrated into the cobble beaches to a depth of four to six feet. Removal of this contamination has included the ecologically disruptive physical cleaning methods such as backhoe and the less intrusive bioremediation techniques.

Thus, from a tragic oil spill there is some hope for the future. There now exists a better pathway for government to bring bioremediation products to the scene in an oil spill emergency. Dr. Michael Griffin, Research Manager for Sybron Chemicals, recalls "Last year it was hard to find any focal point to talk to about applying a bioremediation product to clean up a marine oil spill. NETAC created that focal point, eliminated the confusion, and established a process so that we knew what was required to convince the government to field test our products."

Dr. Clayton Page of Environmental Remediation, Inc. agreed stating that, "NETAC made the process understandable by establishing criteria for vendors to meet." He said, "More importantly, NETAC provided a bridge between industry and the governmental agencies, creating an avenue to move our product to the testing phase more efficiently."

NETAC has established criteria that leave the agencies better prepared to review future bioremediation proposals. EPA plans to institutionalize the lessons learned under this process for future marine oil spills. Protocols have been established that make it easier for alternative bioremediation technologies to be considered in the future.

NETAC was formed to perform a single mission, *to assist in the commercialization of new environmental technologies*. According to Edgar Berkey, President of NETAC, "This project clearly demonstrates the use of NETAC's capabilities to promote new environmental technologies and provide the mechanism for getting technologies from the laboratory into the field."

Full-Scale Bioremediation of Contaminated Soil and Water

Geoffrey C. Compeau, William D. Mahaffey, and Lori Patras

INTRODUCTION

The full scale remediation of contaminated soil and water involves successful interpretation and scale-up of site assessment, material characterization, and laboratory treatability data. The primary goal of the site investigation is to determine the types and quantities of material that are present at a particular site and to identify geographic, geologic and other features that will impact the direction and implementation of remediation. Laboratory characterization and treatability data are developed during evaluation of the feasibility of remedial alternatives. Each of these activities can have an enormous impact on the cost of remediation. The types of laboratory characterization and testing that occur are specific to the remedial technology to be employed during full-scale clean-up operations. In all cases however, this aspect of investigation has the same goals: to accurately determine if the compounds are amendable to treatment and the amount of time and the cost required to treat the contaminants of concern according to the regulated clean-up criteria. The focus of this paper is the integration of laboratory data into full-scale bioremediation projects.

Biological treatment technologies for contaminated soils and groundwater fall into four main categories: 1) solid-phase biotreatment (landfarming); 2) slurry-phase biotreatment; 3) in situ biotreatment and 4) combined biotechnologies with chemical or physical treatment. The appropriate treatment process depends on the physical/chemical nature of the contaminant and the matrix in which it is found. Because of the variability in the source of chemical contamination, the nature of contaminated soil and other concerns, treatability studies of contaminant reduction in specific soils are the most appropriate way of establishing proper treatment time and conditions. Laboratory evaluation of biological remediation involves chemical, physical and biological evaluation of representative soils and waters from the site for

---

Geoffrey C. Compeau and William D. Mahaffey • ECOVA Corporation, Redmond, Washington 98052. Lori Patras • Unocal Corporation, Brea, California 92621.

baseline concentrations of contaminants, nutrients, microbial activity and other parameters critical to implementation of biological remediation. Laboratory-scale tests are performed to evaluate treatments leading to the highest rate of biodegradation. The selection of appropriate treatment options is usually determined by the results of the baseline chemical and microbiological evaluation. A completed study of this type should yield the rate of degradation and the time of clean-up expected with a high degree of certainty. These data can then be used to accurately assess the cost schedule and design of a pilot scale implementation, or if a high enough confidence exists, a full scale implementation.

The purpose of this paper is to present three separate case studies of successful implementation of bioremediation on a full-scale. Each of these remediations was preceded by laboratory treatability testing of materials to be remediated to determine schedule and cost of implementation. These case studies were also chosen to illustrate areas of technical consideration that should be evaluated prior to full-scale biological remediation. Among these technical considerations are the appropriateness and effectiveness of inoculation to increase the rate of bioremediation and the appropriate use of kinetics to determine treatment times.

CASE HISTORY: BIOREMEDIATION OF BUNKER C FUEL HYDROCARBONS

Soil remediation activities are being conducted at a former tank farm facility in southern California. The soil undergoing remediation consists of berm soils and soils underlying a former concrete-lined surface impoundment which was used to store bunker fuel oil. The quantity of soil treated will be in excess of 280,000 cubic yards, and the soil is being treated in eight separate land treatment units (LTUs) at the site. The petroleum contamination contained hydrocarbons in the range of C-10 to C-35 carbon chain length.

Monitoring of Total Petroleum Hydrocarbons

The project encompassed the treatment of over 280,000 cubic yards of soil contaminated with petroleum hydrocarbons in concentrations of up to 30,000 ppm as total petroleum hydrocarbons (TPH). The analytical method used was EPA method 418.1 and the cleanup standard was 1000 ppm TPH. In order to guide the excavation of the soil and facilitate process monitoring of the solid-phase process, a mobil laboratory was placed onsite and staffed with environmental chemists and microbiologists. The lab has analyzed up to 150 samples per day during peak periods of production from the excavation and land treatment units. Over 20,000 samples have been analyzed in the laboratory at this stage in the project.

In one part of the project, a gas chromatograph was brought to the site to guide the remediation of light kerosene-like solvent residues located in a separate disposal area on the site. For this part of the remediation, the analytical protocol was EPA method 8015 and the cleanup standard was 100 ppm TPH. These soils were incorporated into a separate LTU for treatment in several consecutive lifts. The gas chromatograph was also used to qualitatively evaluate the progress of the remediation by determining what fraction of hydrocarbons had been treated and what fraction remained.

The laboratory data management system for TPH as well as nutrient analyses is a PC-based software package. The system provides direct data input for each sample from the time it is taken (via a laptop-mounted computer) through the analysis to the final report. In addition, export utilities allow transfer of selected blocks of data between system modules and/or commercial software packages such as spreadsheet or graphics programs. Data integrity is assured through the use of triple-redundant databases, automatic backup to floppy disk, and a complete audit trail facility. The audit trail facility tracks and records every change made to a sample record. The audit trail database is invisible to, and totally inaccessible by, mobile lab personnel.

Treatment Concentrations

An initial treatability evaluation was conducted to determine the highest concentration of TPH that could be successfully treated to the regulated standard of 1,000 ppm TPH. It was determined that a starting concentration of approximately 4000 ppm total petroleum hydrocarbon would be optimal and that it was potentially possible to treat up to 5000 to 6000 ppm TPH in these soils.

Since the excavation program required continued progress and the sequential stacking of lifts of soil to accommodate the excavation requirements, an area was set aside at the treatment site and an LTU was charged with 5500 cubic yards of soil with an average concentration of 5595 ppm TPH to serve as a pilot study. This pilot treatment required 12 weeks for treatment, confirming laboratory treatability data. The data from this LTU treatment verified the upper limit of TPH concentration treatable considering scheduling requirements.

Nutrient and Biological Monitoring

In addition to contaminant chemistry, the site support laboratory supported the nutrient addition program and monitored biological activity in the LTUs. Ammonia and nitrate nitrogen concentrations as well as phosphate concentrations were routinely analyzed in the LTU soils. To evaluate biological activity, total heterotrophic organisms counts were made in the treatment soils. The microbial analysis program at the site was augmented with plating of soil onto minimal media containing specific hydrocarbons as sole source of carbon for growth. The development trends for the hydrocarbon degrading population were evaluated in this way. Laboratory evaluations of the soil from the remediation and small-scale studies were conducted at the site to more clearly establish the population of organisms involved and the community interactions responsible for the degradation of hydrocarbons. Obvious changes in the microbial population occurred over time in the LTU's. The

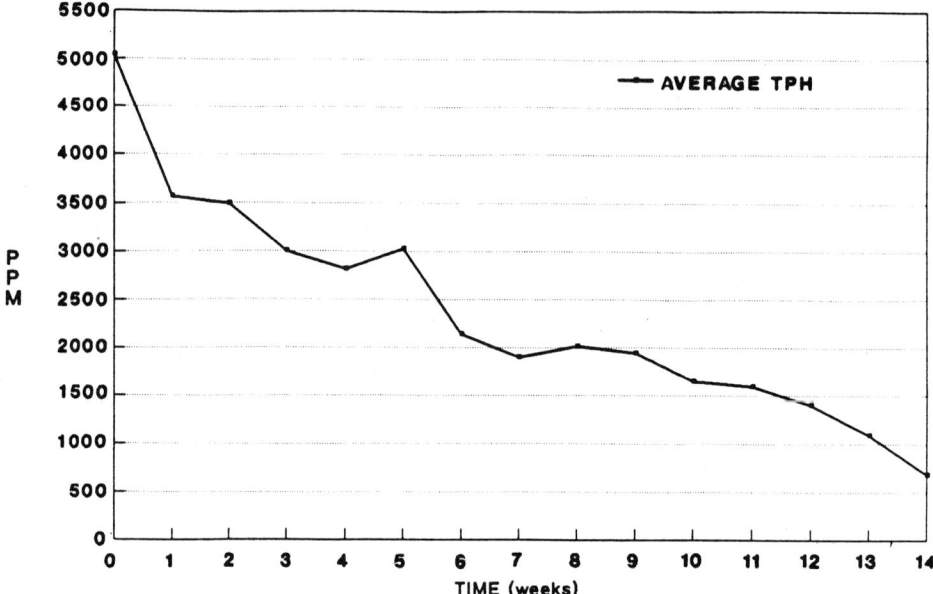

Figure 1. Plot of TPH degradation in soil in land treatment units during full-scale operations. Average of analysis (n ≥ 20).

evaluation of hydrocarbon-degrading activity helped to clearly define the importance of these changes.

Summary of Remediation Data

TPH Degradation. Figure 1 is a representative of TPH data from treatments employed during the remediation. The pattern of degradation presents a similar pattern to that observed in all LTU soils. A high initial rate (55 ppm/day) is followed by a period of reduced rate (10 ppm/day) as the composition of the petroleum hydrocarbon and microbial community changes. After these changes, the rate of remediation increases (37 ppm/day) once again.

There is a critical period of time in the remediation in which the rate slows. This is during the period of 6 - 9 weeks in soils which have a starting concentration of approximately 5,000 PPM. This phenomenon is not observed in LTU soils which have starting TPH concentrations below approximately 3,500 PPM. Changes in nutrient utilization and microbial populations indicate that the changing rate is attributable to changes of the ecology of the degrading population during remediations.

Nutrient and Microbial Monitoring. The significance of the changes in TPH degradation noted above are borne out by the overall changes in nutrient concentrations and the heterotrophic (including petroleum hydrocarbon degrading) microbial

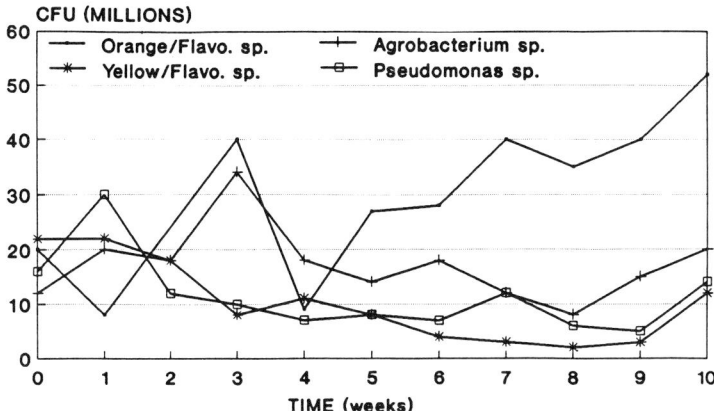

Figure 2. Changes in dominant heterotrophic bacterial populations during TPH degradation. Average of analysis (n ≥ 10).

populations which occur during this period. The analysis of the heterotrophic population indicates that significant changes occur after seven weeks of treatment (Figure 2). Over the final five weeks of the remediation, brightly pigmented bacteria emerge from the population. Heterotrophic organisms in general decrease and the proportion of the hydrocarbon degrading organisms increase. A dominant organism in the remediation, distinguished by a distinctive orange pigment, was identified as possessing the ability to metabolize a wide range of hydrocarbon substrates. Clear evidence of growth by this orange-pigmented organism was demonstrated using pentadecane (C-5), octadecane (C-18), pristane (C-15 branched), docosane (C-20), hexacosane, (C-26) as sole carbon sources. Preliminary study on C-30 hydrocarbons is being undertaken. Control plates which contained no hydrocarbons did not demonstrate growth. More definitive experiments concerning the physiology and metabolism of these organisms are being conducted. The physiology and ecology of this organism may be pivotal to the control of the rate of hydrocarbon degradation in the remediation.

The emergence of the organism is correlated to change in nutrient utilization patterns during remediation (Figure 3). During the first eight weeks of remediation, ammonia and nitrate nitrogen were consumed in a stoichiometry of approximately 1:1, after eight weeks, there was an increase in nitrate nitrogen and dramatic decrease in the soils under going remediation. This is likely caused by an increase in ammonia-oxidizing activity in these soils. Experiments are underway to evaluate methods to control these changes leading to increased rate and shorter treatment time for the TPH contamination.

The Effects of Commercial Inocula on the Degradation Rate of Petroleum Hydrocarbons in Soil. The degradation of hydrocarbons is facilitated by organisms that occur naturally in soils. It has been suggested that the rate of petroleum hydrocarbon degradation can be increased in soils through the addition of strains that can degrade hydrocarbons. In theory, this would artificially raise the number of hydrocarbon de-

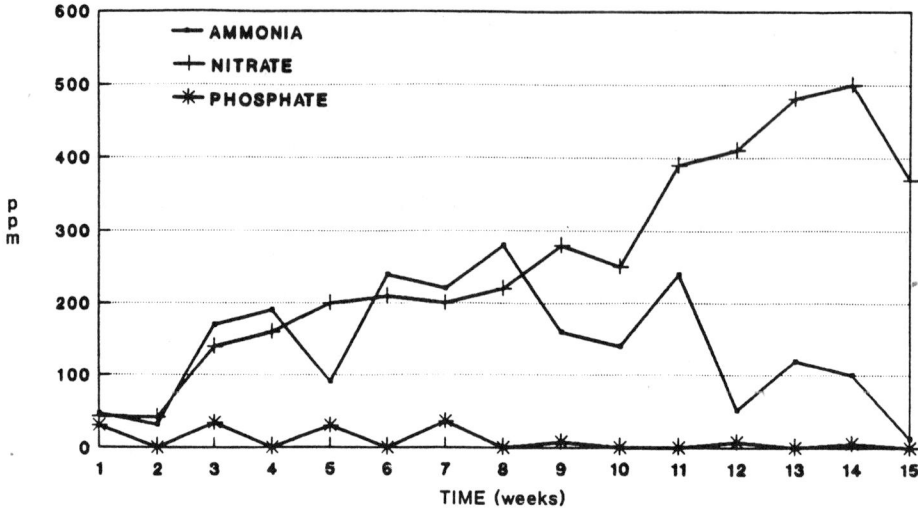

Figure 3. Trends in nutrient utilization during TPH biodegradation. Average of analyses (n ≥ 10).

grading bacteria. However, soils are complex ecosystems, and biological systems develop within the soil to make the maximum use of available resources. In order for introduced organisms to survive and establish themselves, they must compete with the indigenous organisms for resources. The experiments described here were performed to determine the efficacy of introduced microorganisms, not indigenous to treated soils, on the degradation of petroleum hydrocarbons.

The effects of three commercially available hydrocarbon-degrading inocula on the degradation rates of petroleum in soil were examined in soil microcosms. The inocula tested were L-103 and L-104 (Solmar Corporation) and Super-Cee (Microbe Masters, Inc.). Petroleum contaminated soil was obtained from the treatment cells undergoing remedial biological treatment at this site. The effects of the hydrocarbon-degrading inocula were compared to uninoculated controls and also compared to sterilized inocula. Sterilized inocula were tested to determine if any of the other components of the commercial inoculum (ie. nutrients and/or organic additives) stimulated bacterial activity in the soil.

Under the conditions tested, inoculation with the commercially available organisms did not affect the rate of petroleum hydrocarbon degradation. A representative result is shown in Figure 4 for L-104. Taking into account the standard deviation for each set of three composites, the rates are very similar. Over the first four week period, the TPH concentrations in the treated soils decreased from an average (all treatments) of 1266 ppm (standard deviation of 231) to 640 ppm (standard deviation of 82). The concentration of petroleum hydrocarbons was halved in 28 days. None of the living or sterile preparations outperformed the control treatment of adjustment of nutrients, moisture and regular tilling of the soil.

The primary result of the study is that inoculation with these cultures does not lead to an increase in petroleum hydrocarbon degradation as compared to in-

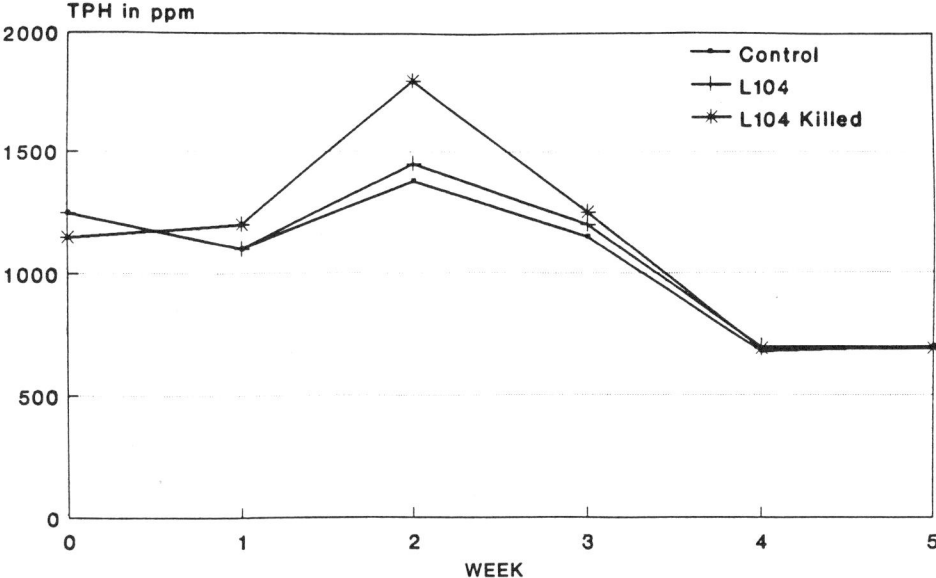

Figure 4. TPH degradation in control (indigenous microorganisms) vs. commercial inocula (Solmar). Average of analyses (n = 6).

digenous microorganisms. The reasons for this may be the inability of the strains to colonize the soil since none of the colony morphologies of the inoculated strains were recognized after four weeks of incubation. The introduced strains may also be outcompeted by the resident microbial strains in the soil. It should be noted also that no negative effect was noticed resulting from each inoculation. Though the inoculated systems, both killed and live, did enable the systems to maintain a higher number of organisms, evidence indicates that this increase in organisms did not affect petroleum degradation rates.

Conclusions

The solid phase remediation program implemented for this site has been extremely successful. Field data has agreed with treatability results. Over 150,000 cubic yards of soil have been treated and removed from the LTUs to date. Approximately 120,000 cubic yards of soil remain to be treated. Understanding changes in the natural ecology of the soil system during remediation may lead to enhancing the rate of TPH degradation of high molecular weight hydrocarbons. Several optimization studies are being conducted on-site during the remediation at an incremental cost to the remediation. The use of commercially available inocula was not successful in stimulating TPH degradation rates in the soil undergoing treatment.

CASE HISTORY: SOLID PHASE REMEDIATION OF WOOD TREATING SITE SOILS

Wood treating operations conducted over a period of 60 years resulted in soil and groundwater contamination with pentachlorophenol and creosote. The 30-acre site is included on the U.S. EPA National Priorities List and the Minnesota Permanent List of Priorities. There is an estimated 30,000 tons of waste to be treated, with concentrations of PCP of up to 3,000 ppm and of total polynuclear aromatic hydrocarbons (PAHs) to 2,000 ppm. Solid phase biological treatment was selected as the technology of choice to reduce pentachlorophenol concentrations to 150 ppm and carcinogenic PAHs to 100 ppm.

The scope of work included the design and construction of an eight acre land treatment unit (LTU) by ECOVA. The design included all necessary provisions for containment of excess water, security, and watering systems, and equipment utilization such that the site could be treated successfully by a two man operating team. Dust and organic vapor levels were monitored throughout the project with readings taken from the perimeters of the site and by individuals working in the LTU.

Treatment Concentrations

The remediation was conducted as an aerobic process. Based on previous experience and treatability study data, concentrations of PCP up to approximately 1000 ppm were considered appropriate to treat in the soils present at this site. Evaluations of microbial numbers and activity clearly demonstrated that the indigenous microorganisms in these soils were capable of PCP degradation. As a precaution, a PCP-degrading inoculum was prepared (detailed below) in case microbial activity was diminished to unacceptable levels during treatment. These events might include precipitous decrease in pH during the release of hydrochloric acid (HCl) as a degradation product or prolonged episodes of rain which might interrupt the aerobic process.

After 17,000 cubic yards of soil were placed in the LTU, a baseline sampling of the LTU soils was conducted and a starting concentration of 410 ppm of PCP and less than 100 ppm of the regulated PAH compounds were detected in the LTU soils. The process monitoring of this remediation therefore focused exclusively on PCP in the treatment soils.

Process Monitoring

The baseline concentration of PCP and PAH compounds was determined by an independent laboratory by EPA Method 8270 and by the ECOVA Project Chemistry Laboratory. The size and duration of the project did not require an onsite laboratory for conducting analyses. During baseline sampling, sixteen subsamples were taken per acre and composited into one sample for analysis per acre this practice was repeated for the verification baseline conducted at the end of the treatment. During process monitoring, however, one discrete sub-sample from each quar-

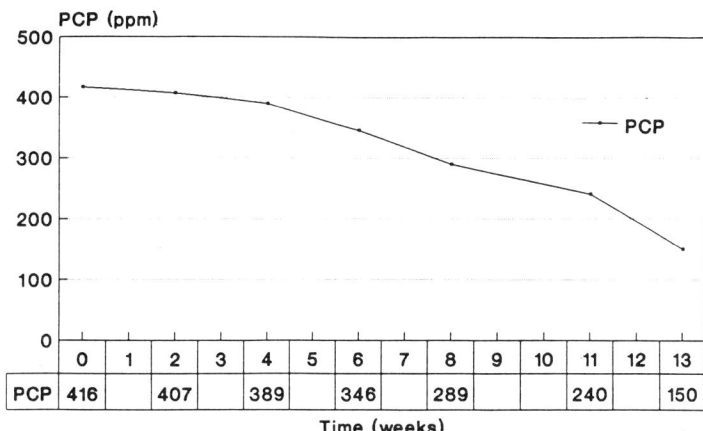

Figure 5. PCP biodegradation rate average of analysis (n = 32).

ter acre was analyzed for PCP (four samples per acre) and on average one sample per acre was analyzed for other process variables.

Process monitoring samples for PCP, nutrient concentrations and microbial activity were sent to the Project Laboratory in Redmond and analyzed within 72 hours of receipt. This approach was critical to the evaluation of project progress, the direction of equipment and the effective management of the biological process for nutrient control. The critical nature of moisture and pH control required that these measurements be conducted by site personnel as part of a program that included perimeter monitoring of air-borne dust.

Summary of Remediation Data

Figure 5 presents the actual site data and progress achieved during the treatment of the first 17,000 cubic yards of soil at this site. The slow onset of the process is attributable to the unseasonal precipitation levels during this period. Generally, the remediation required thirteen weeks. The calculation from the laboratory treatability work indicated a remediation of ten to eleven weeks in duration. Calculations were performed to evaluate the need for alkali addition to these soils during the bioremediation process. These calculations and evaluations of the buffering capacity of the soil indicated that no application of alkaline material would be required during the bioremediation. This proved to be the case during field operations.

The original treatability data was accumulated in a study of degradation kinetics of PCP in soil at a single concentration. Process monitoring field data collected during the actual remediation revealed very important kinetic considerations to be incorporated into the remediation design. The 32 cells composing the eight acre treatment area could be segregated into approximately equivalent groups of cells at initial concentration ranges of from 100–400 ppm, 400–700 ppm, and 700–1000

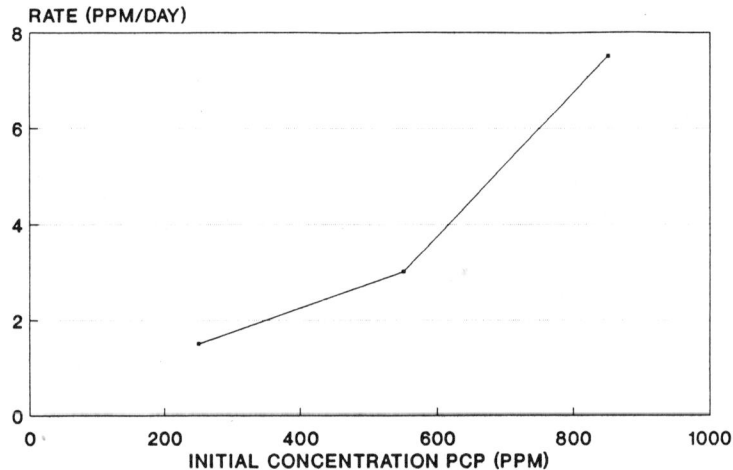

Figure 6. Rate of PCP degradation in soils plotted versus average initial PCP concentration of groups of sampling cells.

ppm PCP. These groups of cells were followed through the remediation to determine the effects of starting concentration of PCP on the rate of degradation.

Degradation rates of organic contaminants are often described by first order kinetic equations with the calculation of "half-life"; the time required for the concentration of a compound to reduce in concentration by 50%. However, in the case of PCP degradation in soil, the rate of degradation is dependent on PCP concentration. Figure 6 presents the rate of degradation versus the initial concentration of PCP in the soil. At 100–400 the average initial rate of degradation is 1.2 ppm/day; at 400–700 ppm the rate is 3.0 ppm/day; and at 700–1000 ppm the rate increases to 7.8 ppm/day. It is clear that within range of PCP concentrations encountered in these soils, the rate of degradation increases as the concentration increases. At some point however, not determined here, PCP will reach an inhibitory and toxic concentration in soil. According to the "half-life" concept, irrespective of starting concentrations, the time required for the compound of concern to reduce in concentration by 50% in constant. During this remediation, the rate of degradation did increase with increasing concentration of PCP. However, at low concentrations, the rate of degradation slows leading to proportionately longer treatment times. At starting concentration of from 200–800 ppm, 50% reduction was achieved in 8–9.5 weeks; however, at 200 ppm, a period of 12 weeks was required to achieve 50% reduction.

Conclusions

The dependence of degradation rate on PCP concentration may occur for a number of reasons. Among these are that the soil components are competing with microorganisms for the PCP in the soil at lower concentrations and the concentration of PCP available to organisms in soil is therefore reduced. Another reason is that the wood treating waste present is composed of petroleum hydrocarbons, PCP and PAHs. The presence of these components effects the solubility, transport and biodegradation rate of the PCP in the soil. A further explanation is that at higher concentrations, PCP induces a greater number of PCP-degrading organisms or increased degradation activity in organisms. What is clear from this data, is that the time required for remediation is best established by evaluating the rates of degradation at a range of loading rates and under actual conditions if possible, not from theoretical kinetic data.

CASE STUDY: BIOREMEDIATION OF PENTACHLOROPHENOL CONTAMINATED SOIL: SOIL WASHING

Bench scale studies were performed to develop a microbial culture and biodegradative process which could treat PCP at higher concentrations than previously reported. Several substrate formulations and culture techniques were evaluated. Ultimately a "self-feeding" continuous culture system (pH auxostat) was used to select for biodegradative activity with PCP as the carbon and energy source. After a period of 50 days, influent PCP concentrations reached 3500 mg/l at a dilution rate of 0.066 h^{-1}. Of the total theoretical chloride which could be released from PCP, 99% was detected as free chloride in the reactor effluent. PCP analysis of the effluent verified complete degradation by the microbial consortium. The reactor was converted to a constant PCP feed. At steady state conditions the dilution rate was 0.05 h^{-1} with an influent PCP concentrations of 1560 mg/l and a biomass yield of 0.18 mg (dry weight) per mg of PCP. Mineralization studies performed with the microbial consortium using (U-^{14}C)PCP indicated that 36.5% of the label was released as ^{14}C-carbon dioxide.

During operations at a PCP formulating facility significant quantities of PCP solutions were spilled resulting in the contamination of an estimated 3400 cubic yards of soil. Representative soil samples from the site revealed PCP concentrations ranging from 2 mg/kg to 9000 mg/kg. Several treatability studies were performed to determine the efficacy of a slurry phase bioremediation process for these soils. Results of the study showed little if any indigenous microbial degradative potential. Inoculation with the PCP degrading consortium resulted in the degradation of PCP to below detectable levels (<1.0 ppm) as measured by HPLC. Data from laboratory treatability studies were used to design and implement the full-scale remediation process. The process involves soil washing to remove PCP from the soils followed by bioremediation of the pregnant wash solution in a slurry-phase bioreactor. The PCP degrading consortium is grown on site and serves as the inoculum for the full scale treatment reactors. Results from the field operations during the first two months of treatment are presented to show proof of process.

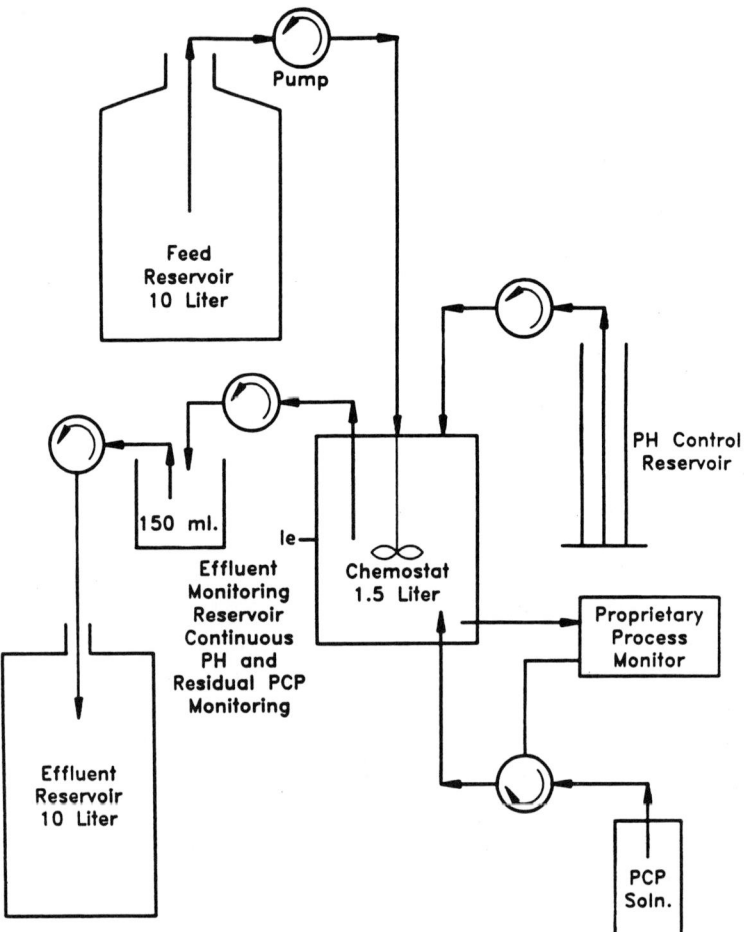

Figure 7. Schematic representation of the bench scale bioreactor for PCP biodegradation studies. The basic unit consisted of New Brunswick BioFlo fermentor equipped with pH controlled feed pumps.

Bench Scale Bioreactor Studies

A schematic representation of the bench scale bioreactor is presented in Figure 7. PCP was fed to the reactor from a separate feed reservoir to facilitate greater control of the feeding rate. Since PCP is quite soluble under alkaline conditions it was always added to the reactor in this form. The Bioflo reactor was operated with a number of substrate formulations in an attempt to increase the PCP degrading biomass by providing cosubstrates. As can be seen in Figure 8, each substrate formulation yielded differences in the extent and rate of PCP degradation as evaluated on the basis of dechlorination activity. The results indicate that formulations 1 through 4 resulted in repression of dechlorination activity as compared to formulations 5 and 6 where PCP was the sole source of carbon.

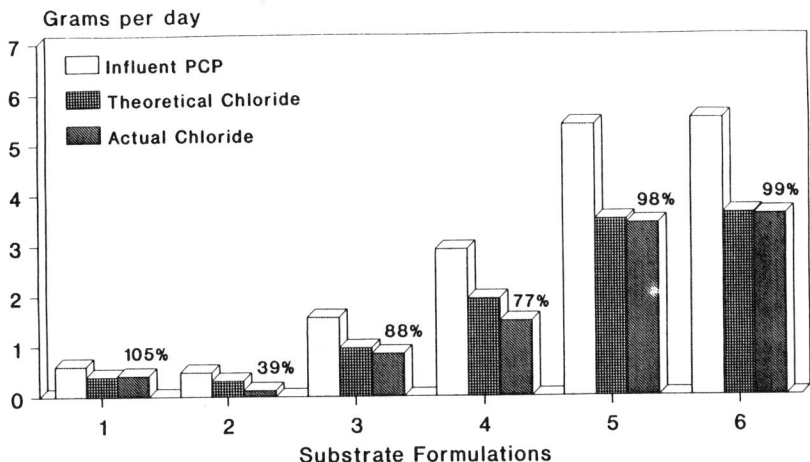

Figure 8. Effects of various substrate feed formulations on the dechlorination of pentachlorophenol in the bench scale continuous flow bioreactor operating in a self-feeding mode.

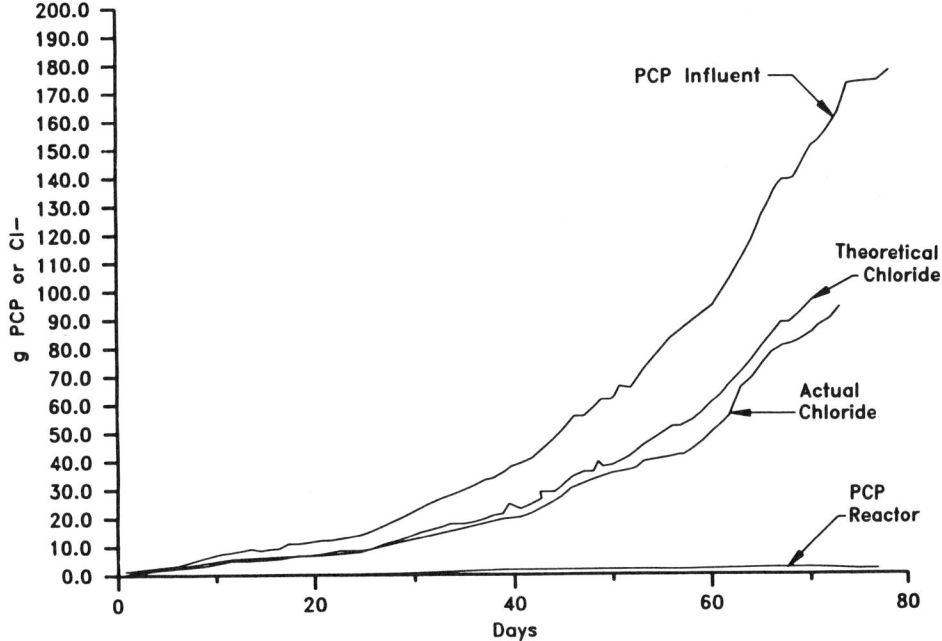

Figure 9. Effect of "self-feeding" operations on the performance of the continuous culture bioreactor in degrading PCP.

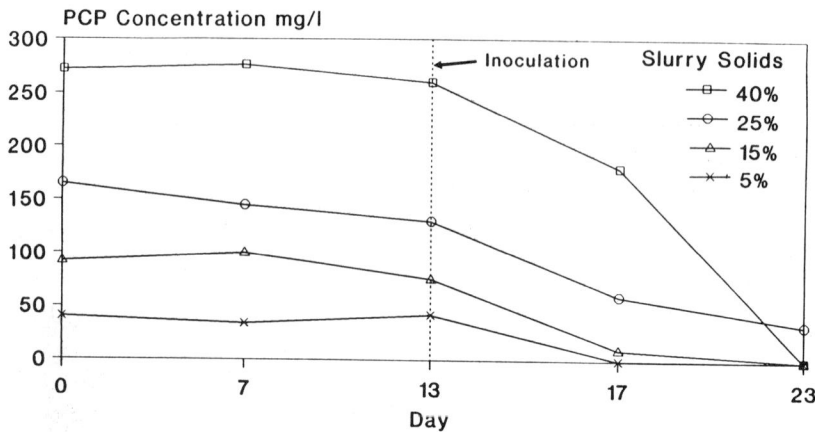

Figure 10. Results of the slurry phase biotreatability evaluation demonstrating the enhanced biodegradation of PCP in contaminated soils by supplementation with a microbial consortium capable of PCP mineralization.

During the evaluations of the various substrate formulations it became apparent that as a result of PCP biodegradation 5 equivalents of HCl could be produced. Therefore the PCP feed was formulated to provide an equimolar amount of sodium hydroxide (NaOH) to neutralize the acid formed. It was hypothesized that the addition of PCP based on the production of acid may be an efficient means of regulating the continuous flow process. This was implemented by interfacing PCP addition to a pH controller which actuates the feed pump when the pH drops below a set point. The addition of the alkaline PCP feed solution neutralizes the pH while providing additional substrate. In this "self-feeding" operational mode the influent PCP loading to the reactor increased over time (Figure 9). In addition, the data suggests that greater than 95% of the PCP was biodegraded as is apparent from the low concentrations detected in the reactor mixed liquor. This is further substantiated by the fact that the levels of free chloride are within 90% of the theoretical value based on PCP loading.

Application of the PCP Consortium to a Bioremediation

Operations at a former PCP formulation facility resulted in the contamination of an estimated 3000 cubic yards of site soil. A biotreatability study was required to prove the feasibility of a bioremediation program specific for PCP. The results of a simple slurry-phase biotreatability study, performed on soils contaminated with high concentrations of PCP, is summarized in Figure 10. During the first two weeks of the study there was little if any PCP biodegradation in any of the various slurries tested (e.g., 5% to 40% soil solids). On day 13 each of the slurries received an inoculum of the PCP consortium to yield approximately 1×10^8 cells/ml of slurry. PCP was biodegraded to less than 2.0 mg/l with the exception of the 25% slurry

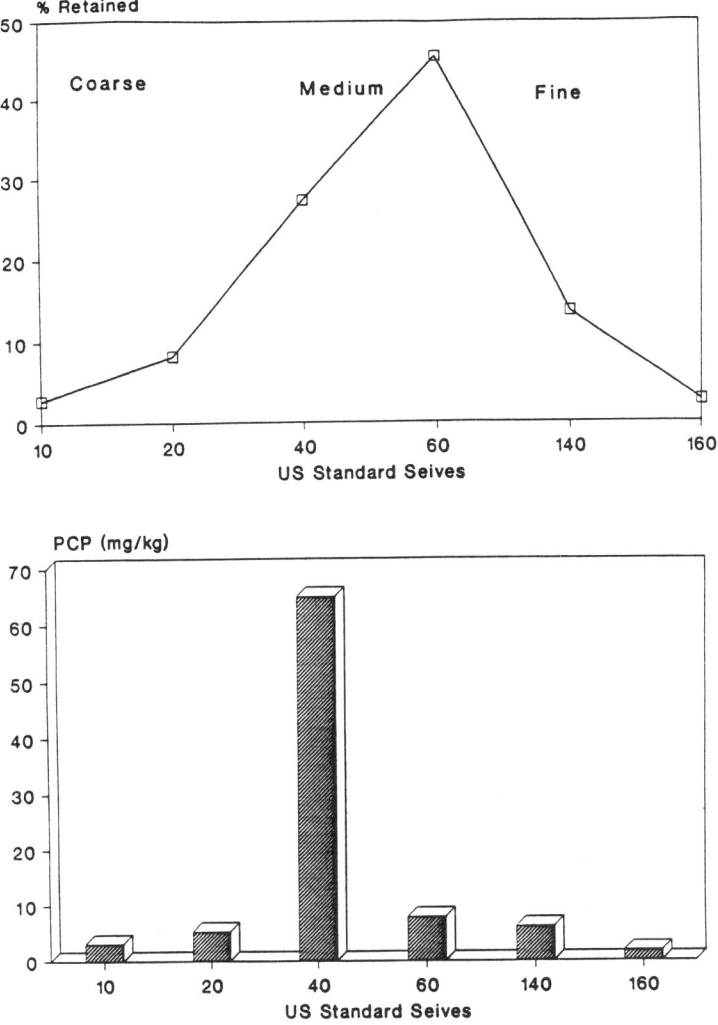

Figure 11. Results of the treatability evaluations to obtain particle size analyses and the distributions of pentachlorophenol on contaminated site soils.

where the levels were only reduced to 48 mg/l. In these particular slurries the pH was not maintained at neutrality (pH 7.5). Rather the pH was allowed to drop in order to evaluate the effects on the rate of biodegradation. By day 17 the pH had decreased to pH 6.5 and by the end of the study had dropped to pH 5.5. These results suggested that the indigenous population was inadequate in degrading PCP and that the PCP consortium may be useful as a supplement for achieving bioremediation of the contaminated site soils.

Additional treatability studies were performed in order to classify site soil on the basis of standard sieve analyses and to determine the distribution of PCP

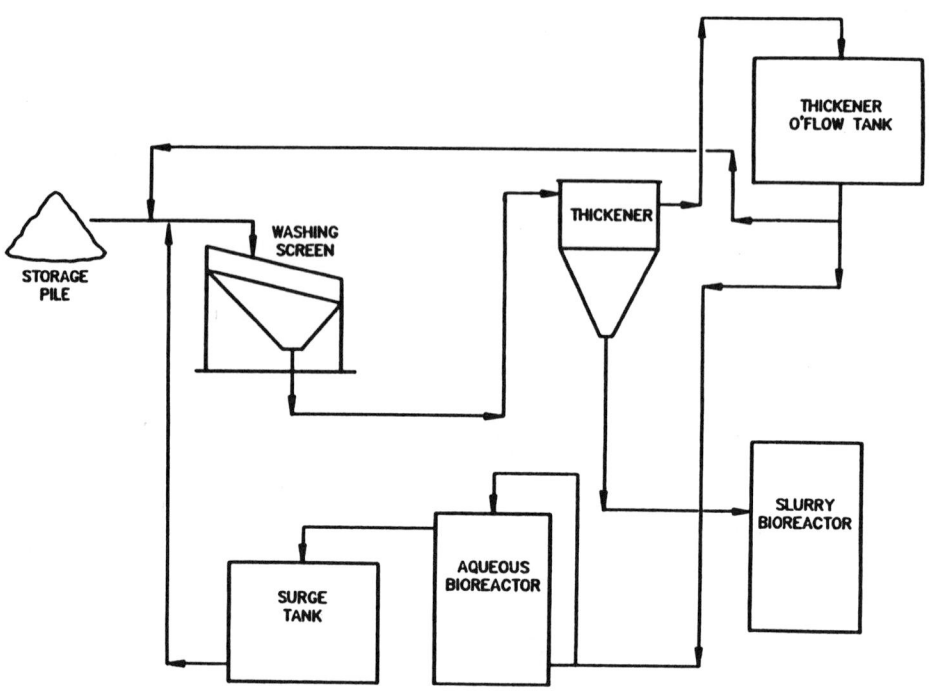

Figure 12. Schematic of field operations for the bioremediation program to clean pentachlorophenol contaminated soil.

in the various size fractions. The results of these studies are shown in Figure 11. The bulk of the PCP contamination is resident on course soil particles greater than 60 mesh in size. This represents material which is difficult to keep suspended as a slurry and also constitutes the bulk of the soil volume. It was decided that it would be more efficient and economical if the PCP could be washed from the soil and the resulting liquor biotreated. The rationale behind this decision was based upon the fact that material greater than 60 mesh constitutes approximately 80% of the total soil volume. A soil washing solution was formulated and found to be highly efficient in removing PCP from the soil particles which are larger than 60 mesh.

Based upon the information developed during bench scale treatability testing Ecova has designed and begun the implementation of a full scale remediation program for PCP contaminated soils. This program is centered on a combination soil washing and screening process which results in the removal of PCP to a cleanup level of 0.5 mg/kg soil. The system is designed to process from 15 to 30 cubic yards of soil per day. The resulting wash solution is a slurry which contains the PCP and the minus 60 mesh soil particles at approximately a 20% solids loading. This material is subsequently treated in on-site slurry phase bioreactors which have been inoculated with the PCP degrading consortium. A general schematic of the field operations is presented in Figure 12.

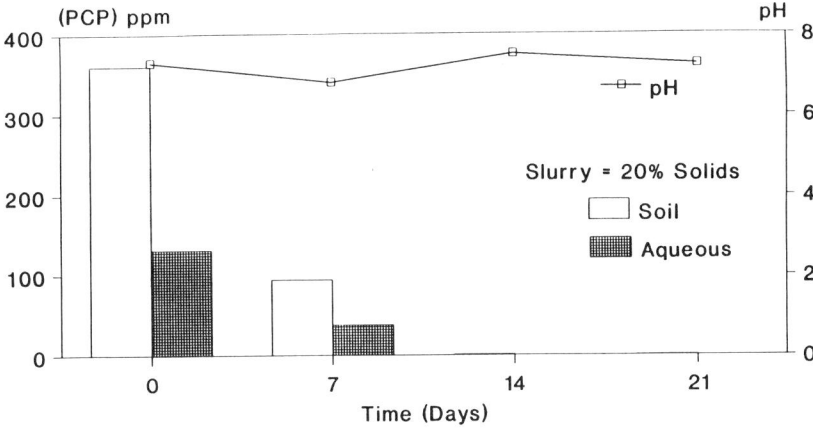

Figure 13. Field data from Tank 1 operations for the bioslurry treatment of solutions generated during soil washing. The system was inoculated with the PCP degrading consortium at time zero to yield approximately 10^7 cells/ml of slurry.

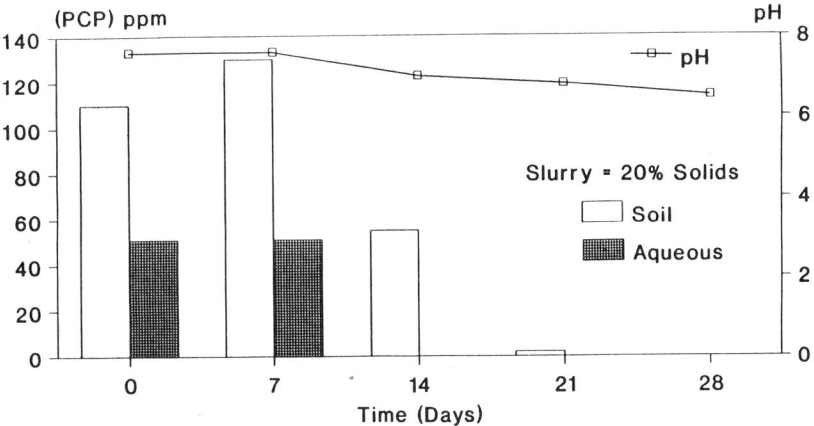

Figure 14. Field data from Tank 2 operations for the bioslurry treatment of solutions generated during soil washing. The system was inoculated with the PCP degrading consortium on day 7 to yield approximately 10^7 cells/ml of slurry.

During startup of site operations, two 25,000 gallon slurry bioreactors were operated in batch mode in order to demonstrate the utility of inoculation with the PCP consortium. Wash solution from the soil washing operations were used to charge each of the slurry bioreactors (Tanks 1 & 2). Substrate formulation number 4 was used to prepare the inoculum in two on-site fermentors of 1000 and 1200 liter capacity. Tank 1 (Figure 13) received an inoculum of 10^7 cells/ml of slurry at time zero, while Tank 2 (Figure 14) received inoculum on day 7. Prior to inoculation the bacterial population density was less than 10^3 cells/ml. Fertilizer was added to achieve a TOC:N ration of 25:1 and a TOC:P ration of 300:1. Low levels (500

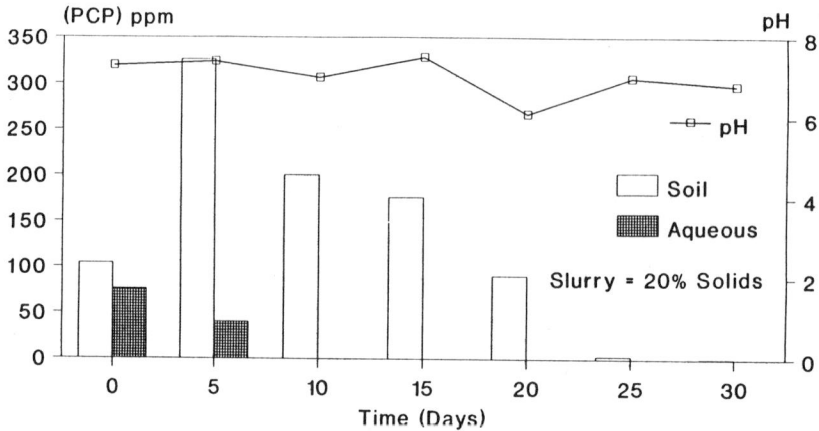

Figure 15. Field data from Tank 3 operations for the bioslurry treatment of solutions generated during soil washing. The system was inoculated with the PCP degrading consortium at time zero to yield approximately 10^7 cells/ml of slurry.

mg/l) of hydrocarbon solvents were detected in both slurry batches. It is apparent from the data in Figure 13, that the PCP consortium was very efficient in degrading PCP from 370 ppm on slurry solids to below the detection limit of 0.5 ppm (mg/kg) over a period of 14 days. In Tank 2 (Figure 14) PCP degradation did not commence until after the inoculum had been added on day 7 which confirms the results observed during the treatability study. It is important to observe that the treatment time for this tank required 21 days of treatment. This may be the result of a lower initial PCP concentration or the fact that the pH of this unit fell below pH 7.0 as a result of a malfunction in the pH controller.

In a parallel field experiment soil, which had not been subjected to the soil washing/screening step, was used to charge Tank 3 at a 20% solids loading rate. This was the maximum loading rate at which the soil could be effectively suspended. After charging the reactor with soil and water an oily solvent (mineral spirits) sheen could be detected at the air/slurry interface. Based on the results of a total petroleum hydrocarbon analysis (Infrared method) the levels of hydrocarbon were 15,000 mg/l, indicating that a hot spot had been excavated from the stockpile. The tank was seeded with an inoculum of the PCP consortium to achieve approximately 10^7 cells/ml of slurry on day 0. During the first 5 days of operation the aqueous PCP levels were reduced by about 40% (Figure 15). However PCP concentrations in the soil increased 3 fold and was correlated with the loss of solvent from the system. Apparently the solvents serve as a reservoir for PCP which partitions onto the soil particles as the solvent is stripped from the system. Subsequently the PCP in the soil particles was observed to undergo biodegradation. The rates of PCP degradation in this system are much slower than those observed in Tanks 1 & 2. The total treatment time required to achieve the target cleanup level of 0.5 mg/kg was 30 days.

Conclusion

In conclusion, a bioprocess for the remediation of PCP contaminated soils has been successfully developed by integrating sound microbiological and chemical research with good materials and process engineering to achieve a solution to a problem. It is our philosophy that only by continuing efforts in which microbiology, chemistry and engineering are integrated in the early stages of research can there by efficient transfer of new technology to the field of hazardous waste remediation.

SUMMARY OF CASE STUDIES

These case histories have demonstrated the use of bioremediation for full-scale clean-up of petroleum hydrocarbon and pentachloroepheneol contaminated soils. Bench-scale studies performed before and during remediation supported the use of inoculation in enhancing the degradation of PCP in slurry reactors. Inoculation was not effective during remediation of petroleum hydrocarbons in soils. Data from the full scale remediation of PCP in soils underscore the problems associated with the strict application of kinetic rate constants (eg. half-life) to degradation of organic compounds in soil. The half-life if PCP in soils increased at low concentrations in soil.

Feasibility and Other Considerations for Use of Bioremediation in Subsurface Areas

Karolyn L. Hardaway, Mark S. Katterjohn, Craig A. Lang, and Maureen E. Leavitt

INTRODUCTION

The movement and persistence of hydrocarbons in soil and aquifer matrices are controlled by physical, biological, and chemical processes, making remediation of subsurfaces complex and difficult. Pump and treat systems constitute the most commonly employed technology used to clean up aquifers, but it is widely recognized that progress is slow. Bioremediation, the process of stimulating microbes to rapidly degrade hydrocarbons within a reasonable time frame, is a possible alternative for remediation of soils and groundwater. However, for *in situ* applications, a multidisciplinary approach is required to extensively characterize and define the biological, physical, and chemical properties of the subsurface and of the hydrocarbons.

BACKGROUND

Texas Eastman Company, established in Longview, Texas in 1950, manufactures chemicals and plastics using olefin feedstocks. During the early years, some of the process waste was held in earthen treatment basins for settling and/or treatment, an acceptable practice at that time. These earthen basins have been emptied, and the sludge has been removed and incinerated, but some underlying hydrocarbon-containing soils and groundwater remain.

Closure and remediation of this site is in progress. Areas of migration and concentration of hydrocarbons in the subsurface were determined during Resource

---

Karolyn L. Hardaway • Texas Eastman Company, Longview, Texas 75607. Mark S. Katterjohn • International Technology Corporation, Austin, Texas 78746. Craig A. Lang and Maureen E. Leavitt • International Technology Corporation, Knoxville, Tennessee 37923.

Address correspondence to: Karolyn L. Hardaway, Texas Eastman Company, Utilities Laboratories, P. O. Box 7444, Longview, Texas 75607.

Conservation and Recovery Act (RCRA) Facility Investigations conducted by Texas Eastman Company and International Technology Corporation. A RCRA-regulated corrective action program was implemented, and in 1986 International Technology Corporation installed a pump and treat system to remediate the site. This corrective action program physically removes and/or contains the hydrocarbons and limits the groundwater flow, thus preventing further spread of the subsurface organics. The current pump and treat system is effective, proven technology, but predicted remediation time at this site has been estimated to be 50 years or longer.

Texas Eastman Company decided to explore alternate technologies that would be effective in cleaning the site and could also reduce the company's costs and time requirements. One such alternative technology, *in situ* bioremediation, offers several advantages over conventional methods: it utilizes indigenous bacteria which can transform the organics to harmless substances or carbon dioxide and water; it does not require physical removal of the soils; and it prevents continued deterioration of the groundwater. However, since this technology is not applicable to every site and all types of hydrocarbons, *e.g.*, more recalcitrant compounds, Texas Eastman Company undertook a study to determine its feasibility at this site.

International Technology Corporation designed a two-part laboratory study to determine the feasibility of remediating the site's subsurface hydrocarbons by *in situ* bioreclamation techniques and also to identify factors existing in the subsurface that would contribute to the success or failure of bioremediation. This feasibility study was designed to run concurrently with an on-going RCRA Facility Investigation to reduce costs in the sampling and geological characterization of the area.

OBJECTIVES

The overall goal of this study was to develop a technical evaluation of the applicability and cost-effectiveness of *in situ* bioreclamation for remediation of subsurface areas at this site. In consideration of this goal, the following objectives were identified:

- To determine the distribution and growth requirements of the indigenous microbial population responsible for hydrocarbon degradation.
- To evaluate the technical feasibility of applying *in situ* bioreclamation considering site hydrogeology, nutrient and oxygen transport, and chemical properties of the hydrocarbons and their distribution.
- To establish site-specific requirements for a field trial, which could be used in the implementation, monitoring, and completion of an *in situ* bioreclamation system.
- To review and identify the technical issues and approaches in combining direct contact treatment, vapor extraction, and *in situ* bioreclamation with other remediation technologies (*e.g.*, pump and treat).
- To evaluate cost differences and time savings which might occur if *in situ* bioreclamation was implemented.

- To assess potential adverse environmental impacts, future human health risks, and the public's perception of the various technologies.
- To review regulatory agencies' requirements, opinions, and applications of *in situ* bioreclamation.

TECHNICAL FEASIBILITY

Accurately defining the technical and scientific problems of bioreclamation requires a multi-disciplinary approach involving input from numerous fields of study including microbiology, chemistry, engineering, and, most importantly for *in situ*, hydrogeology. Major technical considerations for the application of *in situ* bioreclamation for the remediation of subsurface areas must include examination of: microbial populations and growth requirements, soil and groundwater characteristics, hydrocarbon distribution and transport, hydrocarbon biodegradability, and hydrogeological characteristics of the site.

Soil samples were collected at three depths; one in the unsaturated zone and two from the saturated zones. Groundwater samples were also collected from the site. The samples underwent a series of chemical and microbial characterizations to evaluate the potential for *in situ* biodegradation as a remediation tool for this site. Methods employed for most analyses are described in *Standard Methods for Examination of Water and Wastewater*, Seventeenth Edition, 1989. Geology data from logs of soil borings and well data from recovery and monitor wells were used to assay the hydrogeological characteristics of the site and the feasibility of transporting solutions through the subsurface. For each assay discussed, only pertinent data are shown, and, when possible, ranges are given.

Microbial Analyses

The success of *in situ* bioremediation is contingent upon the presence of microbes and requires microbial populations which can be stimulated to rapidly degrade the hydrocarbons in a timely manner. Normally, existing indigenous microbial populations which have adapted to the subsurface environment and existing hydrocarbons are used and nutrients and oxygen are supplemented.

Baseline Enumeration. To determine if sufficient native bacteria were present, all groundwater and soil samples were quantified for the indigenous populations of aerobic heterotrophic bacteria and hydrocarbon-degrading bacteria using Standard Method 9215C. Dilutions of soil and groundwater were plated on nutrient agar for growth of aerobic, heterotrophic bacteria. Hydrocarbon degraders were spread on mineral salts agar and hydrocarbon vapors were supplied as the carbon source. Plates were incubated at 25°C for one week and the number of colonies counted to determine microbial densities.

Overall, soil bacterial populations were low, while most groundwater samples had detectable heterotrophic bacteria (Table 1). In fact, microbial populations in

Table 1. Microbial Densities in Soil or Groundwater Samples

Sample	Total aerobic heterotrophs	Hydrocarbon degraders
Soil	cfu^a x 10^3/g Soil	
667	20	< 3
669	29	< 3
670	< 3	< 3
672	180	< 3
676	< 3	< 3
678	46000	14000
730	< 3	< 3
Groundwater	cfu x 10^3/ml Groundwater	
669	1.4	< 0.3
672	3.3	1.8
678	540	410

^{a}cfu = colony forming units.

two-thirds of the soil samples were below detection limits of 3,000 colony-forming units (cfu) per gram of soil (data not shown). The remaining samples ranged in population from 10^4 to 10^7 cfu/g soil, which are more indicative of typical soil populations. The groundwater samples contained typical numbers of hydrocarbon-degrading bacteria, between 10^3 and 10^5 cfu/ml. Low numbers of microbes in some areas are not limiting to *in situ* treatment since native bacteria may be reintroduced through the injection system.

Stimulation Tests. To determine if the microbial population could recover from the stressful conditions in the soils and grow, microbial stimulation tests were conducted. This test involves using two identical slurries of soil and groundwater. Hydrogen peroxide, an oxygen source, was added to both slurries and one was augmented with nutrients (phosphate and ammonia).

Results of these tests were mixed (Figure 1). In the case of sample 667, the bacterial populations in both the oxygenated and nutrient-augmented slurries increased substantially over the initial enumeration value. This is an indication that oxygen might be the main limiting factor for growth since the increase in both groups was similar. Alternatively, sample 669 displayed substantial increases in the nutrient-augmented slurry, while the population in the oxygenated slurry remained the same. This response is an indication that nutrients were largely responsible for the recovery of the population. In sample 672, little change was detected in either slurry, possibly indicating the presence of an inhibitory substances or environment.

Physical and Chemical Characteristics of Soil and Groundwater

Various soil constituents may interact with the oxygen source and nutrients, limiting their bioavailability to the microbes. To ensure that the subsurface environment is amenable for microbial activity, soil and groundwater samples were characterized to determine their compatibility with the nutrients and oxygen source and to identify factors that may limit availability.

Measurement of Naturally-Occurring Parameters. Soil borings from the unsaturated and saturated zones were analyzed for extractable iron, pH, moisture content, degree of nutrient adsorption, and interference with hydrogen peroxide stability. Conductivity, pH, and concentrations of ammonia, phosphate, chloride,

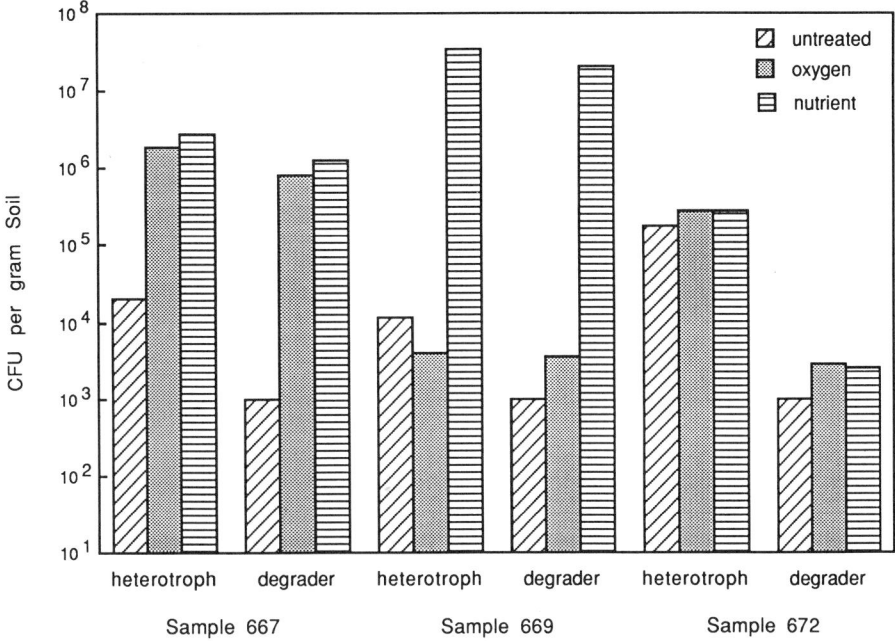

Figure 1. Stimulation of microbial populations. Cross-hatched bars represent untreated soil samples, shaded bars are soils with hydrogen peroxide added, and both hydrogen peroxide and nutrients were added to samples represented by bars with horizontal lines.

magnesium, calcium, and iron were determined for groundwater (gw) samples (Table 2).

Iron catalyzes the degradation of hydrogen peroxide, thus reducing the amount transported to the bacteria. Therefore in soils with high iron content, another source of oxygen, usually nitrate, may be necessary. High concentrations of calcium and magnesium precipitate phosphates, and will reduce the amount available for bacterial consumption. High chlorides may inhibit bacterial activity. Chloride is also used as a tracer to monitor migration of the nutrients injected; however if chlorides are high, another tracer is used.

Results shown in Table 2 indicate that in most of the samples, iron, calcium, magnesium, and chlorides fall within acceptable ranges and should not inhibit microbial activity nor preclude the use of bioremediation. One well had high iron levels, but most samples contained < 200 ppm. The pH values were slightly acidic, but should not pose a threat to biodegradation.

Orthophosphate concentrations were determined by Standard Method 424F and Standard Method 417B was used to measure ammonia nitrogen levels. Phosphate levels ranged from below detection (0.1 mg/l) to 2 parts per million (ppm), while ammonia levels ranged from 2 to 14 ppm. These background levels of nutrients are low and are not adequate to support enhanced biodegradation. These data support the results of the nutrient stimulation test (Figure 1); namely, nutrient aug-

Table 2. Ranges of Naturally-Occurring Parameters in Soil and Groundwater Samples

Parameter[a]	Range
Ammonia	2-14 mg/l
Phosphate	< 0.1-2 mg/l
Magnesium	< 4.2-220 mg/l
Calcium	41-210 mg/l
Chloride	7-22 mg/l
Conductivity	366-3725 μmhos/cm
Total Iron	< 0.5-286 mg/l
Total Iron - soil	9.1-850 mg/g
pH - soil and gw[b]	3.8-6.3

[a] Unless specified, parameters were measured in groundwater samples.
[b] pH ranges in soil and groundwater were the same.

mentation will be required to stimulate the growth of some of the native bacterial populations.

Hydrogen Peroxide Stability. The compatibility and stability of the oxygen source with the soil were tested by mixing 500 ppm hydrogen peroxide with the soil and, at various intervals following the spike, determining the concentration remaining in the soil. If hydrogen peroxide is degraded abiotically by the soil constituents, less is transported away from the injection sites for use by microbes and another oxygen source must be supplied. Most soil samples showed an acceptable hydrogen peroxide stability with slow degradation rates over a 24-hour period (Figure 2). Following a second spike, degradation rates were even slower. This test confirms that hydrogen peroxide can be used as an oxygen source in these soils. The exception was sample 667; the rapid

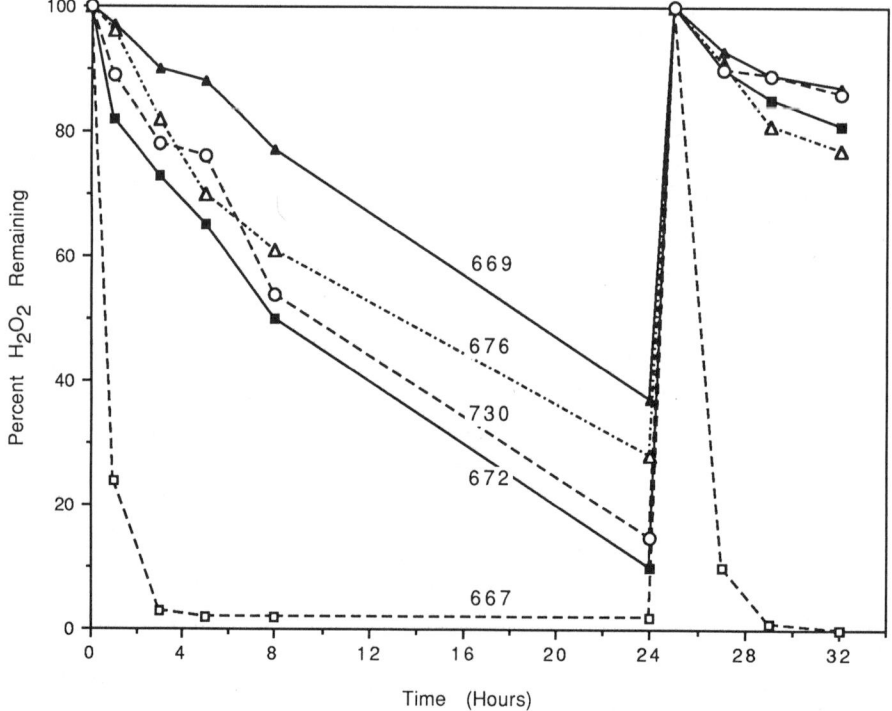

Figure 2. Hydrogen peroxide stability in soil samples 667 (□), 669 (▲), 672 (■), 676 (△), and 730 (○).

Table 3. Nutrient Adsorption Test

Soil sample	Percent adsorbed to soil	
	Ammonia	Phosphate
667	52	100
670	13	53
678	61	96

degradation cannot be explained with available data, since all parameters, including iron, were within acceptable ranges.

Nutrient Binding. Although calcium and magnesium levels in the soils are low, other constituents may bind nutrients. The effect of soil constituents on nutrient availability was analyzed by measuring the percent of nutrients adsorbed to the soil. Soils were slurried in a 1:2 ratio with water containing 500 ppm phosphates and ammonia and shaken for four hours. The amount of nutrients remaining in the supernatant was measured. Relatively high percentages of nutrients were adsorbed to most of the soils (Table 3). However, this can be overcome by increasing nutrient concentrations to maintain adequate concentrations for bacterial use. Normally, optimum bacterial growth occurs when carbon:nitrogen:phosphate ratios are 100:10:1, respectively.

Hydrocarbon Distribution, Transport, and Biodegradability

The solubility and sorbtive properties of a hydrocarbon control how easily it moves through the subsurface. This partitioning between the aqueous and solid (soil) phases determines how much of the hydrocarbon will be present as free product, adsorbed to the soil, or dissolved in the groundwater, and ultimately determines its availability and degradability. Thus, by understanding a hydrocarbon's physical and chemical characteristics and its distribution, one can determine how readily the site can be treated and the time required.

Specific hydrocarbons present, their abundance in the unsaturated and saturated zones, and the extent of hydrocarbon distribution were determined by evaluation of soil borings and groundwater samples for total petroleum hydrocarbons, selected Appendix IX constituents, and specific base neutral extractable and volatile organic compounds. The primary constituents detected are typical of petrochemical plants: benzene, chloroform, chlorobenzene, toluene, xylenes, naphthalene, acetone, styrene, and ethylbenzene. Total petroleum hydrocarbon concentrations ranged from below the detection limit (< 10 ppm) in 60 percent of the soils analyzed, up to 486 ppm in most of the remaining samples. Total organic carbon concentration in the groundwater samples ranged from 8 to 44 ppm.

The hydrocarbons detected in this phase of the study were surveyed for their potential biodegradability. All of the predominant constituents, typically found at most petrochemical plants, have been reported in the literature to be biodegradable in both laboratory and field studies.

Overall, the characterization studies of the groundwater and soil indicated favorable conditions for *in situ* bioremediation. However, the low microbial populations coupled with mixed results from the nutrient stimulation and nutrient adsorption studies warranted continuing with the second, or bench-scale, biodegradation phase of the technical feasibility study.

Bench-Scale Biodegradation Study

This treatability study was designed to examine the effects of nutrients on hydrocarbon degradation, to evaluate the relative efficiencies of nitrate and oxygen as electron acceptors, to develop a preliminary assessment of the extent of degradation, and to establish operating criteria for a bioremediation field trial.

Experimental Conditions. Several sets of subsurface microcosms were established with composites of hydrocarbon-containing soil and freshly collected groundwater at a 1:10 ratio, respectively. One liter glass jars, fitted with Teflon-faced septa, were used for the microcosms, and marbles were added to reduce headspace and to provide better agitation. Inorganic nutrients (ammonia and phosphate) and either hydrogen peroxide (oxygen source) or potassium nitrate, both of which served as electron acceptors, were added to experimental vessels. Controls were treated with mercuric chloride to inhibit microbial activity.

Dissolved oxygen (DO), oxidation-reduction potentials (ORP), pH, and nutrient levels were monitored throughout the study in one set and targeted for optimum levels of 2 ppm DO, +200 mv ORP, and a pH of 7 in the hydrogen peroxide-amended treatments. Vessels were corrected or supplemented if needed. A similar set was used to monitor conditions in the nitrate-amended microcosms. Since an anaerobic, denitrifying atmosphere was maintained in the nitrate- amended microcosms, desirable targets for DO and ORP were 0 ppm and 0 to -200 mV, respectively. ORP values above 0 mV indicated that denitrification was minimal, so nitrate was added.

Microbial Enumeration. The indigenous aerobic and denitrifying bacterial populations showed appropriate increases in most of the samples by the time the study was completed. Both treatment conditions had similar densities of aerobic and denitrifying bacteria, ranging from 10^4 to 10^7 cfu/gram soil. The abiotic controls exhibited no detectable microbes.

Degradation Kinetics. To define the kinetics of biodegradation, indicator compounds (*i.e.*, total petroleum hydrocarbons, dissolved organic carbon, and pentane-extractable organics) were analyzed periodically during the 51-day study and degradation rates of total and specific hydrocarbons were determined using least squares regression. The rate of degradation is expressed as a half-life ($t_{0.5}$) or the time, in days, required to reduce the concentration by one-half. Samples for analysis were obtained from an unopened set of vessels. Total petroleum hydrocarbons were analyzed by EPA Method 418.1 and pentane-extractable organics by a modified version of EPA Method 3510 or Standard Method 6410. The half-lives of specific chemicals in each treatment condition were compared with those seen in the biologically-inhibited controls.

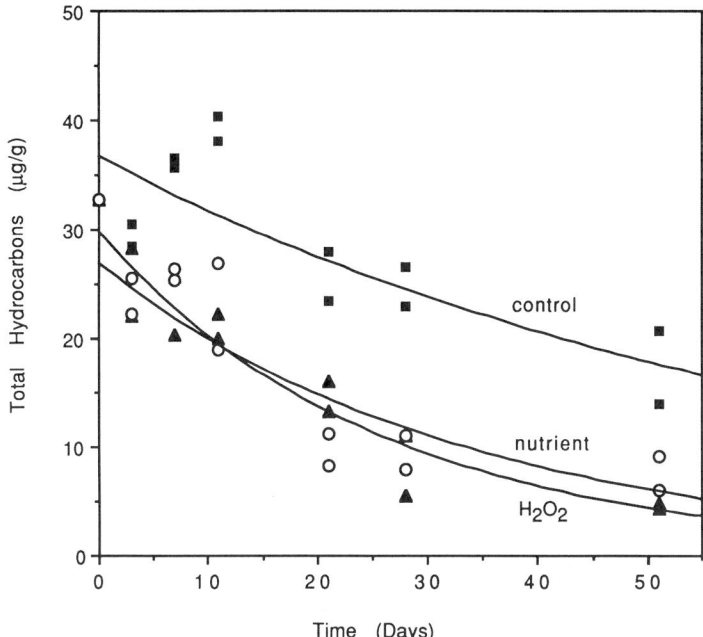

Figure 3. Total hydrocarbon content of hydrogen peroxide-amended microcosms at various times during the study. Half-life ($t_{0.5}$) in days and the correlation coefficient (r) are given for each treatment. Abiotic controls (■), $t_{0.5}$ = 51, r = 0.88. Hydrogen peroxide only (▲), $t_{0.5}$ = 18, r = 0.99. Hydrogen peroxide and nutrients (○), $t_{0.5}$ = 23, r = 0.91.

a. Degradation Rates in Hydrogen Peroxide Microcosms. Results of the aerobic treatability study show an overall reduction in total hydrocarbons (Figure 3). Differences between half-lives in the biologically-inhibited and the biologically-active hydrogen peroxide-amended treatments demonstrated that the compounds can be degraded by the indigenous aerobic bacteria at the site.

Percent reduction in total petroleum hydrocarbons after 51 days was 86 percent in the biologically active, hydrogen peroxide-amended treatment, but only 47 percent in the abiotic control. In fact, certain pentane-extractable organic analytes showed > 99.99 percent disappearance in the biologically active set compared to a 60 percent loss in the biologically inactive control (data not shown).

b. Degradation Rates in Nitrate-Amended Microcosms. Degradation kinetics in the nitrate-amended microcosms showed similar results, although at lower rates than hydrogen peroxide-amended treatments (Figure 4). All the biologically active microcosms exhibited lower half-lives than those of the inactive control.

c. Nutrient Augmentation. Addition of nutrients had little impact on degradation rates of total hydrocarbons in either system, suggesting that the background nutrient concentrations were not initially limiting to biodegradation. Nutrient supplementa-

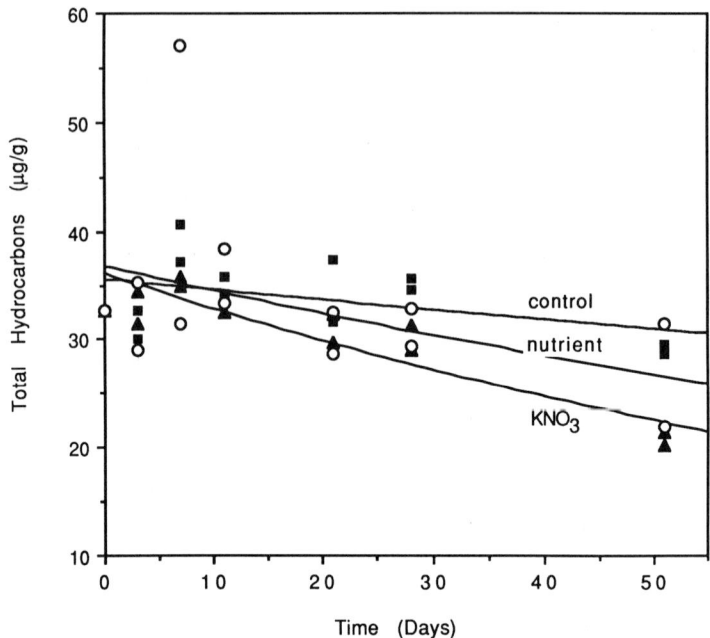

Figure 4. Total hydrocarbon content of nitrate-amended microcosms. Half-life ($t_{0.5}$) in days and the correlation coefficient (r) are given for each treatment. Abiotic controls (■), $t_{0.5}$ = 289, r = 0.46. Nitrate only (▲), $t_{0.5}$ = 77, r = 0.92. Nitrate and nutrients (O), $t_{0.5}$ = 120, r = 0.66.

tion would not be expected to enhance the rate of biodegradation until the background nutrient concentrations have been reduced to limiting levels.

d. Abiotic Degradation. Reduction in total hydrocarbons in the biologically inactive controls was due to abiotic degradation since no detectable microbes were present in these samples after 51 days. Abiotic degradation in the mercuric chloride-inhibited controls could be caused by several factors, *e.g.*, hydrogen peroxide oxidation, volatilization, and/or non-specific abiotic degradation.

e. Half-Lives of Individual Pentane Extractable Organics. Half-lives for individual pentane-extractable organics were combined and the variabilities in half-lives for each treatment condition are shown in Table 4. Rates of degradation are not equal for all chemicals under similar conditions. For example, in the biologically active nitrate-amended treatment, half-lives varied from 9.8 days for one analyte to 82.5 days for a compound more difficult to degrade. The half-life in the abiotic control for the latter chemical was 100.5 days.

The magnitude of some of the pentane-extractable organic analytes' half-lives indicated that little degradation occurred during the 51-day incubation period (Table 4). Chlorinated compounds, found at ppb levels, were not significantly reduced dur-

Table 4. Collective Ranges of Half-Lives for Pentane Extractable Organics

Treatment	Range of $t_{0.5}$ in days
H_2O_2 + $HgCl_2$ - Abiotic	26.1-51.0
H_2O_2 Only	9.2-20.8
H_2O_2 + Nutrients	5.0-22.9
KNO_3 + $HgCl_2$ - Abiotic	46.2-288.8
KNO_3 Only	9.8-82.5
KNO_3 + Nutrients	15.6-119.5

ing the study (data not shown). However, this may be due to the limited treatment period or to the treatment conditions, e.g., aerobic rather than anaerobic.

When compared with abiotic controls, each pentane-extractable organic in the biologically active hydrogen-peroxide amended treatments exhibited a shorter half-life than its control. This relationship is maintained in the other treatment conditions for each pentane-extractable organic analyte. The consistent differences between the analyte half-lives of the abiotic controls and the biologically-active treatments may be taken as evidence that hydrocarbon concentrations within the microcosms were reduced by bacterial degradation.

Hydrogeology

The data thus far indicate that bioremediation is feasible. To promote rapid degradation of subsurface hydrocarbons by native bacterial populations, the metabolic activity of these microbes must be stimulated. This is accomplished by injecting fluids containing nutrients and an oxygen source to bacteria residing in the subsurface. These solutions may solubilize the hydrocarbons so that they are available to the bacteria as a carbon source.

If *in situ* bioreclamation technology is to succeed, the hydrogeology of the site must be well defined and have sufficient permeability to allow effective transport of injected solutions, resulting in hydrocarbon degradation within a reasonable time frame. Therefore, the rate of groundwater movement establishes the maximum rate of *in situ* bioremediation by controlling the rate of nutrient and oxygen transport. If the site hydraulic conductivities and hydraulic gradients are low, *in situ* treatment may not be technically feasible or cost-effective.

The evaluation of hydraulic feasibility focused on three types of data: the sand-body geometry, results of hydraulic testing, and withdrawal rates of existing groundwater recovery systems.

Stratigraphy. The sand-body geometry, or stratigraphy, should be well defined in order to design the most effective pattern for the array of injection/withdrawal wells and/or trenches. The location and distribution of the sediments, particularly the more permeable sand units, should be well defined in order to determine the groundwater flow relationships between contiguous geologic units containing affected groundwater. This distribution is also important in designing the most effective pattern for the array of injection/withdrawal wells or trenches.

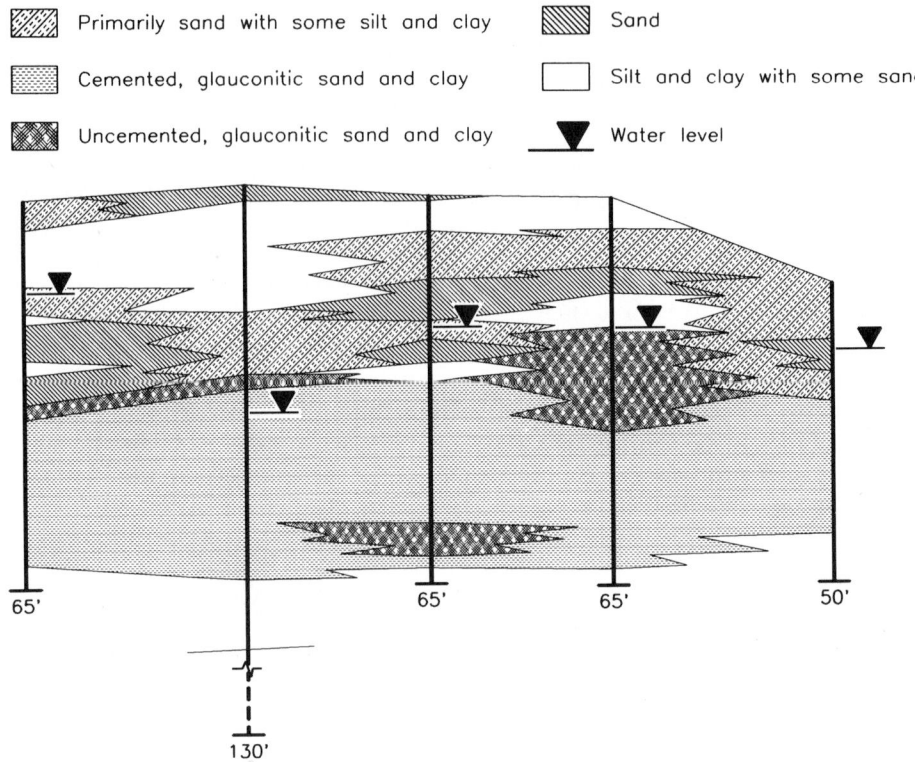

Figure 5. Shallow lithologic cross section.

Data from the logs of soil borings were used to delineate the lithology and stratigraphy of the subsurface. This lithologic cross section from the site shows that the subsurface is composed of interfingering beds of alluvial sand, silt and clay, with a few clay lenses (Figure 5). These boring logs also allowed the delineation of the channel sand deposits beneath the site. The sand bodies are hydraulically connected by the thinner interfingered sand beds.

Results of Hydraulic Testing. Well logs, slug tests, and well recovery and pumping tests of the existing groundwater system were evaluated to determine the hydraulic conductivity of the sediments. Results of these tests yielded sediment hydraulic conductivities that ranged from order of 10^{-5} to 10^{-2} cm/sec (Table 5). The tighter soils with high clay and silt contents typically have the lower permeabilities on the order of 10^{-5} to 10^{-6} cm/sec. Numerous clayey and silty sand units are present within the shallow subsurface which have hydraulic conductivities typically on the order of 10^{-4} cm/sec. Sands present in the channel sand deposits represent the highest hydraulic conductivities of 10^{-3} cm/sec and higher.

Transmissivities and flow rates are fair to marginally acceptable for *in situ* treatment (Table 5). Those soils which have low hydraulic conductivities below levels

Table 5. Ranges of Hydrogeologic
Characteristics Within the Site

Characteristic	Range
Hydraulic conductivities	10^{-5} to 10^{-2} cm/sec
Transmissivities	6 to 3500 gpd/ft
Average flow rate/well	<0.1 to 2 gpm
Conceptual trench flow	55 gpd/linear trench foot

generally accepted for bioremediation (10^{-4} cm/sec) also have slow water movement and fortunately little hydrocarbon contamination migration. *In situ* treatment is possible, but more difficult in these lower permeability ranges.

The bulk of the site has sufficient permeability and apparent hydraulic interconnection to enable circulation of biorestoration fluids. Hydrogeological characteristics of the site do not preclude *in situ* bioremediation, but they are marginal in areas and will be the rate-limiting factor in *in situ* treatment.

PROPOSALS

Well Array for *In Situ* Treatment

Results of the microbial, soil, and groundwater characterizations, laboratory treatability studies, and hydrogeologic evaluation indicate that *in situ* bioremediation can be used to degrade the bulk of hydrocarbons at this site. However, because of the low hydraulic conductivities, a large number of wells would be needed to achieve adequate flow rates necessary to transport nutrients and oxygen away from the injection point. Rather than wells, series of injection/withdrawal trenches may be more cost-effective and overcome the low flow rates (Figure 6).

The trenches would increase the flow rates from an average of 0.6 gpm/well to approximately 40 gpm/1000 feet of trench, providing good flow and transport rates for nutrients and oxygen sources. The arrangement of the trenches should take into account the orientations of the sand bodies, since the sands allow greater rates of groundwater flow through the matrix. By identifying favorable locations for injection/withdrawal wells or trenches, it should be possible to maximize coverage of the sand bodies, while maintaining depressed hydraulic heads in areas where inward directed gradients can aid in moving fluids from the less permeable units to the sands.

Pilot Test

A field pilot test is the next recommended technical project phase to aid in estimating the potential for the success of *in situ* bioremediation on a large scale. The pilot study data will be used to more precisely estimate the number of wells

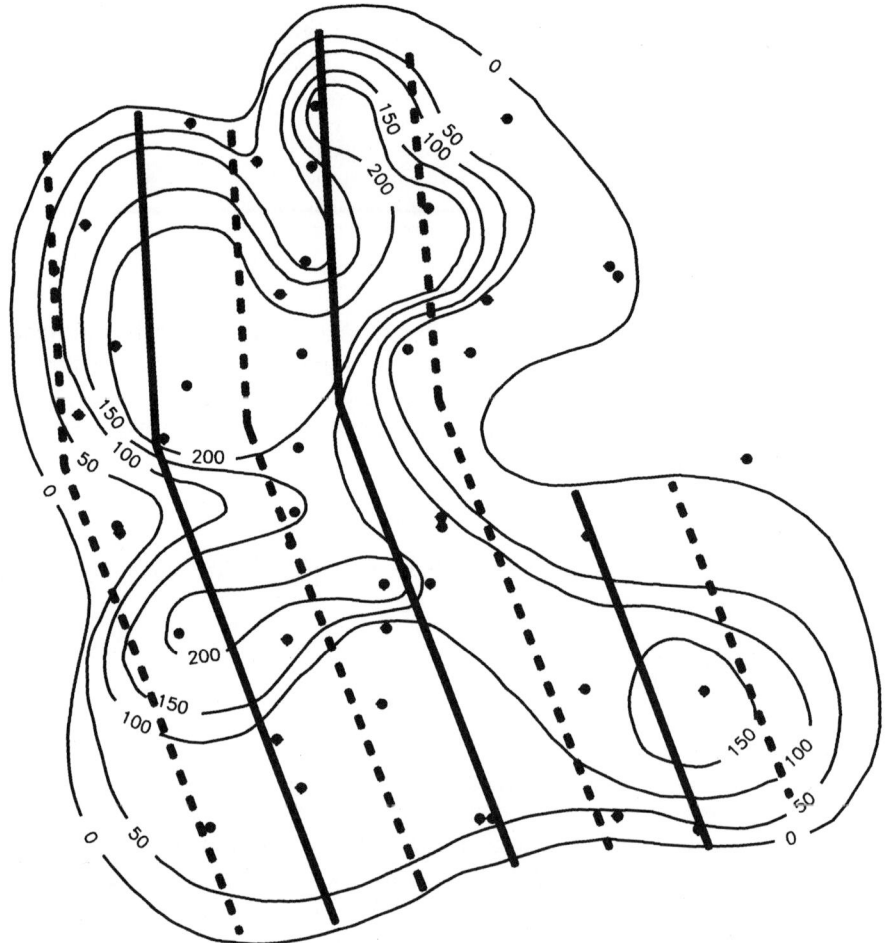

Figure 6. Conceptual extraction/injection trench system. Dashed lines (- - -) represent extraction trenches, solid lines (——) the injection trenches, and circles (●) show the existing wells. Isopleth contour (thin lines) based on PID readings (ppm).

or trenches required, the number of pore volumes to be pumped, and the real time transport and amounts of nutrients and peroxide consumed by biological and non-biological factors.

OTHER SELECTION CRITERIA

In addition to the above, there needs to be an evaluation of some of the non-technical aspects of the technology, such as costs and regulatory requirements.

The cost-effectiveness of bioremediation must be compared with the cost of the present pump and treat remediation program. Because time required for clean-up influences costs, one should ask how long it will take to remediate the site and if another method would be as effective in removing the hydrocarbon. Since the hydrogeology of the site is marginal, other technologies may be more feasible. Also, potential environmental impacts and the public's perception of the technology should be considered.

More often, EPA is allowing the use of bioremediation as an alternate technology to clean a site. Site-specific regulatory requirements for remediation methodologies and targets, *e.g.*, "how clean is clean" or "when do we stop", are negotiated between industry and the regulatory agencies. Currently, endpoints are not clearly defined and can be identified as leachable concentrations, total residual concentration, amounts of specific chemicals, some other surrogate parameters, or based on risk analysis.

Flexibility is needed in determining the closure of a particular site. For example, some aspects of RCRA closures may inhibit further biodegradation by blocking migration of water, oxygen, and nutrients to the bacteria. These type of issues need to be carefully explored and the best solution determined for the specific site.

SUMMARY

In summary, successful implementation of an *in situ* bioreclamation project requires a technical understanding of the site's microbial populations and their growth requirements, the physical and chemical characteristics of the soil and groundwater, the hydrocarbon distribution, transport, and biodegradability, and most importantly, the hydrogeology of the site. Even if the above factors indicate it is technically feasible, other factors must also be considered. These are: cost-effectiveness of the technology, potential impacts on human health and the environment, the public's perception, and regulatory requirements, both current and anticipated future regulations. Bioremediation may be a desirable alternative for a portion of this site, but it may have to be augmented or done in combination with other technologies. Information obtained in the field can be used to predict the potential for successful *in situ* treatment. Further evaluation of the cost, the proposed trench system, and other technologies, as well as input from the regulatory agencies are necessary before implementing a full-scale system.

ACKNOWLEDGEMENTS

We thank T. E. Bumpass, W. T. Edmonds, and T. W. McAninch for their valuable discussions and editorial comments; and R. S. Ford, D. A. Graves, and C. M. Renwick for their excellent technical assistance.

Integration of Biotechnology to Waste Minimization Programs

Godfred E. Tong

PERSPECTIVES ON WASTE MINIMIZATION PROGRAMS OF THE CHEMICAL INDUSTRY

This paper focuses on the waste minimization programs in the USA chemical process industries and the role of biotechnology in these programs. The role of biotechnology in waste minimization is reviewed from an R&D perspective. To obtain a good perspective on the waste minimization activities of the chemical industry in the last 5 years, one needs to answer the following two questions:

- How has the processing industry responded to the increased environmental regulations and social consciousness on waste generation?
- What waste minimization programs were implemented during this period and what are planned for the future?

The immediate positive action taken by the processing industry was to identify and quantify the sources of wastes being generated. This was done by extensive "audits" in each of the chemical processing plant facility. These "audits" pinpointed the sources of waste discharged, enabled characterization of the nature of the waste, and ultimately allowed the quantification of these wastes. The results of these quantitative "audits" on waste generation enabled individual companies to define the waste minimization targets and the resources to meet these targets (Ref. 1).

The waste minimization programs that were implemented after the completion of these audits can be classified under four categories:

Source Reduction Programs. Examples of projects implemented under these programs included more aggressive preventive maintenance projects to stop leaks in pumps, flanges, valves etc., implementation of back-up systems to prevent accidental spills, and minor process modifications such as changing from solvent-based to water-based carriers to coat tablets thereby reducing solvent emissions and tighter process controls to reduce production of "off-spec" products. These projects not

---

Godfred E. Tong • Monsanto Corporate Research, Monsanto Company, St. Louis, Missouri 63167.

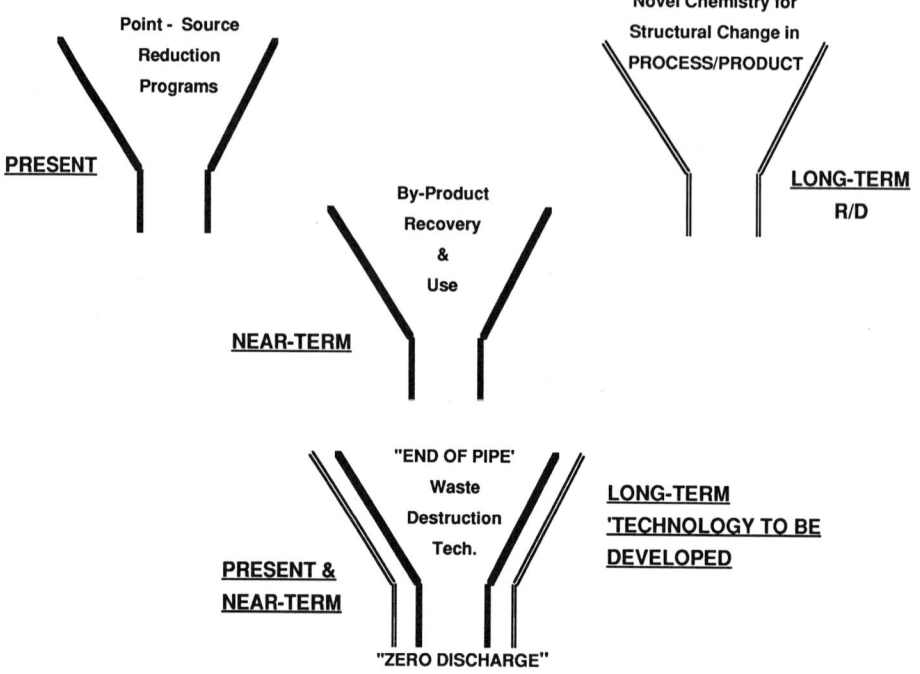

Figure 1. Funnelling of resources for waste minimization.

only significantly reduced the wastes generated but significantly reduced manufacturing costs due to lower raw material usage and waste treatment costs (Ref. 2) . For instance, thirteen major US ethylene oxide producers focussing on ways to stop process leaks in their source reduction program achieved a 74% reduction of ethylene oxide fugitive gas emissions (Ref. 2) .

As indicated in Figure 1, point-source reduction programs were the near-term projects implemented and the bulk of the waste minimization resources have been focused on this area of waste reduction. From an R/D perspective, these point source reduction programs did not involve the use of new technologies and equipment systems.

By-product Recovery and Utilization Programs. Examples of projects implemented under these programs included solvent recovery, evaporative concentration, and upgrading of recovered by-products to meet product specifications for new markets for the recovered chemicals (Ref. 3).

Unlike point-source reduction programs which were the immediate actions taken by the processing plants, by-product recovery and utilization programs require greater effort to develop the technologies to upgrade by-products and to develop markets for these recovered chemicals. The more readily implemented byproduct recovery programs involved projects where markets or uses for the recovered by-

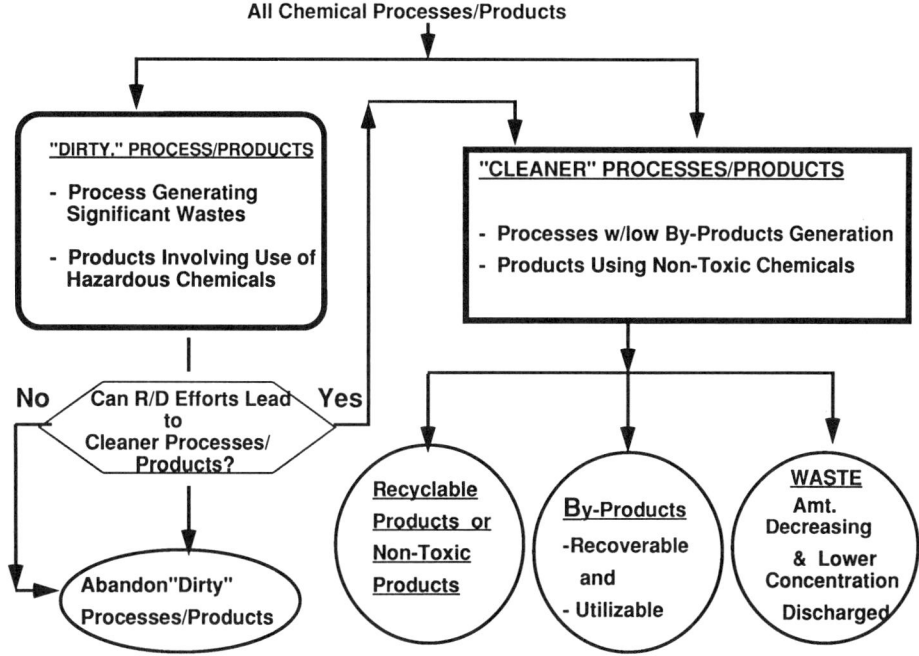

Figure 2. Long-term view of "zero-discharge" goals for chemical processes.

product were internally consumed such as the use of the recovered organics as energy sources.

"End-of-Pipe" Treatment of Wastes. These "end-of-pipe" treatment involves either physical, thermal, chemical, and/or biological treatment technologies for wastes. These treatment technologies are designed to destroy wastes, converting them to carbon dioxide, water and/or minerals. Relative to point-source reduction and by-product recovery projects, these "end-of-pipe" treatment programs are, in general, less economically attractive. Therefore from an R/D perspective, there is an increasing need to develop lower cost and more efficient waste destruction technologies.

While environmental biotechnology is often viewed in association with bioremediation of wastes disposed in non-active process plant sites or with end-of-pipe biotreatment, this paper treats environmental biotechnology from a broader perspective so as to include the application of this type of technology to waste minimization programs. A more detailed discussion of this broader view of biotechnology is given in the next section.

"Structural" Changes in Environmentally-Unsound Processes or Products ("dirty" processes or products). The net result of the quantitative "audits" on waste generation coupled with the more stringent environmental regulations has enabled the individual companies to define the economic incentives to allocate R&D resources to

replace these unsound environmental processes or products. As indicated in Figure 2, the majority of these R/D efforts to structurally change or abandon these "dirty" processes will require novel chemistries and will require substantial resources and time to achieve these longer term goals. They are, however, an integral part of the "zero discharge" goals that many processing companies are committed to pursue.

As costs for end-of-pipe treatment of wastes increases for the manufacturers using inefficient processes, the successful development cleaner processes will eventually replace these less efficient, environmentally-unsound processes. Thus, Figure 2 explains the long-term view that "zero-discharge or near zero discharge", while more difficult to achieve, does make economic sense in the 1990s.

Major trends implied by the long-term movement of the chemical process industry toward "zero-discharge" include the following:

a. The amount of waste per unit product will continue to decrease with time.
b. The amount of unrecoverable waste generated will be of a more dilute in concentration.

This latter waste minimization trend, toward increasingly dilute waste, is of significance, as biotechnology is the more appropriate and effective technology in most of the instances when one does a comparative cost evaluation of the treatment alternatives, that is, thermal, physical, chemical and biological treatment technologies, wastes are dilute.

BIOTECHNOLOGY ROLES IN WASTE MINIMIZATION PROGRAMS

The first section of this paper identified the aspects of the waste minimization programs relevant to biotechnology. This section defines the roles that biotechnology plays in the broadest context of waste minimization programs. One can examine biotechnology applications under two categories:

- Location within the processing plant where biotechnology is applicable,
- Type of biocatalysts that can be used in these waste minimization applications.

Biotechnology applications to waste minimization programs can be seen at three locations within a typical process plant (see Figure 3):

- in the process plant and off-site utilities,
- in by-product utilization
- in "end-of-pipe" treatment

In the Process Plant and Off-Site Utilities

Biotechnology applications are still limited but are emerging as a result of environmental regulations. The two biotechnology applications to be cited here are not well known.

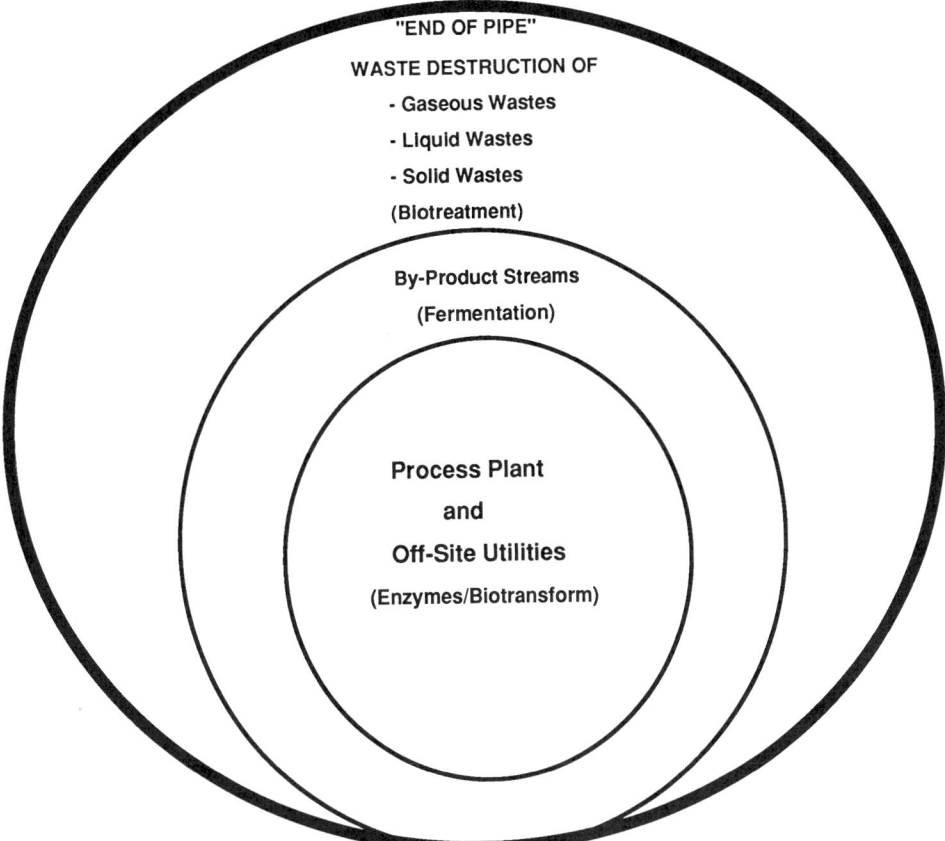

Figure 3. Biotechnology roles in waste minimization.

1. The application to be described is a biotechnology version of the "point-source" reduction program. The use of water-treatment chemicals for cooling water systems in processing plants is undergoing a "structural" change. Use of controversial corrosion control chemicals such as chromates is beginning to be phased-out as more states ban its discharge to the environment. Chromates are being replace by more environmentally acceptable additives such as phosphates which are nutrients for microbial (source of biocorrosion and biofouling problems) and algae (source of biofouling problems). Biocide chemical usage to control biofouling and biocorrosion in these high volume cooling water systems is being reduced to meet more stringent discharge guidelines (Ref. 4). These "structural" changes are resulting in the need for biotechnologists to

- develop and apply effective bioassay systems to monitor and control microbes responsible for biocorrosion and biofouling.

- apply principles of effective microbial growth inhibition systems to minimize the use of chemicals in these cooling water systems.

2. Specific examples of enzyme applications in the food processing, animal feed, textile and leather treatment and detergent industries have been well documented (Ref. 5). Enzyme applications involving aqueous process streams in other related industries such as the paper and pulp processing are beginning to emerge. Hemicellulase and cellulase are being tested to hydrolyze hemicellulose in pulp in order to reduce the consumption of bleaching chlorine in the pulping process and thereby reduce the generation off chlorinated organics wastes in bleached pulp production.

Biotechnology in By-Product Utilization

Most of the potential biotechnology applications in the area of by-product utilization for the chemical process industry are speculative. One can, however, cite a number of existing food-processing industry by-products currently used as raw materials for the fermentation industry; viz., cane/beet molasses, cheese whey for protein and ethanol production, and corn steep liquor for antibiotic production.

The application of biotechnology to chemical processing by-products is likely to be limited to the conversion of by-products to energy products such as methane, ethanol, acetone, and organic esters. A potential exception to this is the application of enzymes or biotransformation to racemic mixtures of agricultural chemicals such as insecticides, pesticides and herbicides. As research to find the more active isomers and the more readily degraded isomers progresses, manufacturers will be able to focus on technologies to produce purer, more active isomers of agricultural chemicals (Ref. 6).

The use of biotransformation or specific enzymes to achieve this goal is one of the potential technologies that can be applied to reduce environmental pollution by the less active isomers which may also be slower to degrade in field applications.

Biotechnology in "End of Pipe" Treatment (Waste Destruction Technology)

One can list three different types of biotreatment for wastes from process plants. Microbial catalyzed, green plant catalyzed, and combinations of green plant and microbial catalyzed biotreatment. More detailed discussions of each of the specific biotreatment applications are given in the next section.

Biotechnology roles in waste minimization can classified according to the types of biocatalysts one can use for each application. Figure 4 shows the hierarchy of biocatalyst systems that one can be used both in the by-product use and in the "end-of-pipe" treatment areas. One can make a generalized comment on the use of the higher forms of biocatalysts. As one progresses to higher forms of biocatalysts (enzymes to microbes to green plants) the reaction rates decrease. From the point of view of end-of-pipe biotreatment, the treatment rates will be slowest and will require larger "bioreactors" to achieve equivalent treatment. Despite this

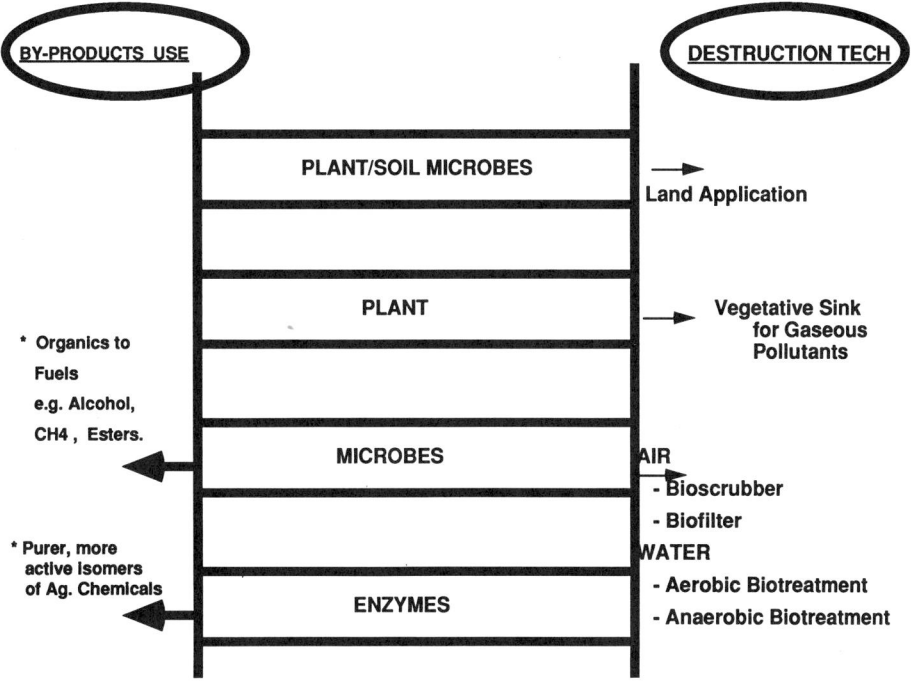

Figure 4. Hierarchy of biocatalyst systems.

inherent disadvantage of using higher forms of biocatalysts such as green plants, their role in biotreatment will become increasingly important as the chemical industry approaches "near-zero" discharge where the concentrations of organics and inorganics in waste streams becomes increasing dilute.

TECHNICAL ISSUES OF BIOTECHNOLOGY APPLICATIONS

A number of existing and potential biotechnology applications in waste minimization are briefly discussed in the context of the technical issues which are the sources of new discoveries and development for R/D.

Biotechnology in By-Product Use—Technical Issues

The anaerobic microbial fermentation of organics in aqueous waste to methane is a biotechnology application that can potentially be applied to other chemical byproduct streams. This application is currently limited by the low fermentation rates particularly at high concentrations (> 1 wt% of organics in aqueous solutions).

A second potential application for solids is in the area of "disposable" plastic starch (Ref. 7) . The development of starch with improved physical and chemical properties to substitute for "disposable" plastics can potentially provide a source of inexpensive raw material for the production of ethanol for liquid fuel use. At present, grain ethanol production is limited by the high cost of its raw material (starch cost is 60-70% of manufacturing cost). For this application to be feasible, developments in two areas are needed—successful product development of starch plastics and the development of segregation and collection centers for the disposable starch plastic wastes.

While this is speculative, fermentable organic solid wastes such as starch can be converted to fuels (methane, ethanol, butanol, acetone and organic esters) using existing conversion technology.

Biotechnology in "End of Pipe" Treatment—Technical Issues

The "end-of-pipe" biotreatment applications can be viewed according to the types of wastes encountered—gaseous or air and aqueous liquid. This review will focus on those biotreatment technologies that will become increasingly important as one starts to address the dilute concentration wastes of the chemical process industry.

Gaseous Wastes. Control and removal of odors from gas wastes can be achieved by biotreatment. While alternative technologies such as carbon absorption and catalytic oxidation exist, biological systems such as soil bed, peatbed (Ref. 8) or other forms of biofilters or bioscrubbers (Ref. 9) are becoming competitive treatment systems particularly in applications involving high gas flow rates (> 5000 m^3/h). In addition to their use in odor control, biofilters and bioscrubbers have been used to treat waste gases containing low levels of aromatics such as toluene, esters such as ethyl and butyl acetates, alcohols such as ethanol and butanol, and ketones such as acetone.

The technical issues associated with biofilter applications include selection of improved microbial strains to degrade chlorinated hydrocarbons such as vinyl and methylene chloride and amides such as dimethyl formamide (Ref. 9), slow treatment rates thus requiring land area, reliability of process controls, and other operational issues such as ease of changing the biofilter beds.

Aqueous Liquid Wastes. In the area of biotreatment of aqueous wastes, there are two components in industrial aqueous wastes that makes biotreatment inefficient: high salt content (above 3% wt) and high organic nitrogen content. High salt contents in chemical waste streams produce significant reductions in biotreatment rates. There is a need for the development of heterotrophic and osmiophilic biocatalysts such as algae or microorganisms. High organic nitrogen content in industrial wastes will cause high levels of ammonia which are inhibitory to microbial activity to be produced in biotreatment systems. Also, the microbial metabolism of nitrogen results in the generation of high levels of microbial sludge which are costly to handle.

As end-of-pipe treatments are applied to chemical process wastes, the discharge from these streams become increasingly dilute in organics and inorganics. For dilute organics, a form of "tertiary" biotreatment that is applicable to such stream is land application. Land application or land treatment involves mixing or dispersion of wastes into the upper zone of the soil-plant system with the objective of microbial stabilization, adsorption, immobilization, selective dispersion or crop recovery, leading to an environmentally acceptable assimilation of the waste. There are many issues that need to be resolved to make this an environmentally sound application. These include the appropriate selection of compatible waste stream, soil type and crops (grasses, pines etc.).

For dilute inorganic waste streams, biosorption is a technology being considered. Biosorption involves the use of inactivated, nonliving microbial biomass as a sorbent for the concentration of heavy metals. Biosorption can be viewed as part of a well-integrated effluent handling system for aqueous streams containing dilute heavy metals (Ref. 10).

SUMMARY AND CONCLUSIONS

Increasingly effective waste minimization programs will continue to be implemented. Biotechnologists have a definite role in every phase of waste minimization including

- point-source reduction
- by-product utilization
- end-of-pipe treatment

Existing as well as potential future biotechnology applications in each of the above areas of waste minimization programs were identified. A broad range of biocatalysts from enzymes to microbes to green plants that can function to minimize waste was discussed. Chemical industry movement towards zero-discharge goals and the resulting need for the treatment of dilute wastes was highlighted as an opportunity for increasing the application of biotechnology.

Land application of dilute non-toxic organics and aqueous salts, biofilters for dilute organics in waste gases, and biosorption of heavy metals were identified specifically as the emerging biotechnologies relevant to waste minimization programs.

More speculative biotechnology applications involving the utilization of potentially recoverable liquid and solid by-products were likewise identified. These included the following:

- potential conversion of liquid organics and solid by-products (such as plastic starch) to liquid or gaseous fuels.
- the use of biotransformation or specific enzymes to produce purer, more active isomers of agricultural chemicals in order to reduce environmental pollution.

REFERENCES

1. Parkinson, G., *"Reducing Wastes Can Be Cost-Effective"*, Chem. Eng'g., July, 1990, pp. 30-33.
2. Berglund, R., et al., *"Fugitive Emissions From the Ethylene Oxide Production Industry"*, Environmental Progress (Vol. 9, #1), pp. 10-17, 1990.
3. Osborne, F. T., Chi, C. T., Lester, J. H., *"Recovering Value from Process Wastes"* Int. Conference on Pollution Prevention: Clean Technologies and Clean Products, Washington D. C., June 10-13, 1990.
4. *"Consolidation Squeezes Water Treatment Industry"*, Chem. Business, July/August, 1990, pp. 13-15.
5. *"Promising New Markets Emerging for Commercial Enzymes"*, Chem. and Eng'g. News, September 24, 1990, pp. 17-18.
6. Armstrong, D. et al., *"Separations of Optical Isomers"*, Anal. Biochem. 167 (2), 261-4, 1987.
7. Miller, B., *"All Starch Polymer Enters Biodegradable Derby"*, Plastics World, March, 1990, p. 12.
8. Valentine, F., *"Making Chemical Process Plants Odor-Free"*, Chemical Eng'g, Jan, l990, pp. 112-119.
9. Ottengraf, S. P., *"Exhaust Gas Purification"*, in Biotechnology, Rehm, H.J., and Reed, G., Editors, Vol. 8, pp. 425-452 (1986).
10. Volesky, B., *"Biosorption of Heavy Metals"*, CRC Press, Boca Raton, Florida, 1990, pp. 51-64.

Bioremediation of Explosives Contaminated Soils (Scientific Questions/Engineering Realities)

Craig A. Myler and Wayne Sisk

INTRODUCTION

The organic compounds trinitrotoluene (TNT), hexahydro-1,3,5-trinitro-1,3,5-triazine (RDX) and 1,3,5,7-tetranitro-1,3,5,7-tetrazocine (HMX) have all been demonstrated as subject to biological attack to some degree. In fact, RDX and HMX are routinely treated in wastewaters from production facilities by a conventional anaerobic biological process. The presence of the subject compounds in soil, however, has proven a more difficult removal problem for biological processes, especially in the case of TNT, for which biological treatment was not considered possible before 1975. This paper will discuss the origins of soil contamination by explosives and the current efforts to reduce the treatment costs of these soils using biological methods.

Source of the Problem

Production of explosives has, quite naturally, been concentrated during times of military conflict. Likewise, research and development on explosives waste treatment and disposal followed the same general history. In the early 1900's research was conducted to determine the effects of TNT on human health. Animal and human studies concentrated on ingestion or adsorption of TNT with by-products of TNT metabolism typically being analyzed in the bodily fluids and tissues of the test subjects. Metabolism studies continued in the 1940s and 1950s (Channon, et. al. 1944, Lemberg and Callaghan, 1944) but additional research was conducted on effluent treatment from production and processing operations. The results of studies evaluating the treatment of TNT production waste streams concluded that biological treatment was not feasible, due principally to the presence of sodium salts of the nitrotoluene sulfonic acids in the wastes. (Schott, et. al., 1943) In general, waste dilution through discharge into streams was considered acceptable, even at production waste concentrations as high as 1000 ppm of the combined waste made up of cooling water, acid wash waters and red water streams. In the past, waste streams from ordinance filling operations contained the purified explosives only and were either mixed with existing sewage streams or allowed to settle in basins or lagoons. (Ruchhoft, et. al., 1945, Wilkinson, 1945) Removal of TNT in domestic sewage treatment plants was acknowledged as being too ineffective to be practically feasible as a sole means of disposal. Shell loading wastes were understood to present a toxicity risk to aquatic organisms and were therefore typically discharged to settling ponds or percolation basins. Biological treatment was only investigated as far as feasibility in existing sewage treatment facilities and only standard tests were conducted.

Pure Explosives as Wastes

In the early 1970s a new requirement for research presented itself in the form of pure explosives requiring disposal and/or treatment. These wastes were either from off-specification production or unserviceable munitions. They were typically disposed of by open burning/open detonation or through ocean dumping. Increased environmental awareness and action through legislation made these options undesirable, and new technologies were sought. (Brown, 1976) Biological treatment of raw explosives was investigated in earnest. (Osmon and Andrews, 1978, Hoffsommer, et. al., 1978) During this period the degradation of explosives compounds, particularly TNT, RDX and HMX was established. RDX and HMX were demonstrated to be degradable by anaerobic bacteria while TNT was shown to degrade aerobically and anaerobically. While conventional waste treatment methods were investigated, it was at this time that composting was first considered as a process for treatment of explosives. It is also in this period that engineers and scientists began to disagree on the results from various experiments. Laboratory data collected on TNT composts carefully monitored in the laboratory did not match that which came from pilot scale studies.

Explosive Contaminated Soil

In 1982 serious efforts were initiated to evaluate composting as a technology for remediation of explosives contaminated soils. Laboratory studies addressed various aspects of using microorganisms as a means to render explosives contaminated soils safe for land disposal. Additional studies addressed the metabolism occurring in these systems. As the requirement for remediation was established through regulatory action and the magnitude of the problem became clear, research in biological treatment of explosives contaminated soils has accelerated.

EXPLOSIVES CONTAMINATED SOIL REMEDIATION

As mentioned previously, waste streams from the explosives industry were often discharged to settling basins or lagoons as an accepted treatment process.

As these compounds are typically solids at ambient temperature and only sparingly soluble they have remained, for the most part, in the soil, only slowly migrating into groundwater. Currently, 26 active or inactive army ammunition production and processing properties are on the National Priorities List (NPL). The only implemented technology available for remediation of explosives contaminated soils is incineration, a process which costs approximately $300 per ton of contaminated soil. If each of the 26 NPL sites on average contained 200,000 tons of soil to be treated (this estimate is only hypothetical at this time as the extent of contamination is only now being determined) the total cost of remediation would exceed 1.5 billion dollars. This range of projected cost warrants investigation of alternative treatment methods, biological treatment standing out as the most promising for cost reduction.

Objective of Biological Remediation

The objective of biotreatment of explosives contaminated soils has been confused by many researchers with a requirement to create carbon dioxide and water from the principle contaminants. While the latter is a desirable outcome, it should be clear that the objective of remediation is to reduce to an acceptable level, the risk associated with the contaminants present. There are actually two risks associated with explosives contaminated soils. The first is an acute risk and comes from a detonation of the explosives present and a propagation of that detonation. The level of explosives required for explosive propagation in soils is 12 percent by weight. (Balasco, 1987) This does not preclude any explosion if explosives content is below 12 percent but describes a limit below which propagation will not occur. The second risk is more a chronic problem which is associated with the toxicity of the explosive compounds. Concentrations acceptable in soil and in water are dependent on many factors and are negotiated on a site by site basis. The relative insolubility of explosives is grounds for negotiating relatively large residual soil concentrations as leachability and transport are small in most cases. With the two risks in mind, treatment must first demonstrate the removal of the explosive hazard. This is accomplished by reducing the explosives concentration to below 12 percent (120,000 ppm). Second, the treatment must reduce the potential toxicity risk to acceptable levels. This includes reduction of the primary toxic constituents and by establishing, through acceptable protocols, the toxicity of the finished product.

COMPOSTING

The most advanced biological treatment technology for explosives contaminated soil remediation is composting and related solid phase systems. This is due in part to unsuccessful aqueous treatment system research described earlier and to the significant composting effort conducted since 1975.

Composting Background

While composting is considered an aerobic, thermophilic system by many, its use in the current discussion will include various solid phase systems including windrowing, aerated static pile, and in-vessel systems. Landfarming is also considered a solid phase treatment system. A literature review on explosives contaminated soils composting was conducted (Woodward, 1990) which identifies key works since 1982. Early workers attempted to fully characterize composting on a laboratory scale (Kaplan and Kaplan, 1982, Isbister, et.al., 1984), typically utilizing 10 to 20 grams of compost. These studies were felt to be definitive as the latest analytical techniques were used in an attempt to fully characterize the compost system. The best known of these studies was performed by Kaplan and Kaplan (1982) in which a simulated compost system produced metabolic products previously identified by McCormick, et. al. (1976). In fact, many of the identified products had been identified in the animal and human metabolism studies previously conducted. While laboratory studies seemed to indicate potential problems with composting explosives contaminated soils, pilot scale systems demonstrated feasibility of meeting the objectives mentioned previously. Osmon and Andrews (1978) concluded that natural biodegradation or land farming techniques resulted in the formation of compounds as described by Kaplan and Kaplan, but, they also concluded that composting did not produce any of these compounds at the completion of a 115 day compost test. Other laboratory and pilot scale composts of explosives and contaminated soils (Doyle, et. al., 1986, Isbister, et. al., 1982, Klausmeir and Jamison, 1980) demonstrated the avoidance of undesirable by-product accumulation and a general reduction in overall toxicity. Williams, et. al. (1988) demonstrated composting on a field scale as an acceptable means of reducing explosives concentrations in soils. Results of one experimental compost pile, shown in Figure 1, demonstrate significant explosives reduction as well as the extinction of unwanted by-products. Follow on analysis of the product from the field scale experiments (Griest, et. al., 1990) provided indication that leachates from explosives contaminated soil compost exhibit low toxicity. In Greist, et.al., 1990, aqueous leachate from thermophilic explosives contaminated soil compost was subjected to Ames testing. The compost leachate results showed 26 and 128 revertants per plate in strains TA 98 and TA 100 while the control showed 30 and 147 respectively. While not comprehensive, these results clearly indicate potential for using composting to reduce toxicity to acceptable levels.

Full Scale Composting of Explosives Contaminated Soils

A full scale design estimate was prepared (Lowe, et. al., 1989) which described various composting options including various sized aerated static pile facilities and mechanically agitated in-vessel systems. Essentially, worst case scenarios were used in formulating the facilities and operations designs. Kinetics data from the field demonstration (Williams, et.al., 1988) were used. Assumptions made included a 50 dollar per ton compost amendment, 90 day composting period and a requirement for hardened structures in which composting would be performed. Figure 2 presents

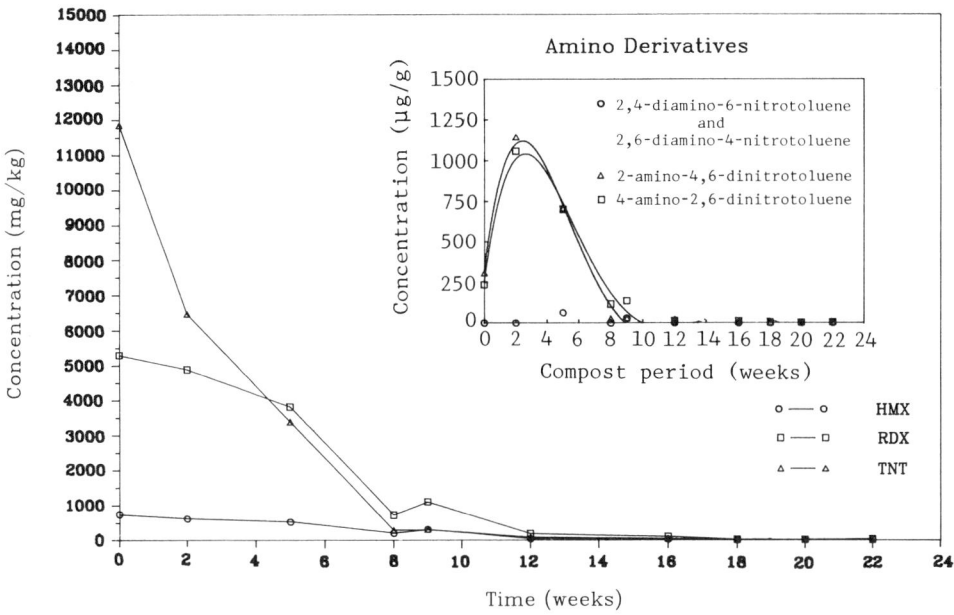

Figure 1. Explosives degradation in demonstration of composting (from Williams, et al., 1988).

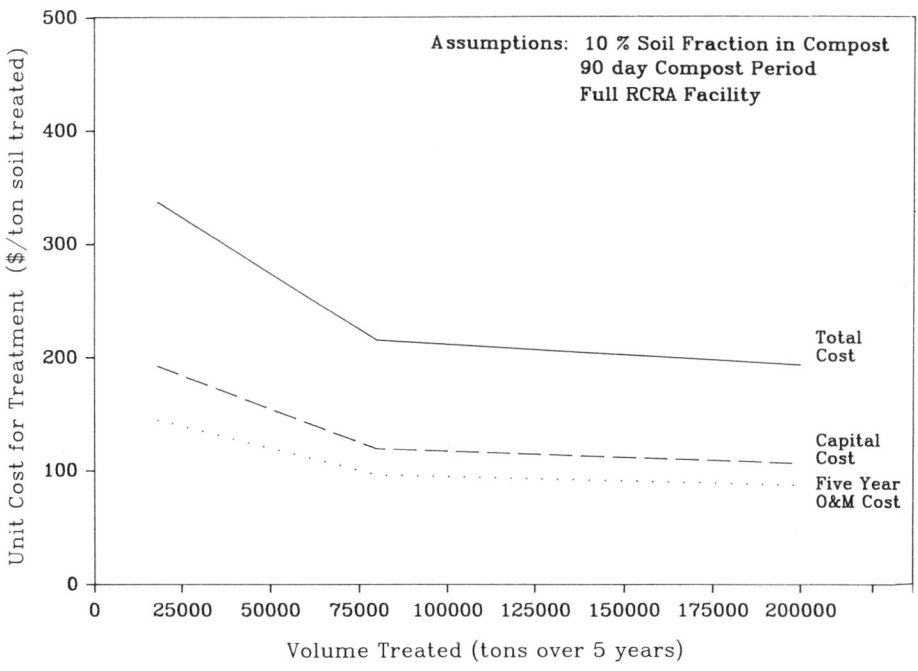

Figure 2. Cost estimate for explosives contaminated soil compost operation (from Lowe et al., 1989).

costs estimated for both capital and operations of a compost facility. The sensitivity to volume of soil treated is primarily due to capital costs. Operations costs are dependent on both the amount of soil treated per batch as well as the kinetics of degradation.

Sampling, Analysis and Control

Compost piles are non-homogeneous by design. Typically, a mixture of materials is made which exhibits the proper carbon, nitrogen and phosphorus levels as well as the appropriate moisture content, pH and porosity. When contaminated soil is introduced into the compost mixture the basic requirements for microbial growth do not change. During operation of the compost pile, differences in temperature, aeration, moisture, and pH occur throughout the compost mass. Non-homogeneity impacts sampling, analysis and control operations.

Sampling. Discrete samples taken from the compost mass may exhibit large variation in contaminant levels, especially at the beginning of the composting process. These variations tend to decrease as the compost process proceeds. For regulatory purposes the beginning and end points of the process are the only required sampling points for hazard analysis and should be performed using strict quality assurance/quality control procedures. Intermediate sampling is done to assure the compost process is able to proceed and includes temperature and moisture measurements. Averaging of temperatures throughout the compost mass is performed as wide variations can exist.

Analysis. There are currently no specific methods for determining explosive compounds in compost. Methods approved for soil analysis have been used to determine explosives levels in compost. One technique which is currently being used is to air dry the compost sample prior to shipment for analysis. Drying inhibits continued microbial action. Continued microbial action may skew analytical results. A dry compost sample can also be homogenized by milling. Homogenization reduces intra-sample errors in analysis.

Control. Little data have been reported on control of compost systems used for soil remediation. Temperature is by far the most frequently measured parameter used to effect process control in these systems. Typically, an average compost temperature is measured and aeration or mixing is performed to maintain a set point temperature. For explosives contaminated soils, both mesophilic (35 °C) and thermophilic (55 °C) temperatures have been effective but the thermophilic range demonstrated better explosives degradation performance. (Williams, et. al., 1988) Current development efforts incorporate relative humidity and oxygen measurements to better control moisture as well as temperature in the soil composts.

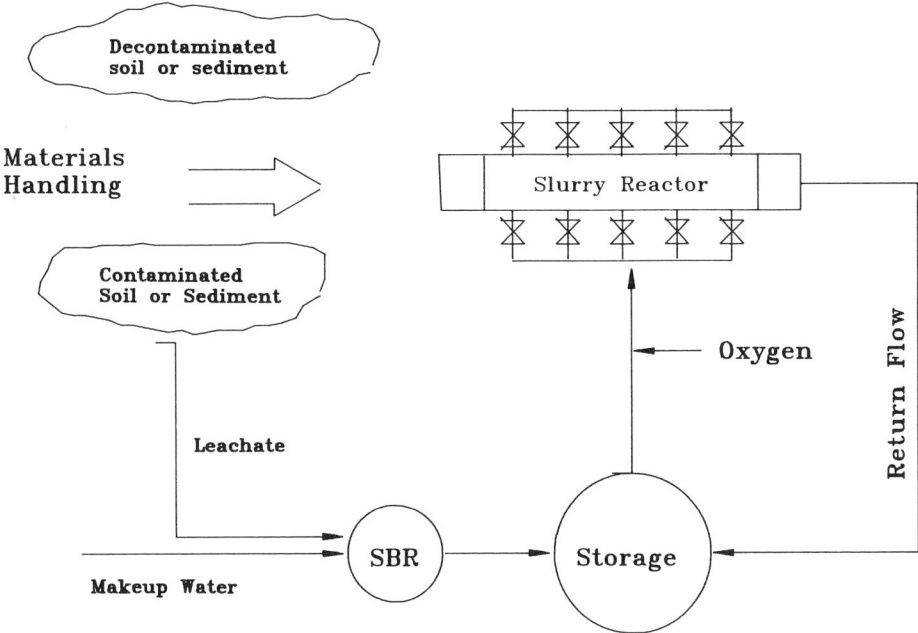

Figure 3. Conceptual slurry reactor for explosives contaminated soil remediation (from Montemagno and Irvine, 1989).

SLURRY REACTORS

Recently, aqueous soil treatment systems have been proposed for treatment of hazardous wastes. These systems have been described as slurry reactors (USEPA, 1989) and their use in hazardous waste treatment is in its infancy. Such reactor systems are under consideration for explosives contaminated soils. The potential benefits of slurry reactors for explosive contaminated soils are a reduction in handling sensitivity due to the aqueous phase, no volume increase once the soils have been dewatered, and decreased treatment costs. Bench scale studies are currently underway and preliminary results are promising. A conceptual full scale explosives contaminated soil treatment system (Montemagno and Irvine, 1989) is shown in Figure 3. The primary concerns in these systems are the metabolic products of degradation as well as system operability over extended time frames and climactic conditions.

INOCULATION

While both solid phase biological reactors and aqueous phase reactors have demonstrated potential using indigenous bacteria and fungi, increased rates of reaction and more positive control are the goals of innoculant, or prepared biological seed cultures. Only recently have isolated cultures capable of degrading explosive compounds been demonstrated. (Unkefer et.al., 1990, Fernando, et.al., 1990) Continued efforts are being pursued to field these microorganisms. Unfortunately, methods of inoculating soils contaminated with hazardous wastes are not well established. The conditions for growth of these cultures in the laboratory may not translate easily to a field situation.

SCIENTIFIC QUESTIONS/ENGINEERING REALITIES

The acceptance of biological means for explosives contaminated soil remediation is slow in coming. The apparent recalcitrance of the explosive TNT to biological dissociation is the primary obstacle to scientific endorsement. Reports and papers describing the results of limited early studies have incorrectly indicated that it was infeasible to biodegrade TNT contaminated soil by composting or other biological means. Although recent developments in isolating, identifying and culturing TNT degrading microorganisms have done much to stimulate continued research, basic knowledge gaps between laboratory and field studies remain. Other basic scientific information germane to biological treatment is also lacking. Simple adsorption/desorption phenomena for explosives on soils in the presence of microbial consortium is not available. Standard analytical methods for potential metabolic by-products have not been developed and standard analytical reference materials for these by-products have not been prepared. Finally, determination of a remediation end point has not been defined for biological systems. What is known about explosives biological treatment is enough to continue development efforts designed to implement biological treatment of contaminated soils on a full scale. The engineering realities of such an implementation are concentrated not on scientific unknowns but on physical and economic demands such operations will require. While domestic sewage and waste treatment systems utilizing biological methods are commonplace, hazardous waste treatment case studies of full scale implementations are scarce. Facilities requirements are only estimates at this point based on best guess scenarios. Operations costs, control requirements, system capabilities, operational ranges and other information must be determined from experiment. In effect, what the engineers require for design factors are not obtained from basic knowledge research but from implementation pilot studies.

CONCLUSION

Explosives contaminated soils pose a significant problem at hazardous waste sites and many of these sites are expected to require remediation. Incineration of these soils would be expensive (approximately $1.5 billion). The use of biological processes are under development to provide an alternative to incineration and thus, reduce these costs. While all the scientific questions have not been answered, the real obstacle to implementation of biotechnology has centered around a lack of engineering design data. Studies are continuing on both fronts to develop the necessary information for full scale implementation of bioremediation for explosives contaminated soils.

REFERENCES

Balasco, A.A., 1987, Testing to Determine the Relationship Between Explosive Contaminated Sludge Components and Reactivity, Final Report, U.S. Army Toxic and Hazardous Materials Agency, Aberdeen Proving Ground, MD, Report No. AMXTH-TE-CR-86096.

Brown, J.A., 1976, The Incineration Properties of Surplus Military PEPs, Final Report, Naval Surface Weapons Center, Crane, IN, Report No. N60921-76-M-E946.

Channon, H.J., Mills, G.T., and Williams, R.T., 1944, The Metabolism of 2:4:6-trinitrotoluene (a-T.N.T.), *Biochem J.*, 38:70-85.

Doyle, R.C., Isbister, J.D., Anspach, G.L., and Kitchens, J.F., 1986, Composting Explosives/Organics Contaminated Soils, Final Report, U.S. Army Toxic and Hazardous Materials Agency, Aberdeen Proving Ground, MD, Report No. AMXTH-TE-CR-86077.

Fernando, T., Bumpus, J.A. and Aust, S.D., 1990, Biodegradation of TNT (2,4,6-Trinitrotoluene) by phanerochaete chrysosporium, *Appl. Env. Micro.*, 56(6):1666-1671.

Greist, W.H., Stewart, A.J., Tyndall, R.L., Ho, C.H., and Tan, E., 1990, Charaterization of Explosives Waste Decomposition Due to Composting, Final Report, U.S. Army Medical Research and Development Command, Ft. Dietrick, MD, Report No. DOE IAG 1016-8123-A1.

Hoffsommer, J.C., Kaplan, L.A., Glover, D.J., Kubose, D.A., Dickinson, C., Goya, H., Kayser, E.G., Groves, C.L., and Sitzman, M.E., 1978, Biodegradability of TNT: A Three-Year Pilot Study, Final Report, Naval Surface Weapons Center, Crane, IN, Report No. NSWC/WOL TR 77-136.

Isbister, J.D., Anspach, G.L., Kitchens, J.F., and Doyle, R.C., 1984, Composting for Decontamination of Soils Containing Explosives, *Microbiologica*, 7:47-73.

Isbister, J.D., Doyle, R.C. and Kitchens, J.F., 1982, Composting of Explosives, Final Report, U.S. Army Toxic and Hazardous Materials Agency, Aberdeen Proving Ground, MD, Report No. DRXTH-TE.

Kaplan D.L. and Kaplan, A.M., 1982, Thermophilic Biotransformations of 2,4,6-Trinitrotoluene Under Simulated Composting Conditions, *Appl. Env. Micro.*, 44(3):757-759.

Klausmeir, R.E. and Jamison, E.I., 1982, Composting of TNT: Airborne Products and Toxicity, Final Report, U.S. Army Armament Research and Development Command, Dover, NJ, Report No. ARLCD-CR-81039.

Lemberg, R. and Callaghan, J.P., 1944, Metabolism of Symetrical Trinitrotoluene, *Nature*, 154:768-769.

Lowe, W., Williams, R. and Marks, P., 1989, Composting of Explosive-Contaminated Soil Technology, Final Report, U.S. Army Toxic and Hazardous Materials Agency, Aberdeen Proving Ground, MD, Report No. CETHA-TE-CR-90027.

McCormick, N.G., Feeherry, F.E., and Levinson, H.S., 1976, Microbial Transformation of 2,4,6-Trinitrotoluene and Other Nitroaromatic Compounds, *Appl. Env. Micro.*, 31(6):949-958.

Montemagno, C.D. and Irvine, R.L., 1989, Feasibility of Biodegrading Trinitrotoluene (TNT) Contaminated Soils, Proceedings of the 14th Annual Army Environmental R&D Symposium, Nov 14-16, 1989.

Osmon, T.L. and Andrews, C.C., 1978, The Biodegradation of TNT in Enhanced Soil and Compost Systems, Final Report, U.S. Army Armament Research and Development Command, Dover, NJ, Report No. ARLCD-TR-77032.

Ruchhoft, C.C., LeBosquet, M., and Meckler, W.G., 1945, TNT Wastes from Shell-Loading Plants-Color Reactions and Disposal Procedures, *Ind. Eng. Chem.*, 37:937-943.

Schott, S., Ruchhoft, C.C. and Megregian, S., 1943, TNT Wastes, *Ind. Eng. Chem.*, 35(10):1122-1127.

Unkefer, P.J., Alvarez, M.A., Hanners, J.L., Unkefer, C.J., Stenger, M. and Margiotta, E.A., 1990, Bioremediation of Explosives, Presented at the JANNAF Special Session on Alternatives to Open Burning/Open Detonation, Tyndall Air Force Base, FL, March 1990.

U.S. Environmental Protection Agency, 1989, The Superfund Innovative Technology Evaluation Program: Technology Profiles, EPA/540/5-89/013.

Wilkinson, R.W., 1945, Treatment and Disposal of Sewage and Waste Waters from Shell-Filling Factories, Institue of Sewage Purification Journal and Proceedings, Part 1, 145-150.

Williams, R.T., Ziegenfuss, P.S. and Marks, P.J., 1988, Field Demonstration-Composting of Explosives-Contaminated Sediments at Louisiana Army Ammunition Plant (LAAP), Final Report, U.S. Army Toxic and Hazardous Materials Agency, Aberdeen Proving Ground, MD, Report No. AMXTH-IR-TE-88242.

Woodward, R., 1990, Evaluation of Composting Implementation: a Literature Review, Final Report, U.S. Army Toxic and Hazardous Materials Agency, Aberdeen Proving Ground, MD, Report No. TCN 89363.

Practices, Potential, and Pitfalls in the Application of Biotechnology to Environmental Problems

Carol D. Litchfield

INTRODUCTION

The application of biotechnology to environmental problems is not a new concept. Although not generally recognized as examples of biotechnology, waste treatment plants (WTP) and composting certainly fit the definition of using biology and technology together. WTP's have been the subject of much study by engineers who have developed numerous designs for dealing not only with domestic waste but with industrial wastes too. In fact, the DuPont Company has a commercial enterprise that treats thirty-five million gallons of hazardous waste per day from throughout the United States! Perhaps this type of process has not been recognized as an example of biotechnology because of the complexity and only partial microbiological understanding of WTP's. However, the information gained from these plants is being directly applied to treating many of the environmental problems of today, and these systems should be considered as prototypes for today's applications of biotechnology to environmental problems.

This paper will review some of the current *applications* of biotechnology to environmental problems emphasizing on site and *in situ* bioremediation as field tests and full scale efforts, describe a few of the potential areas for development or field testing, and mention some of the perceived as well as real problems in applying biotechnology in the environment.

FIELD APPLICATIONS OF BIOREMEDIATION

Types of Bioremediation

There are basically five types of bioremediation: above ground bioreactors (AGB), solid phase treatment, composting, landfarming, and *in situ* treatment. All of these technologies have their uses and are currently being applied to bioremediate plant effluents, oily sludge bottoms from refineries, contaminated soils, and groundwater. Examples from the literature will be described for each technology.

When to Use Bioremediation

Several factors must be considered in evaluating whether to use biotreatment for a particular site. Bioremediation is appropriate whenever you can be certain that no toxic by-products will be formed, there are microorganisms which can degrade the contaminants, and it is economical. To be certain that more toxic by-products are not formed, it is essential to know the metabolic pathways that have been described based on studies of pure cultures and single compounds. These data are then confirmed using the natural mixed consortium or pure culture that will be employed at the site. Also, the laboratory studies must involve testing the mixture of contaminants at the *in situ* concentrations. Optimization studies performed on only two or three compounds from a list of 15 will not allow the engineers to design the best treatment system. It is also not possible to predict from pure compound studies how mixtures will interact or which materials will be degraded first. This is site and mixture specific.

Another factor to consider in determining whether to use bioremediation is the availability of the waste. For example, nonpoint-source contamination from agricultural runoff does not lend itself to biotreatment, while feed lot wastes certainly could be biotreated before reaching streams or groundwater.

Finally, the economics of the various bioremediation options must be evaluated. Some of the factors which are important here include: time to develop the necessary data base, length of time for the remediation, costs of construction, location of the contamination in the soils or in soils and groundwater, the hydrogeology of the site, location of the contamination in relation to site boundaries and surface structures, and space for the treatment system.

Applications of Bioremediation Techniques

Above Ground Bioreactors. Above ground bioreactors (AGB) have been designed to deal with contamination in either the liquid phase or the solid phase, present as a slurry, and require much the same technology as used in fermentors and WTP's. The AGB concept is most useful for soils when the soils have already been excavated, because the cost of excavation is a significant portion of the total costs for a soil or sludge slurry bioreactor.

The preferred AGB designs use suspended microbial growth or growth on a fixed solid support. The suspended support medium can be activated carbon, plastic spheres, glass, diatomaceous earth, etc. Two advantages of having the microbes attached to a surface in a biofilm are a reduction in the amount of sludge produced and a more rapid biodegradation rate probably due to the greater surface area. Different media such as granular activated charcoal may provide the best biofilm support for anaerobically operated digestors. This is also true for the DuPont aerobic industrial waste treatment plant mentioned earlier (Hutton, personal communication) which pioneered the powdered activated carbon treatment process. A word of caution, though, has been raised by Tsezos and Bell (1988) who noted that non-biodegraded contaminants can be sorbed onto the biofilm and the support medium, especially charcoal, thereby making the sludge that is formed a potentially hazardous material.

The inoculum can come from the contaminated materials, an activated sludge treatment plant, pure cultures, or potentially genetically engineered microorganisms (GEM). AGB systems lend themselves to the application of GEM technology because of the enclosed reactor design which allows for greater control of the GEM. Pharmaceutical companies use GEM in fermentation, so the most likely first application in the environment will be in AGB systems.

Depending on the chemical nature of the contaminants, the AGB may be operated in an aerobic or anaerobic mode, though, to date, most have been aerobic operations. It is not unreasonable to consider two bioreactors in series with the first one operated in the anaerobic mode (for example, to dehalogenate a compound) and the other reactor in the aerobic mode (to mineralize the resulting metabolic by-products).

One of the early applications of AGB technology to groundwater remediation was the work of Frick, et al. (1988) who reported the bioremediation of pentachlorophenol (PCP) at a wood-treating plant in Minnesota. These investigators had previously isolated a flavobacterium capable of degrading PCP and used it as the inoculum in their AGB. During four weeks, 70% of the PCP in contaminated surface water was removed, and greater than 99% of the PCP in groundwater was removed during single passage through the bioreactor. BioTrol has subsequently designed and installed a large groundwater and soil treatment system using immobilized bacteria to degrade PCP and a range of polynuclear aromatics (PNA) (Pflug and Burton, 1988).

Using the indigenous microbial population, ECOVA designed and installed an AGB to treat methyl ethyl ketone, isopropanol, ethylene glycol, and methyl isobutyl ketone. A consortium of microorganisms was found to be involved in these degradations: *Acinetobacter calcoaceticus* var. *anitratus*, *Pseudomonas putida*, *P. paucimobilis*, *P. testosteroni*, and *Geotrichum candidum*. After a 10-day acclimation period, the effluent met the targeted concentrations with most of the contaminants at the nondetectable level (St. John and Sikes, 1988).

We are currently installing an AGB system in Florida to treat groundwater contaminated with acetone, isopropanol, methyl ethyl ketone, chlorotoluenes, dioctyl phthalate, and xylenes. While microbial numbers in the groundwater and soil are low, use of an AGB followed by recirculation of the water through the sandy aquifer will stimulate the indigenous bacteria. The inoculum will be site soils and groundwater. Laboratory testing demonstrated that these materials could be degraded, Table 1, even at very high initial concentrations. Efforts are also under way to purify and identify the cultures that grew on the vapor-phase plates for future use as inocula, if needed.

Another example of an immobilized inoculum and AGB technology applied to contaminated groundwater is the work by Portier, et al. (1989). In this effort

Table 1. Degradation by Aquifer Microorganisms of Very High Concentrations of Mixed Organic Wastes in Sealed Microcosms

Contaminant	Initial conc. (ppm)[a]	Final conc. (ppm)[b]	Percent reduction
Acetone	7165	893 (4)	88
Methyl ethyl ketone	62700	19835 (8)	68
Xylenes	25300	2610 (8)	90
Isopropanol	5630	5218 (12)	7
Chlorotoluenes	73	46 (3)	37
Dioctyl phthalate	1060	830	22

[a]Concentrations added to each sealed microcosm.
[b]Number in parentheses is number of replicates.

the bioreactor was operated in a plug flow mode using specific bacteria immobilized on diatomaceous earth. An *Acinetobacter* sp. was the primary organism involved in organochlorine removal, while another group of bacteria was responsible for the degradation of the organophosphate pesticides. Removal rates during the field trial averaged over 50%.

In a final example, two air-lift bioreactors were developed in Germany with immobilized pseudomonads and arthrobacters. More than 90% of the aromatics for the tar refinery wastes were biodegraded, and in another bioreactor greater than 99% of naphthalene sulfonic acid was destroyed. While this process has not been commercially applied, the pilot plant scale test is encouraging the company to proceed to a full scale system.

Sequencing batch reactors (SBR) are often chosen because they permit one to operate over a larger range of contaminant concentrations (2-3 orders of magnitude) and irregular flow rates. Although more sludge is generated in an SBR, this design is much less susceptible to upsets than the previously described designs. Municipal and chemical landfill leachates are ideal situations for treatment in an SBR. The system developed by Chem Waste Management uses activated sludge and normally a 48 hr cycle including anaerobic-settling time. This plant is removing 81% of the COD from the leachate with a feed/decant volume of 52,500 gallons per cycle (Daley and Leary, 1989).

Wick and Pierce (1990) reported the results of Occidental Chemical Company's treatment of the Hyde Park Landfill leachate. Bacteria were isolated from the site and inoculated into SBR's in which over 90% of the TOC and 70 to 90% of the TOX were removed. The major compounds monitored were: chlorendic acid, phenol, benzoate, and o-, m-, p-chlorobenzoic acid (Wick and Pierce, 1990).

Solid Phase. When contaminated soil or treatment lagoon sludge has been excavated, it can be incinerated, possibly moved to a landfill, or bioremediated. An example of biotreating a hydrocarbon contaminated viscous sludge was described by Stroo, et al. (1989). They first reviewed the principles of liquid:solid treatment systems, with emphasis on the importance of the liquid:solid ratio, the mixing rate, and nutrient addition rates. After 2 to 8 weeks, the oily sludges contained less than 10% of the original oil and grease while 30 to 80% of the carcinogenic PAH's were removed, and the noncarcinogenic PAH's were at nondetectable levels. The material from sludge pits, however, had considerably longer half-lives (12 to 16 weeks) with 10 to 30% of the noncarcinogenic PAH's still present after eight weeks.

Another way of treating soils is to mound them either in an enclosure or on top of a liner and percolate water and nutrients through the mound. This type of treatment can also be very effectively combined in a treatment train with vacuum extraction of any volatile contaminants.

Composting. Other approaches that have also been used to treat solids (soils and sludges) include composting and landfarming. Composting has been widely used by farmers, mushroom growers, and suburban gardeners for years. It has recently been tried by Williams, et al. (1989) on the bioremediation of explosives: TNT (2,4,6-trinitrotoluene), RDX (hexahydro-1,3,5-trinitro-1,3,5-triazine), HMX (octahydro-1,3,5,7-tetranitro-1,3,5,7-tetraazocine), and Tetryl (N-methyl-1,2,4,6-tetranitroaniline). In the thermophilic pile, maintained at 55 C, greater than 90% of the explosives had been biodegraded or transformed within 80 days. From a starting concentration of 18,000 mg/kg, the final levels were 74 mg/kg after about 150 days. Obviously, composting should be considered more often as a means of remediating high concentrations of difficult to degrade wastes as are often found in sludges, tank bottoms, and aged soils.

Landfarming. Another on-site treatment that has been widely used in the last 30 years by the petroleum industry is landfarming. Bartha and Bossert reviewed the available literature on this subject in 1984. They discussed the importance of loading rate, moisture, fertilization, etc. to the proper design and functioning of a landfarm. Of course, for this technique to be applicable, the water table must be deep, there should be a thick clay layer under the landfarm area (either natural or constructed), and monitoring wells must be installed to insure that no groundwater is contaminated by the leachate. In a recent field demonstration, special plots were constructed, and oily lagoon wastes or buried oily wastes were applied at different loading rates. Over a 35-day period, approximately 70% of the aromatic compounds, and 90% of the aliphatic compounds were biodegraded (Bianchini, et al., 1988). Thus landfarming or modifications such as constructed plots are also viable techniques for complex, highly contaminated solid wastes.

***In Situ* Bioremediation.** *In situ* bioremediation (ISB) has been practiced since 1972 when Raymond first applied the concept to a gasoline spill in New Jersey. The site had sustained a 3200 barrel spill up gradient of the municipal well field. After free product was recovered, the addition of ammonium sulfate and phosphates and sparged air over a one-year period, led to the degradation of approximately 1080 barrels of gasoline (Raymond, et al. 1975). However, it was not until the late 1980's that *in situ* bioremediation gained favor with remediation firms and regulatory agencies.

The basic concepts of ISB are deceptively simple. The primary concept is that there are indigenous bacteria in groundwater and subsurface soils. Next, is that these microorganisms have adapted to the contaminants and to their physical/chemical environment. Finally, these adapted organisms will naturally degrade the organic contaminants until some nutrient reaches a limiting concentration. This is most often oxygen, but nitrogen and phosphate are also frequently involved. Thus, ISB is a natural, on-going process which is simply stimulated by the external addition of the limiting nutrients. For this reason many refer to ISB as *enhanced* ISB.

Since its first applications in the early 1970's, enhanced ISB has been used to clean up many sites contaminated with gasoline, diesel fuel, and JP-4 aviation fuel. The EPA and Coast Guard co-sponsored a field test of this technology at the Traverse City U. S. Coast Guard station (Ward, et al. 1989). At the three monitoring locations reported by Ward et al., there was an increase in the BTEX concentrations after a short lag period followed by a 50 to 100% reduction during the remainder of the study. This initial increase in contaminant concentrations is commonly seen, and it is believed to be due to the production of surfactants by the growing microorganisms (Raymond, personal communication). Seven months after nutrient addition began, the BTEX concentrations were less than 0.3 mg/kg aquifer material (Wilson and Kampbell, 1989).

At another leaking underground storage tank site where enhanced ISB was applied, Litchfield and co-workers noted over two orders of magnitude fluctuations in the microbial biomass during weekly sampling of the monitoring wells. This could not be directly correlated with nutrient levels or oxygen availability. Following 18 months of hydrogen peroxide addition as the oxygen source, however, the groundwater had gone from over 3000 ppm TPH to less than 1 ppm in the immediate vicinity of the former storage tank (Litchfield, et al. 1988).

Jhaveri and Mazzacca used enhanced ISB to clean up a site in New Jersey which had been contaminated with methylene chloride, n-butyl alcohol, dimethylaniline, and acetone (1985). After three years of operation, over 90% of the contaminants had been removed from the soils and groundwater.

Currently, a chlorobenzene (CLBZ) contaminated site is under going enhanced ISB. Preliminary laboratory testing demonstrated that bacteria were present which could degrade chlorobenzene, Table 2, and further testing revealed a complex consortium of both gram positive and gram negative bacteria (Litchfield and Belcher, 1990. Some strains could mineralize CLBZ to CO_2, while other strains could only metabolize CLBZ to 3-chlorocatechol. The former bacteria, however, could also mineralize 3-chlorocatechol, so this intermediate did not accumulate in the groundwater. It is expected that the project will take three to five years to complete.

Finally, the chlorinated aliphatic solvents are of great interest because of their wide spread distribution in groundwater. Methane augmentation of aquifers may induce methane monooxygenase mediated solvent degradation. This enzyme has broad specificity and allows the organisms to co-metabolize trichloroethylene (TCE) and dichloroethylene (DCE), although in the field methane solubility may be the rate limiting factor in this type of bioremediation effort.

Table 2. Degradation by Groundwater Bacteria of Chlorobenzene under Various Nutrient Amendments

Nutrient mixture	Percent removal	
	Sample 1	Sample 2
NH_4Cl	> 99.9	87
NH_4Cl+acetate	> 99.9	84
NH_4Cl+acetate+H_2O_2	85	99.9
NH_4Cl+H_2O_2	99	81
$NaNO_3$	> 99.9	N.T.[a]
$NaNO_3$+acetate	>99.9	N.T.
$NaNO_3$+acetate+H_2O_2	N.T.	> 99.9
HCO Killed Control	64	63
No Nutrient Control	N.T.	70
Total amount Chlorobenzene (ppm)	150	150

[a]N.T.= Chemical analyses not performed.

In other work, Bourquin and co-workers have field tested a pseudomonas species with the ability to mineralize TCE when an inducer compound such as toluene or phenol is present (Nelson, et al., 1987). They have recently been granted a patent on a more environmentally acceptable inducer tryptophan (Nelson and Bourquin, 1990).

Litchfield et al. have also completed a field test of the enhanced ISB of these compounds with the result that 90% of the perchloroethylene and 70% of the TCE were degraded within three months, but DCE accumulated even though this was an aerobic system. There was no appearance of vinyl chloride. This would seem to indicate that another pathway for the biodegradation of these compounds is possible when mixed natural consortia are involved (Litchfield, et al. in preparation).

SOME POTENTIAL APPLICATIONS OF BIOTREATMENT/ BIOREMEDIATION TO ENVIRONMENTAL PROBLEMS

Waste Minimization

Given the wide availability of small AGB's and our ever increasing knowledge about degradative pathways, a logical next step is the application of this information to waste minimization programs. This could come in the form of process stream bioreactors to reduce the by-products before they become mixed with other waste streams. An even better use of AGB is to utilize this concept to convert the by-products to new raw materials or products.

An example of the former potential application is the work reported by Heitkamp, et al. (1990) showing biodegradation of a synthetic p-nitrophenol (PNP) waste stream by pseudomonads immobilized on diatomaceous earth. Industrially relevant concentrations were tested with 90 to over 99% removal of the PNP. This is not a unique system, but it certainly needs to be utilized more widely.

A patent was granted in 1990 to Meyer, et al. for using microorganisms to degrade the acetonitrile from HPLC effluents. Another patent was issued to Yoshizawa, et al. in 1989 to treat food-processing wastes. Perhaps, the idea will catch on.

Agricultural Prospects

Two areas of concern in agricultural pollution are pesticide use and specific waste treatment. A great deal of research is underway to evaluate the use of endophytic bacteria to control plant pests. This again builds on years of research in the control of root diseases, reviewed by Gould (1990). Much of the work is of a proprietary nature, but the initial tests look very promising for drastically reducing the use of xenobiotic pesticides within the next 10 years.

Bioreactors for the treatment of agricultural wastes is also not a new idea. They were being tested by Hobson, et al. in 1984. Simplifying the design, reducing the costs, and stabilizing the microbial population so that an "off the shelf" package could be bought would make this type of system usable by the farmer.

Heavy Metals and Desulfurization

Heavy metal removal from groundwater or waste water has been based on the physical precipitation of the materials or the use of ion selective membranes. Recently, Henderson and his co-workers reported using attached sulfate-reducing bacteria to precipitate the heavy metals from synthetic waste streams (Henderson, et al., 1990). Litchfield and Jewell had also demonstrated the efficacy of this technique using expanded bed anaerobic bioreactors where over 99.99% of the lead, nickel, and 90% of the chromium could be removed by these organisms (data not shown).

The removal of sulfur compounds from coal and waste water, on the other hand, has been widely tested in the laboratory. Isbister et al. reviewed the information available in 1987 and described the field test at Atlantic Research on five different coals. Bioleaching of the inorganic sulfur occurs naturally as the result of ferric iron production which then oxidizes sulfide minerals into the soluble ions and elemental sulfur. This then provides the substrate for the action of *Thiobacillus* to oxidize the sulfur further and produce acid mine runoff. Organic sulfur reductions ranged from 19 to 30% with retention times of 9 to 48 hr (Isbister et al., 1988).

Also using the thiobacilli and aerobic processes, a patent was granted to the Rijkslandbouwuniverseit Wageningen in 1989 to remove sulfides from waste water using *Thiobacillus* and *Thiomicrospira* (Netherlands Patent NL 88 01,009). In both cases, only the field implementation of such concepts remains to make these technologies generally available.

PITFALLS IN THE APPLICATION OF BIOTECHNOLOGY

There are really two types of pitfalls present: perceived and real. In many cases the perceived problems are more difficult to deal with because field testing or full scale remediation efforts will not dispel them.

Perceived Pitfalls

Many of the perceived pitfalls result from an inability of the scientist and engineer to communicate with the public, client, and regulators. There is a distrust of microbial processes partially because of the "Andromeda Strain" image, that somehow monster bacteria will be produced with extensive uncontrolled degradative capabilities. The other fear is that pathogenic bacteria will be unleashed due to biotreatment above ground or *in situ*. There does appear to be less concern with above ground technologies than with ISB, because the one is in a confined container and can be controlled, while ISB is often portrayed as a black box in the unidentified and uncontrollable subsurface.

The other major perceived obstacle occurs because the extensive history of biotechnology and environmental problems is forgotten. Composting, waste treatment plants, fermenters have been around a long time, but we tend to forget that fact in the excitement of being considered novel, new, and innovative. What is innovative is the hundreds of applications of the "old" biotechnology to today's problems and we should emphasize that.

Real Problems

There are real limitations to the application of biotechnology that result from the heterogeneous nature of the waste, where it is located, and the complexity of the wastes and process streams. For the former, a site may contain not only contaminated soils and water, but also concrete, process tanks, building debris, etc. At such times we must be imaginative in combining other technologies into a treatment train. Bioremediation will not always be the solution for all problems.

A very real impediment to applying biotechnology can be the developmental time and costs. At a contaminated site frequently deadlines have been established and must be met. These will not allow one to isolate the organism(s), develop the best immobilization medium, and design the best bioreactor. Again, using off the shelf components, building on previous work, and limiting testing to confirmatory studies and not research can all help to reduce this lead time.

Also, because remediation affects the bottom line in a negative way, clients often want the quickest (in terms of implementation) and cheapest treatment. However, this attitude can be changed through communication and demonstration that microbial treatment is usually the most rapid and most complete treatment system if contaminant mineralization is possible.

SUMMARY

The limitations of applying biotechnology to environmental problems are mainly our imaginations and the willingness of engineers and microbiologists to really communicate. Both field tests and full scale demonstrations of the biotreatment of hazardous wastes have been performed over the past 10 years, and the time is ripe to engineer these systems into lower cost, off the shelf units so that biotreatment will become the first technology considered, not the last.

REFERENCES

Anonymous, 1990, Tar Refinery Waste Degradation Biotechnical Process, *Eur. Biotechnol. Newsl.* 89 p2.

Bartha, R and Bossert, I., 1984, The Treatment and Disposal of Petroleum Wastes, in: *Petroleum Microbiology* (R. M. Atlas, ed.) MacMillan Pub. Co. New York, pp 553-577.

Bianchini, M. A., Portier, R. J., Fujisaki, K., Templet, P. H., and Matthews, J. E., 1988, Determination of Optimal Toxicant Loading for Biological Closure of a Hazardous Waste Site, in: *Aquatic Toxicology and Hazard Assessment: 10th Volume*, ASTM STP 971 (W. J. Adams, G. A. Chapman, and W. G. Landis, eds.) American Society for Testing and Materials, Philadelphia, pp. 503-516.

Daley, P. S., O'Leary, K, and Arand, J., 1989, Commercial Treatment of Municipal and Chemical Landfill Leachates by Sequencing Batch Biological Processing, in: *Hazardous Waste Treatment: Biosystems for Pollution Control*, Air and Waste Management Association, Pittsburgh, PA, pp. 125-140.

Frick, T. D., Crawford, R. L., Martinson, M., Chresand, T., and Bateson, G., 1988, Microbiological Cleanup of Groundwater Contaminated by Pentachlorophenol, in: *Environmental Biotechnology Reducing Risks for Environmental Chemicals through Biotechnology* (G. S. Omenn, ed.), Plenum Press, New York, pp 173-191.

Gould, W. D., 1990, Biological Control of Plant Root Diseases by Bacteria, in: *Biotechnology of Plant-Microbe Interactions* (J. P. Nakas and C. Hagedorn, eds.), McGraw-Hill Publishing Co., New York, pp 287-317.

Heitkamp, M. A., Camel, V., Reuter, T. J., and Adams, W. J., 1990, Biodegradation of *p*-Nitrophenol in an Aqueous Waste Stream by Immobilized Bacteria, *Appl. Environ. Microbiol.* 56:2967-2973.

Henderson, W. D., Bewtra, J. K., St. Pierre, C. C., and Biswas, N., 1990, Removal of Heavy Metals from Wastewater using Sulfate Reducing Bacteria in Attached Growth Systems, in: *Proceedings Eighth International Biodeterioration and Biodegradation Symposium*, Elsevier, Essex, England, in press.

Hobson, P. N., Summers, R., and Harries, C., 1984, Single- and Multi-stage Fermenters for Treatment of Agricultural Wastes, in: *Microbiological Methods for Environmental Biotechnology* (J. M. Grainger and J. M. Lynch, eds.), Academic Press, New York, pp. 119-138.

Isbister, J. D., Wyza, R. E., Lippold, J., DeSouza,A, and Anspach, G., 1988, Bioprocessing of Coal, in: *Environmental Biotechnology Reducing Risks from Environmental Chemicals through Biotechnology* (G. S. Omenn, ed.), Plenum Press, New York, pp. 281-293.

Jhaveri, V. and Mazzacca, A. J., 1985, Case History I. Bio-reclamation of Ground and Ground Water by In-Situ Biodegradation and II. Bio-reclamation of Ground and Ground Water, Privately published by: Groundwater Decontamination Systems, Inc. 140 Route 17 North, Paramus, New Jersey, 07652, pgs.52.

Litchfield, C. D. and Belcher, L. A., 1990, Microbial Degradation of Chlorobenzene in Groundwater, in: *Proceedings Eighth International Biodeterioration and Biodegradation Symposium*, Elsevier, Essex, England, in press.

Litchfield, C. D., Erkenbrecher, Jr., C. W., Matson, C. E., Fish, L. S., and Levine, A., 1988, Evaluation of Microbial Detection Methods and Interlaboratory Comparisons During a Peroxide-Nutrient Enhanced *In Situ* Bioreclamation, in: *Proceedings International Conference on Water and Wastewater Microbiology* (B. H. Olson and D. Jenkins, eds), Vol. 2, pp. 52-1 - 52-6.

Meyer, O, Sander, A., and Bambang, S., 1990, Process and Microorganisms for Biodegradation of Acetonitrile in HPLC effluents. German Patent DE 3,831,396.

Nelson, M. J. K. and Bourquin, A. W., 1990, Method for Stimulating Biodegradation of Halogenated Aliphatic Hydrocarbons, United States Patent No. 4,925,802.

Nelson, M. J. K., Montgomery, S. O., Mahaffey, W. R., and Pritchard, P. H., 1987, Biodegradation of Trichloroethylene and Involvement of an Aromatic Biodegradative Pathway, *Appl. Environ. Microbiol.* 53: 949-954.

Pflug, A. D. and Burton, M. B., 1988, Remediation of Multimedia Contamination from the Wood-Preserving Industry, in: *Environmental Biotechnology Reducing Risks from Environmental Chemicals through Biotechnology* (G. S. Omenn, ed.), Plenum Press, New York, pp 193-201.

Portier, R. J., Nelson, J. A., Christianson, J. C., Wilkerson, J. M., Bost, R. C., and Flynn, B. P., 1989, Biotreatment of Dilute Contaminated Ground Water Using an Immobilized Microbe Packed Bed Reactor, *Environ. Progress* 8: 120-125.

Raymond, R. L., Jamison, V. W. and Hudson, J. O., 1975, *Beneficial Stimulation of Bacterial Activity in Ground Waters Containing Petroleum Products*, Final Report American Petroleum Institute, Project OS 21.2, Washington, D. C. p. 141 plus attachments.

Reikslandbouwuniversiteit Wageningen, 1990, Removal of Sulfides from Wastewaters Using Sulfur-oxidizing Bacteria. Patent Netherlands Appl. NL 88 01,009.

St. John, W. D. and Sikes, D. J., 1988, Complex Industrial Waste Sites, in: *Environmental Biotechnology Reducing Risks from Environmental Chemicals through Biotechnology* (G. S. Omenn, ed.), Plenum Press, New York, pp. 237-252.

Stroo, H. F., Smith, J. R., Torpy, M. F., Coover, M. P., and Kabrick, R. M., 1989, Bioremediation of Hydrocarbon-Contaminated Solids Using Liquid/Soils Contact Reactors, in: *Superfund '89 Proceedings of the 10th National Conference*, Hazardous Materials Control Research Institute, Silver Springs, Maryland, pp. 332-337.

Tsezos, M. and Bell, J. P., 1988, Significance of Biosorption for the Hazardous Organics Removal Efficiency of A Biological Reactor, *Wat. Res.* 22: 391-394.

Ward, C. H., Thomas, J. M., Fiorenza, S., Rifai, H. S., Bedient, P. B., Wilson, J. T., and Raymond, R. L., 1989, In Situ Bioremediation of Subsurface Material and Ground Water Contaminated with Aviation Fuel: Traverse City, Michigan, in: *Hazardous Waste Treatment: Biosystems for Pollution Control*, Air and Waste Management Assn., Pittsburgh, PA, pp. 83-96.

Wick, C. B. and Pierce, G. E., 1990, An Integrated Approach to Development and Implementation of Biodegradation Systems for Treatment of Hazardous Organic Wastes, in: *Developments in Industrial Microbiology*, Vol. 31 (J. J. Cooney, V. P. Gullo, A. I. Laskin, O. Sebek, J. C. Hunter-Cevera, and C. H. Ward, eds.), Elsevier, New York, pp. 81-96.

Williams, R. T., Ziegenfuss, P. S., Mohrman, G. B., and Sisk, W. E., 1989, Composting of Explosives and Propellant Contaminated Sediments, in: *Hazardous Waste Treatment: Biosystems for Pollution Control*, Air and Waste Management Assn, Pittsburgh, PA, pp. 269-281.

Wilson, J. T. and Kampbell, D. H., 1989, Challenges to the Practical Application of Biotechnology for the Biodegradation of Chemicals in Ground Water, Preprints of Papers Presented the 197th ACS National Meeting, pp. 74-76.

Yoshizawa, K., Tadenuma,M., Sato, S., Iefuji, H., Shimoii, H., Suzuki, O., Hamazaki, K., and Nitta, Y., 1989, Treatment of Wastewaters from Food-Processing Plants. Japanese Patent Jpn. Kokai Tokkyo Koho JP 01,224,012 (89,224,012).

What Is the K_m of Disappearase?

Ronald Unterman

INTRODUCTION

There are many reasons why the bioremediation of hazardous wastes holds great promise. (1) It is a technology that has already been demonstrated for easier targets, such as gasoline and simple pesticides. (2) Researchers have discovered many microorganisms in the last few years that are capable of degrading what were previously thought to be highly recalcitrant molecules. A good example of this is trichloroethylene (TCE), which only ten years ago was thought not to be biodegradable. In the intervening years, both anaerobic and aerobic biotransformations have been demonstrated encompassing a wide variety of biochemistries. These include aerobic microorganisms which degrade toluene, phenol, methane, ammonia and propane. Likewise, many groups have isolated and characterized polychlorinated biphenyl (PCB)-degrading microorganisms representing many microbial genera. (3) We now have available the tools of genetic engineering to manipulate the genetic material that encodes degradative enzymes and pathways and thereby create vastly superior microorganisms. Of course, there are regulatory and public concerns about the use and release of these microorganisms, but these will in time be resolved. (4) Where bioremediation is applicable, it is one of the most cost-effective approaches available. (5) Bioremediation of hazardous chemicals is a natural process essentially identical to the breakdown conversions that nature catalyzes continuously to recycle our planet's carbon. Indeed, the problem of recalcitrant chemicals in our environment is, in some cases, a product of our attempt to make nature's compounds more stable, for example, by adding chlorine to aromatic compounds to provide chemical or thermal stability. Our goal is now to reintroduce these chemicals back into the carbon cycle.

There are several challenges that bioremediation faces. First, laboratory discoveries need to be transitioned to commercial products; a "flask-to-field" conversion. Another problem clearly faced by this industry is that bioremediation has been oversold. It is not a panacea. It is only one technology among many for treating

---

Ronald Unterman • Envirogen, Inc., Princeton Research Center, Lawrenceville, New Jersey 08648.

hazardous wastes. Other approaches such as incineration, vitrification, solvent extraction and chemical destruction must all be considered and utilized for the complex problems that face our society. Biology will have its niche, but it will only be one of many solutions.

The last challenge I will address is the subject of this paper. The bioremediation industry has a reputation similar to that of the snake-oil salesman. For too many years companies have been selling "foo-foo dust" — a magic elixir for destroying a client's most serious chemical waste problems. The challenge we now face is to set technical standards which can withstand the most critical technical review. It is our responsibility as an industry to prove that our technology truly destroys the targets to which it is directed.

It is critically important that throughout biodegradation process development research (and ultimately commercial marketing), investigators clearly demonstrate that their bacterial decontamination results are unequivocally due to biological activity. Too often, the results of some studies have not been able to quantitatively account for the disappearance of the target substrate. In some cases, highly hydrophobic contaminants are redistributed in reactors or sorbed to unsampled locations in these reactors. In other studies with, for example, volatile organics, vigorous aeration has potentially volatilized the target instead of biodegrading it. Therefore, biodegradation studies and demonstrations must be designed to determine as best as possible a mass balance of the target substrate and hopefully demonstrate the products of this transformation (ideally, carbon dioxide and water).

INSOLUBLE TARGETS

One such pitfall that can be encountered in biodegradation studies can be illustrated by research that the author and his co-workers have been conducting on the biodegradation of PCBs, although the same arguments can be made about other insoluble substrates, such as PAHs. This family of compounds includes over 200 different congeners differing in the number and position of chlorine atoms on the biphenyl nucleus. These compounds are highly insoluble and sorb to soil and sediment particles.

Studies in many academic and industrial laboratories have now clearly demonstrated that PCBs are biodegradable by many microbial species. However, this activity is often limited to the mono-, di-, and trichlorobiphenyls. As originally formulated, the Aroclor series generally contains from di- through octachlorobiphenyls. Although there have been several reports of aerobic degradation of highly chlorinated PCBs such as in Aroclor 1260, these have not been substantiated. Even some reports of biodegradation of lower-chlorinated Aroclors such as Aroclor 1242 have not been well-documented. There have been numerous claims by commercial researchers purporting to demonstrate PCB disappearance mediated by bacterial cultures, however, a critical review of these studies often illustrates the activity of the ubiquitous enzyme "Disappearase".

It is possible that an observed PCB congener depletion in a "biodegradation" process is actually due to physical loss of the PCB and not the true biological degradation. With Aroclor studies, these processes can easily be distinguished because biodegradation results in depletion of specific congeners yielding gas chromatograph (GC) profiles that are distinctly different from those of the original Aroclor mixtures. Physical depletion, on the other hand, results in uniform depletion of all congeners (e.g., adsorptive loss) or depletion of lower chlorinated congeners due to their higher volatility (e.g., evaporative loss). The production of PCB metabolites is, of course, another unequivocal method for demonstrating the biological basis of PCB depletion.

In an attempt to distinguish these different depletion mechanisms and to demonstrate this pitfall, we established a mock, non-biological "biodegradation" process. PCB-contaminated soil was incubated with stirring in the *absence* of bacterial inoculum and with a constant stream of *inert gas* (argon). Following 19 days of incubation, this reactor was analyzed for PCBs in all locations. Although the samples taken from the middle of the soil slurry throughout the 19 days did show greater than 90% depletion of PCBs, the final mass balance was able to account for all of the PCB in the reactor by analyzing all of the physical components of the reactor (glass walls, stirrer, etc.) as well as coalesced PCB droplets in the bottom of the reactor.

PCBs *are* biodegradable; however, ill-conceived experimental protocols cannot be tolerated as a demonstration of this activity. This is not too much to expect from a company marketing a product. The FDA requires a demonstration of efficacy for a new drug, so too should this requirement be made of bioremediation processes. The client, the regulators, and the public all expect and deserve this type of quality control.

VOLATILE TARGETS

Volatile organic compounds pose a different set of challenges than the insoluble chemicals described above. This group includes the chlorinated solvents such as TCE and many lower molecular weight hydrocarbons such as benzene, toluene, xylene, etc. Too often claims have been made that these compounds were biodegraded using traditional, open-vessel, aerated bioreactors. Indeed, some researchers have reported that they observed increased rates of biodegradation as the air flow was increased. Quite obviously, the loss of these chemicals due to volatilization must be considered as a major, if not primary, mechanisms for their disappearance. Again, the activity of Disappearase is evident.

Studies in our laboratory have now demonstrated that toluene- and phenol-degrading microorganisms can biodegrade TCE from concentrations of 2 ppm to less than 2 ppb. However, like other researchers, how does one prove these claims? Generally, the best approach is to utilize sealed systems and compare biodegradation results to uninoculated or sterile controls. In our laboratory, small batch experiments are done in crimp-sealed serum vials with Teflon septa. Samples are removed by syringe and directly analyzed by injection into a gas chromatograph. Thus, a measure of biodegradation is determined relative to sterile controls.

For larger studies, the bioreactor vessels are sealed and samples taken directly from the headspace by direct plumbing to a GC using aluminum or Teflon tubing. Such studies, when done under these conditions, do clearly demonstrate that compounds such as TCE can be biodegraded and not substantially volatilized. However, in the final configuration of full-scale bioreactors, the control of aeration will be a critical parameter, as well as monitoring of the bioreactor offgas.

SUMMARY

It is imperative that for bioremediation to be accepted as a safe, cost-effective technology, researchers must unequivocally demonstrate that the loss of the target substrate is due predominantly to biological activity and not some other, non-biodegradative effect. This is particularly critical with insoluble and volatile substrates as discussed above. It is too easy to conclude that they are depleted by the action of Disappearase. Insoluble substrates sorb tightly to solid particles and are often hard to analyze. Volatile substrates are often stripped by the aeration necessary for aerobic growth.

When presented with claims of biodegradation, it is important for the reviewer (client or regulator) to clearly understand the protocol used in these studies and often to obtain the raw data. The most unequivocal proof that a biodegradation process is real is to observe the production of metabolic products, ideally CO_2. If this is not possible, and only depletion of the target is observed, then other forms of unequivocal data must be obtained. For example, 1) a comparison to a sterile or uninoculated control; 2) the nondepletion of an added internal standard; 3) the nondepletion of an internal standard which is a component of the mixture (e.g., a non-biodegradable PCB congener in an Aroclor mixture), and; 4) a total analysis of the complete bioreactor.

The challenge is now ours. Biochemical targets, even the more recalcitrant, are often biodegradable. However, poorly-designed feasibility studies and specious claims will only serve to further support the observation that the K_m of Disappearase is strictly dependent on the rate of bubbling, the size of your pipette, the shape of your reaction vessel, or your astrological sign! It is time to heat inactivate Disappearase by well-conceived, controlled, and executed experimentation. As we go from the "flask-to-the-field", there is no place in environmental biotechnology for *"Pseudomonas foo-foo-dustus"*.

Use of Treatability Studies in Developing Remediation Strategies for Contaminated Soils

Michael J. McFarland, Ronald C. Sims, and James W. Blackburn

INTRODUCTION

Treatability studies are laboratory or field tests designed to provide critical data needed to evaluate and, ultimately, implement one or more treatment technologies. Remediation of contaminated soils is generally accomplished by using one of the following three types of treatments systems:

1. In Situ
2. Prepared Bed
3. Bioreactor (e.g., slurry reactor, compost unit)

Examples of each of these treatment systems may be found in Sims et al., 1990. The discussion in the present text will focus only on the unsaturated soil treatment systems (e.g., in situ, prepared bed, and composting).

According to the EPA Guidance document, Guidance For Conducting Remedial Investigations and Feasibility Studies Under CERCLA (1989), as many as three tiers of treatability testing may be undertaken for evaluation of a remediation technology. These tiers include: 1) laboratory screening, 2) bench-scale testing, and 3) pilot-scale testing.

Laboratory screening studies yield data that can address a technology's potential to meet performance goals and can identify parameters for investigation during bench- or pilot-scale testing. Screening studies generate little, if any, design or cost data. Bench-scale testing can verify that the technology can meet expected cleanup goals and can provide information in support of a particular remedy. Additionally, bench-scale testing may provide some cost and design information. Pilot-scale testing is intended to provide quantitative performance, cost, and design information for operation of the full-scale unit. Although this general approach to treatability studies

---

Michael J. McFarland and Ronald C. Sims • Utah Water Research Laboratory, Utah State University, Logan, Utah 84322. James W. Blackburn • Energy, Environment and Resource Center, University of Tennessee, Knoxville, Tennessee 37996.

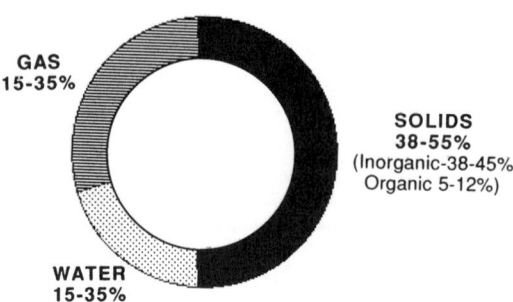

Figure 1. Typical volumetric composition of soil.

is valid for evaluation of most remediation technologies, soil remediation presents some unique design concerns which are related to mass transfer and partitioning of nutrients and/or contaminants within the soil matrix.

In general, laboratory treatability studies may represent optimum conditions with respect to mixing or contact of solid materials with contaminants and microorganisms or amendments. Moreover, homogeneous conditions normally exist throughout the soil reactor (microcosm). Due to these "optimal" conditions, laboratory treatability studies can only provide information concerning "potential levels" of soil treatment achievable. To evaluate a particular treatment technology, information from laboratory treatability studies must be combined with information about mass transfer and partitioning of contaminants and/or nutrients so the limitations of the technology can be determined. Bench-scale treatability studies, together with laboratory screening data, should address these mass transfer concerns to identify possible pathways of contaminant movement and limitations to degradation. Finally, because of the increased limitations to mass transfer at field-scale, pilot-scale treatability studies may be the only type of testing that can provide the critical information regarding full-scale design and operation. Since pilot-scale testing is the most expensive and time consuming of the three tiers of treatability analyses, identification of potential transport pathways and limitations of the treatment alternative should be identified at the laboratory and bench-scale levels.

SOIL BASED PROCESSES

As shown in Figure 1, the soil is a complex system consisting of four phases: (1) soil gas, (2) soil water, (3) inorganic solids, and (4) organic solids. By volume, gases and water comprise about 50 percent of the pore spaces in soil. An organic contaminant, depending upon its solubility and its tendency to volatilize, will distribute itself in certain proportions in these phases. In addition to the fluid phases, the organic contaminant may distribute (i.e., partition) itself between the various soil solid phases. The inorganic solid phase may contain highly reactive charged surfaces that impact both mobility and degradation of the contaminant. Similarly, the surfaces of the organic solid phase may aid in retaining organic contaminants within a soil system, making them more or less susceptible to transformation. Successful remediation of contaminated soil will depend on a thorough characterization and evaluation of pathways of movement and potential removal mechanisms in a soil at a specific site. For each chemical or chemical class, specific characteristics important for evaluating remediation of contaminated soils include: 1) characteristics related to potential leaching, e.g., water solubility, octanol/water partition coefficient; 2) characteristics related to potential volatilization, e.g., vapor pressure, Henry's Law Constant; 3) characteristics related to potential biodegradation, e.g., half-life, transformation; and 4) characteristics related to chemical reactivity, hydrolysis half-life, and soil redox potential.

Under field conditions, the rate and extent of remediation is generally limited by environmental factors such as accessibility and rate of mass transfer of chemical substances (oxygen, nutrients, contaminants etc.) to the soil environment. Moreover, contaminants may partition into one or more phases within the soil matrix. Evaluation of the fate of an organic contaminant in soil requires identification and measurement of the distribution of the contaminant among the physical phases that comprise the system as well as the mechanisms by which the contaminant may be chemically altered in the soil. For example, calculation of the rate of decrease of a parent compound would not account for the possibility that the contaminant may have been transferred from one phase to another or that it was chemically altered so that its properties were altered (e.g., humification).

Once the interphase transfer potential, pathways of escape, and chemical reactions have been identified by laboratory treatability studies and simulation modeling, containment requirements in the full-scale treatment system can be determined. For example, if the major pathway of transport is volatilization, containment with respect to volatilization control is required (e.g., inflatable dome, or closed reactor system). If leaching has been identified, consideration should be given to the temporary removal of soil from the site during which a plastic or clay liner may be placed under the site and the soil returned for treatment or the soil may be placed in a closed reactor system.

The fundamental approach to soil treatability studies which allows for estimation of mass transfer effects within the context of evaluating rates and extent of contaminant removal is the chemical mass balance. A thorough chemical mass balance performed early in the remediation design process (e.g., laboratory or bench-scale) will significantly save both cost and time in development of the pilot-scale treatability study and full-scale remediation system.

MASS BALANCE APPROACH TO SOIL-BASED TREATABILITY STUDIES

A conceptual framework based on the chemical mass balance has been applied to aqueous treatment systems (Blackburn et al, 1985; Blackburn, 1987, DiGrazia, et al., 1990) and has been adapted to treatability studies for contaminated soils (Sims et al., 1990). The concept of a chemical mass balance (i.e., the tracking of transport [gaseous and/or liquid], partitioning, and degradation/transformation of

Table 1. Materials Balances and Mineralization Approaches to Degradation Assessment

Biodegradation approach	Process examined
Materials balances	Recovery of parent compound in the air, soil water, and soil solids (extractable). Recovery of transformation products in the air, soil water, and solids (extractable).
Mineralization	Production of carbon dioxide, and/or methane from the parent compound. Release of substituent groups, e.g., chloride or bromide ions.

contaminants of interest in an environmental system) provides a rational and fundamental basis for asking specific questions and obtaining specific information necessary for determining fate and behavior, evaluating and selecting treatment options, and monitoring treatment effectiveness in laboratory-scale, bench-scale, and pilot-scale treatability studies. While a mass balance is routinely conducted on above-ground liquid treatment processes, a mass balance approach has generally not been applied to the evaluation of soil treatability studies under CERCLA or RCRA corrective action programs. The information needed to construct a mass balance for contamination at a site, or in a treatability reactor, simultaneously addresses the evaluation of potential transport and transformation processes and mechanisms as well as evaluation and selection of a remediation technique.

During the performance of a laboratory treatability study, transformation, detoxification, mineralization, and partitioning (immobilization) processes can be evaluated. The following describes how these soil-based processes are examined within the laboratory soil treatability study.

EVALUATION OF DEPLETION AND TRANSFORMATION

To assess the potential for depletion of organic chemicals at a specific contaminated site, the use of treatability studies incorporating either or both mass balance and mineralization approaches to determine the environmental fate and behavior of the contaminants in the specific soil is recommended (Table 1, as adapted from Rochkind, Blackburn, and Sayler, 1986). A rate of overall depletion or transformation can be calculated either by measuring the loss of parent compound or the production of carbon dioxide with time of treatment. Either rate can be reported as half-life which represents the time required for 50 percent of the compound to be removed, based upon a first-order kinetic rate equation or as a suitable rate constant based on the selection of a transformation rate equation (e.g., zero order, first, order, monod, etc.).

Calculation of the rate of decrease of parent compound or the rates and extent of mineralization (unless mineralization is 100%), however, does not provide necessary and sufficient information. Further information is required to understand whether a contaminant is simply transferred from one phase to another through a

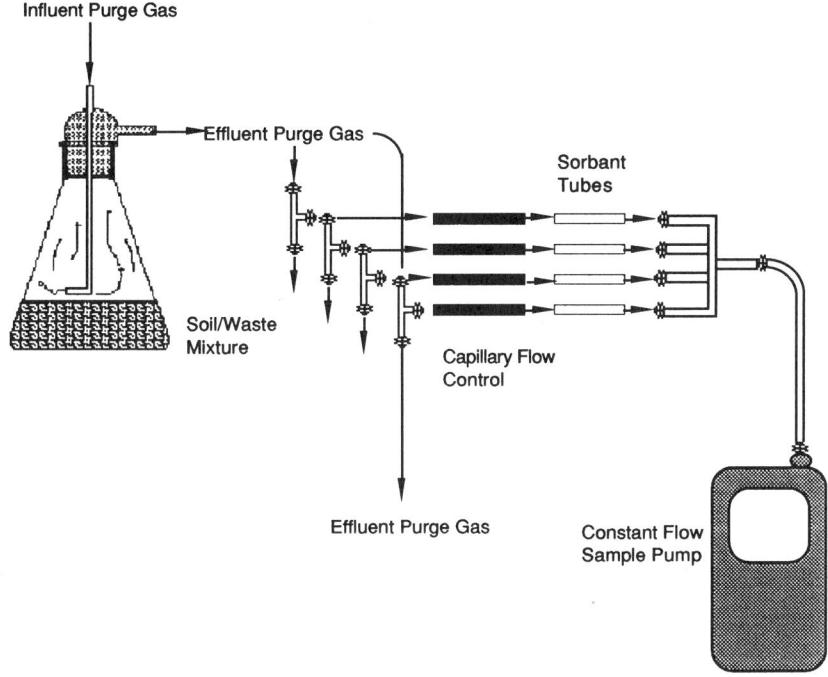

Figure 2. Laboratory flask apparatus used for mass balance measurements.

process of interphase transfer or is chemically altered so that the properties of the parent compound are destroyed. Evaluation of the fate of contaminants in soil requires identification and measurement (or prediction using validated models) of the distribution of the contaminants among the physical phases that comprise the system as well as differentiation of the mechanisms by which the constituents may be chemically altered in a subsurface system.

Studies that have as an objective the determination of the biotic or biotransformation rates, usually must depend upon an indirect approach where all advection and abiotic mechanisms can be measured or predicted and the mass unaccounted for is attributed to biological processes. Thus, accuracy in the measurement or prediction in advection and abiotic processes is important to yield a meaningful biotic rate estimate. It is the biotic rate estimate that is central to process scale-up and application.

Examples of apparatus that can be used as microcosms (i.e., laboratory reactors) to measure interphase transfer, degradation/transformation potential, and effectiveness of proposed remedial treatment technologies are illustrated in Figures 2 through 5.

A description of the use of the microcosm shown in Figure 2 illustrates a typical treatability study methodology. Methodologies used for the other types of microcosms are similar, emphasizing the use of the chemical mass balance approach for determining fate and behavior of waste constituents in a contaminated soil. Con-

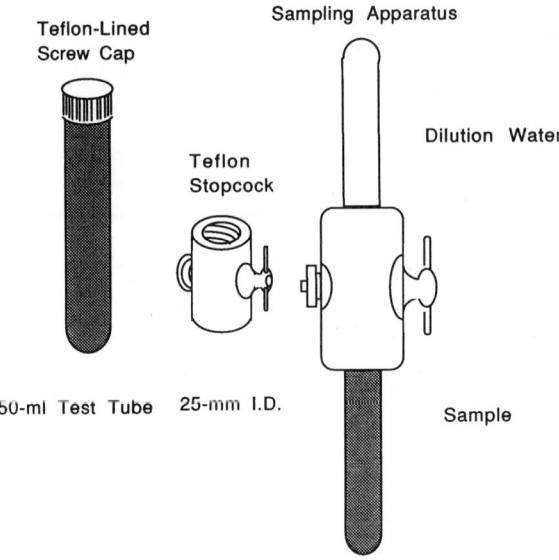

Figure 3. A simple microcosm to study biotransformations of pollutants which are highly volatile or sorb strongly to rubber or plastic.

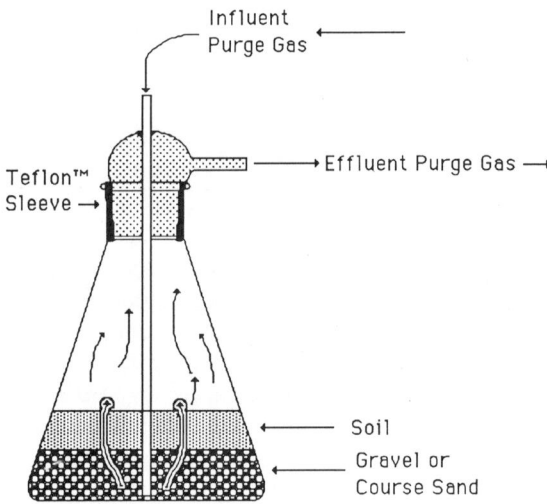

Figure 4. Laboratory flask apparatus used for soil venting evaluation.

taminated soil and treatment amendments are placed in a flask which is then closed and incubated under controlled conditions for a period of time. During the incubation period, air is drawn through the flask and then through a sorbent material. Volatilized materials are collected by the sorbent and are measured to provide an

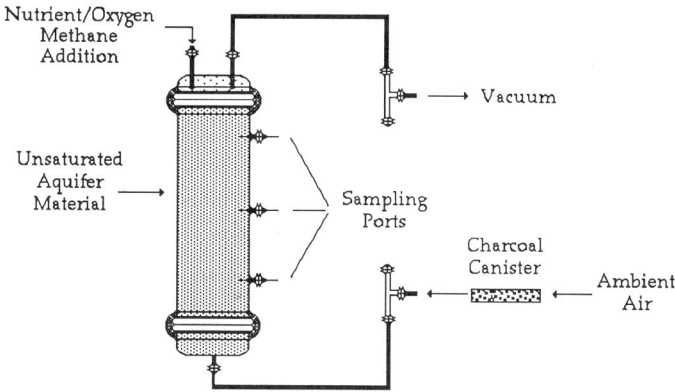

Figure 5. Large scale microcosm system used for the unsaturated zone studies.

estimate of volatilization loss of the contaminants. At the end of the incubation period, a portion of the contaminated soil is treated with an extracting solution to determine the extent of loss of the contaminant in the soil matrix. This loss can be attributed to biodegradation and possible immobilization within the soil matrix. Selection of an appropriate extracting solution is necessary to maximize contaminant recovery from the soil. Another portion of the soil is leached with water to determine the leaching potential of the remaining contaminants. Abiotic processes involved in removal of the parent compound are also evaluated by comparing microbially active soil with soil that has been treated with a microbial poison (e.g., mercuric chloride, formaldehyde, propylene oxide, etc.). By conducting a mass balance on a known amount of contaminant which is added to the soil, the fate and behavior of the chemical can be evaluated as it moves through the multiple phases of the soil system. Abiotic controls must be conducted as kinetic experiments and results compared to biotic systems.

Biotransformation refers to the conversion of a compound or its intermediates to the next product in the biochemical pathway. Intermediate products may be less toxic or more toxic than the parent compound; therefore, the rate and extent of detoxification of the contaminated soil should be evaluated. Microcosm samples generated from the different phases can be either analyzed for intermediate biotransformation products (normally requiring a labeled parent compound) or used in bioassay studies to provide information concerning transformation and detoxification processes.

EVALUATION OF TOXICITY IN SOIL TREATABILITY STUDIES

Bioassays to quantify toxicity measure the effect of a chemical on a test species under specified test conditions. The toxicity of a chemical is proportional to the severity of the chemical on the monitored response of the test organism(s). Toxicity assays use test species that include rats, fish, invertebrates, microorganisms, and seeds. The assays may use single or multiple species of test organisms. The use of a single bioassay procedure does not provide a comprehensive evaluation of the toxicity of a chemical in a subsurface/waste-impacted system. Often a battery of bioassays is used that may include measurements of effects on general microbial activity (e.g., respiration, dehydrogenase activity) as well as assays relating to the activity of subgroups of the microbial community (e.g., nitrification, nitrogen fixation, cellulose decomposition). Bioassays using organisms from different ecological trophic levels may also be used to determine toxicological effects. However, use of a single assay as a screening test to identify relative toxicity in the environment is common in treatability studies. Microorganism assays are often used because of their speed, simplicity, ease in handling, cost effectiveness, and number of statistically significant test organisms required to detect the effects of potentially toxic materials in the environment.

Two bioassays that have been used as part of treatability studies to evaluate wastes in subsurface soil systems are the Ames Salmonella typhimurium mammalian microsome assay and the Microtox™ test system. The Ames assay is a measure of the mutagenic potential of hazardous compounds (Ames et al., 1975) and has been widely used to evaluate environmental samples (Sims et al., 1984, 1985, 1986, 1987, 1988; Sims, 1990). A high correlation has been shown between carcinogenicity and mutagenicity, where about 90 percent of known carcinogens were shown to be mutagenic in the Ames assay (McCann et al., 1975).

The Microtox™ assay is an aqueous toxicity assay that measures the reduction in light output produced by a suspension of marine luminescent bacteria in response to an environmental sample (Bulich, 1979). Bioluminescence of the test organism depends on a complex chain of biochemical reactions. Chemical inhibition of any of the biochemical reactions causes a reduction in bacterial luminescence. Therefore, the Microtox™ test considers the physiological effect of a toxicant, not just the mortality. Matthews and Bulich (1984) described a method of using the Microtox™ assay to predict the land treatability of hazardous organic wastes. Matthews and Hastings (1987) developed a method using the Microtox™ assay to determine an appropriate range of waste application loading for soil-based waste treatment systems. Symons and Sims (1988) used the assay to assess the detoxification of a complex petroleum waste in a soil environment.

Two additional assays that are routinely used for treatability studies are fish and daphnia aquatic bioassays, conducted according to procedures in Standard Methods for the Examination of Water and Wastewater (APHA, AWWA, WPCF 1989).

EVALUATION OF IMMOBILIZATION IN TREATABILITY STUDIES

Immobilization refers to the extent of retardation of the downward transport (e.g., leaching potential) and upward transport (e.g., volatilization potential) of contaminants. Interphase transfer potential for a contaminant among soil phases (e.g., water, air, and solid phases) is affected by the relative affinity of the waste constituents for each phase and is quantified through the calculation of partition coefficients. Partition coefficients are calculated as the ratio of the concentration of a chemical in the soil, oil, or air phase to the concentration of a chemical in the water phase; these are expressed as K_o (oil/water), K_h (air/water), and K_d (solid/water). Partition coefficients are often determined in batch laboratory studies or calculated using structure-activity relationships. Recent investigations on nonequilibrium adsorption of organic compounds have resulted in the development of empirical models used to predict adsorption rate from equilibrium sorption coefficients (Brusseau and Rao, 1989). Such models provide insight to the dynamic behavior of interphase contaminant transfer.

USE OF MATHEMATICAL MODELS

To estimate the effects of transport and biotransformation in pilot- and full-scale systems, data from treatability studies are used as input information for comprehensive unsaturated zone mathematical models. Modeling provides a tool for: (1) integrating degradation, transport, and partitioning information generated in treatability studies with soil and waste/chemical specific characterization; (2) simulating the behavior of contaminants in a soil; (3) predicting the pathways of migration through the contaminated area and, therefore, pathways of exposure to humans and to the environment; (4) identifying constituents that will require treatment and monitoring in the air (volatile) liquid or leachate, and solid (soil or aquifer material) phases; and (5) approximating and estimating rates and extent of treatment that may be expected at the field-scale using selected remedial treatment technologies.

The Regulatory and Investigative Treatment Zone (RITZ) Model developed at the U.S. EPA Robert S. Kerr Environmental Research Laboratory by Short (1986), based upon an approach by Jury et al. (1983), is an example of a model that has been used to describe the potential fate and behavior of organic waste constituents in a contaminated soil system (U.S. EPA, 1988). An expanded version of RITZ, the Vadose Zone Interactive Processes (VIP) Model, has been developed and can be used in treatability studies that use unsaturated zone materials and conditions. The VIP incorporates predictive capabilities for the dynamic behavior of organic constituents in unsaturated soil systems under conditions of variable precipitation, temperature, and waste loadings (McLean et al. 1988, Stevens et al., 1988, 1989, Symons et al., 1988, U.S. EPA 1986). RITZ and VIP models simulate vadose zone processes, including volatilization, transformation, sorption/desorption, advection, and dispersion (Grenney et al., 1987).

CASE STUDY

Laboratory treatability studies were conducted on soil previously contaminated with fossil fuel waste. Toxicity of the contaminated soil, as well as contaminated soil mixed with different amounts of uncontaminated soil, were evaluated using the Microtox™ assay. Based on the results of toxicity testing and the guidance for loading rates based upon experience with the Microtox™ assay (Matthews and Bulich

Table 2. Acclimation of Soil to Complex Fossil Fuel Waste

PAH compound	Unacclimated soil		Acclimated soil	
	Initial soil concentration (mg/kg-dry wt)	Reduction in 40 days (%)	Soil concentration[1]	Reduction in 22 days (%)
Naphthalene	38	90	38	100
Phenanthrene	30	70	30	83
Anthracene	38	58	38	99
Fluoranthene	154	51	159	82
Pyrene	177	47	180	86
Benzo(a)anthracene	30	42	40	70
Chrysene	27	25	33	61
Benzo(a)pyrene	10	40	12	50

[1]After first reapplication of waste (after 168 days incubation at initial level), (mg/kg-dry wt).

1984, Matthews and Hasting 1987, and Symons and Sims 1988), contaminated soil was incorporated into uncontaminated soil at 1:19 weight/weight basis. Laboratory treatability studies also indicated that mixing of contaminated soil with a fertile sandy loam in a prepared bed mode, together with addition of manure and pH adjustment, resulted in stimulation of degradation of polynuclear aromatic hydrocarbon (PAH) compounds in the contaminated soil (Table 2).

If a soil has been previously exposed to similar or the same type of contamination, the soil microbial population may become acclimated to the waste, and waste degradation may occur at a faster rate. In laboratory testing of the contaminated fossil fuel waste, a greater reduction in concentration of PAH compounds was achieved in 22 days in an acclimated soil sample, compared to the reduction observed in 40 days in an unacclimated soil (Table 2) (Sims, 1986).

A field scale pilot study was designed based on results of laboratory treatability testing described above. The fossil fuel contaminated soil was thoroughly mixed into a clean sandy loam soil in a prepared bed mode across a one-half acre site. Sampling of soil cores was performed at 10 feet intervals across 100 feet rows. Data presented in Table 3 are composite values from the sampling efforts. In all cases, concentrations of PAH compounds in the soil were greatly reduced after 91 days. Data quality was poorer at the 91 day sampling period, as measured by the coefficient of variation (CV), which is the mean value divided by the standard deviation. The poorer data quality was attributed to variability in the biotransformation mechanisms and to increased analytical difficulties when levels of constituents are measured near detection limits.

Based on results of the field-scale pilot study, a full-scale implementation of prepared bed remediation was undertaken. Results of chemical analysis for PAH compounds and toxicity analysis using the Microtox™ assay indicated that remediation of the site to background levels of PAHs and to "non toxic" levels of leachate samples was achieved in approximately one year.

Table 3. Field Results for Soil Treatment of PAHs in Coal Gasification Wastes

Compound	C_o* (mg/g)			C after 91 days (mg/g)		
	AVG	SD	CV (%)	AVG	SD	CV (%)
Naphthalene	186	68	37	3	1.8	61
Acenaphthene	729	276	38	1	1.8	157
Phenanthrene	78	28	36	2.6	0.6	23
Benz(a)anthracene	86	42	49	2	0.8	38
Dibenz(a,h)anthracene	52	36	69	ND	–	–

*C_o = Initial Soil Concentration.

ACKNOWLEDGEMENTS

The authors would like to thank the U.S. Environmental Protection Agency, Office of Exploratory Research, Grant No. R-814475-01, Office of Research and Development, Washington, D.C. (Mr. Donald Carey, Project Officer) for information used in the development of this manuscript. The authors would like to acknowledge the assistance of the U.S. Environmental Protection Agency's Robert S. Kerr Environmental Research Laboratory, Ada, Oklahoma. The authors would like to thank Ms. Ivonne Cardona Harris for preparation of this manuscript and Mr. Paul Jarvis for his assistance in development of the figures.

REFERENCES

Ames, B.N., McCann, J., and Yamasaki, E., 1975, Methods for detecting carcinogens and mutagens with the Salmonella/mammalian-microsome mutagenicity test, *Mutat. Res.* 31:347-364.

APHA, AWWA, WPCF, 1989, *Standard Methods for the Examination of Water and Wastewater*, 17th ed., Washington, DC.

Blackburn, J.W., 1987, Prediction of organic chemical fates in biological treatment systems. *Environmental Progress* 6.

Blackburn, J.W., Troxler, W.L., Traong, K.N., Zink, R.P., Meckstroth, S.C., Florance, J.R., Groen, A., Sayler, G.S., Beck, R.W., Minear, R.A., Breen, A., and Yagi, O, 1985, *Chemical fate prediction in activated sludge treatment processes*, EPA/600/S2-85, NTIS PB 85-247674.

Brusseau, M.L, and Rao, P.S.C., 1989, The Influence of sorbate-organic matter interactions on sorption nonequilibrium. *Chemosphere* 18:1691-1706.

Bulich, A.A., 1979, Use of luminescent bacteria for determining toxicity in aquatic environments. (L.L. Markings and R.A. Kimerle, eds.) In *Aquatic Toxicology*, ASTM 667, Amer. Soc. for Testing and Materials, Philadelphia, PA, 98-106.

DiGrazia, P.M., Blackburn, J.W., Bienkowski, P.R., Hilton, B., Reed, G.D., King, J.M.H., and Sayler, G.S., 1990, *Development of a systems analysis approach for resolving the structure of biodegrading soil systems*. Applied Biochemistry and Biotechnology, 24/25:237-252.

Grenney, W.J., Caupp, C.L., Sims, R.C., and Short, T.E., 1987, A mathematical model for the fate of hazardous substances in soil: Model description and experimental results. *Hazardous Wastes & Hazardous Materials* 4:223-239.

Jury, W.A., Spencer, W.F., and Farmer, W.J., 1983, Behavior assessment model for trace organics in soil: Model description, *J. Environ. Qual.* 12:558-564.

Matthews, J.E., and Bulich, A.A., 1984, A toxicity reduction test system to assist predicting land treatability of hazardous wastes. (J.K. Petros, Jr., W.J. Lacy, and R.A. Conway, eds.) In *Hazardous and Industrial Solid Waste Testing: Fourth Symposium*, STP-886, American Society of Testing and Materials, Philadelphia, PA, pp. 176-191.

Matthews, J.E., and Hastings, J.,1987, Evaluation of toxicity test procedure for screening treatability potential of waste in soil, *Toxicity Assessment: An International Quarterly* 2:265-281.

McCann, J.R., Choi, R., Yamasaki, E., and Ames, B.N., 1975, Detection of carcinogens as mutagens in the Salmonella/microsome test: Assay of 300 chemicals, *Proc. Natl. Acad. Sci.* 72:5135-5139.

McLean, J.E., Sims, R.C., Doucette, W.J., Caupp, C.R., and Grenney, W.J., 1988, Evaluation of mobility of pesticides in soil using U.S. EPA methodology, *J. Environ. Eng., Am. Soc. Civil Eng.* 114: 689-703.

Short, T.E., 1986, Modeling processes in the unsaturated zone, in *Land Treatment: A Hazardous Waste Management Alternative,* Water Resources Symposium No. 13, (R.C. Loehr and J.F. Malina, Jr. eds.), Center for Research in Water Resources, The University of Texas at Austin, Austin, TX., pp. 211-240.

Rochkind, M.L., Blackburn, J.W., and Sayler, G.S., 1986, Microbial Decomposition of Chlorinated Aromatic Compounds. EPA/600/2-86/090, Hazardous Waste Engineering Research Laboratory, U.S. Environmental Protection Agency, Cincinnati, OH.

Sims, R.C., 1990, Soil Remediation Techniques at Uncontrolled Hazardous Waste Sites—A Critical Review, *Journal of the Air and Waste Management Association*, 40:703-732.

Sims, R.C., 1986, Loading Rates and Frequencies for Land Treatment Systems, in *Land Treatment: A Hazardous Waste Management Alternative*. Water Resources Symposium No. 13, (R.C. Loehr and J.F. Malina, eds.), Center for Research in Water Resources, The University of Texas at Austin, Austin, TX., 151-170.

Sims, R.C., Sims, J.L., and Dupont, R.R., 1984, Human health effects assays, *J. Water Pollut. Control Fed.* 56:791-800.

Sims, R.C., Sims, J.L., and Dupont, R.R., 1985, Human health effects assays, *J. Water Pollut. Control Fed.* 57:728-742.

Sims, R.C., Sims, J.L., and Dupont, R.R., 1986, Human health effects assays *J. Water Pollut. Control Fed.* 58:703-717.

Sims, R.C., Sims, J.L., and Dupont, R.R., 1987, Human health effects assays, *J. Water Pollut. Control Fed.* 59:601-614.

Sims, R.C., Sims, J.L., and Dupont, R.R., 1988, Human health effects assays, *J. Water Pollut. Control Fed.* 60:1093-1196.

Sims, J.L., Sims, R.C., and Matthews, J.E., 1990, Approach to bioremediation of contaminated soils, *Hazardous Materials* 7:117-149.

Stevens, D.K., Grenney, W.J., and Yan, Z., 1988, *User's Manual: Vadose Zone Interactive Processes Model*, Dept. of Civil and Environ. Eng., Utah State Univ., Logan, UT.

Stevens, D.K., Grenney, W.J., Yan, Z., and Sims, R.C., 1989, *Sensitive Parameter Evaluation for a Vadose Zone Fate and Transport Model*. EPA/600/2-89/039, Robert S. Kerr Environmental Research Laboratory, U.S. Environmental Protection Agency, Ada, OK.

Symons, B.D., and Sims, R.C., 1988, Assessing detoxification of a complex hazardous waste, using the Microtox^{TM} bioassay, *Arch. Environ. Contamination Toxicol.* 17:497-505.

Symons, B.D., Sims, R.C., and Grenney, W.J., 1988, Fate and transport of organics in soil: Model predictions and experimental results, *J. Water Pollut. Control Fed*, 60: 1684-1693.

U.S. EPA, 1986, *Permit Guidance Manual on Hazardous Waste Land Treatment Demonstrations,* EPA-530/SW-86-032, Office of Solid Waste and Emergency Response, U.S. Environmental Protection Agency, Washington, DC.

U.S. EPA, 1988, *Interactive Simulation of the Fate of Hazardous Chemicals during Land Treatment of Oily Wastes: RITZ User's Guide*, EPA/600/8-88-001, Robert S. Kerr Environmental Research Laboratory, U.S. Environmental Protection Agency, Ada, OK.

U.S. EPA, 1989, *Guide for Conducting Treatability Studies Under CERCLA*, Interim Final EPA/540/2-89/058 Office of Research and Development Cincinnati, Ohio and Office of Emergency and Remedial Response Washington, D.C.

Biodegradation of Mixed Solvents by a Strain of *Pseudomonas*

J. C. Spain, C. A. Pettigrew, and B. E. Haigler

INTRODUCTION

Substituted aromatic compounds are used extensively as solvents, synthetic intermediates, pesticides, and fuels. They are released in the environment in tremendous quantities and can pose a considerable human health hazard, particularly in groundwater. The most common groundwater contamination problem in the United States is caused by gasoline components such as benzene, toluene, ethylbenzene, and xylenes leaking from underground storage tanks. These and other petroleum components in gasoline are readily biodegradable and bioremediation is often the treatment of choice for cleanup. Indigenous mixed microbial communities are typically able to carry out the treatment process if supplied with oxygen and mineral nutrients. Biotreatment has been successful for petroleum hydrocarbons in groundwater and soil (Thomas and Ward, 1989) and in landfarming (Bartha and Bossert, 1984).

In contrast to the record of successful biotreatments of natural products such as petroleum, relatively few xenobiotic (manmade) chemicals have been the subject of bioremediation efforts (Frick et al. 1988; Morgan and Watkinson, 1989). Most simple substituted aromatic compounds are biodegradable and the degradation pathways have been studied in the laboratory with pure cultures of bacteria. Unfortunately, the information obtained from work with single compounds and pure cultures cannot be extrapolated to the field because most pure cultures degrade only a narrow range of related compounds. In contrast, bioremediation often involves complex mixtures of contaminants and undefined, mixed populations of microorganisms. Mixed cultures offer greater flexibility but can require long acclimation periods when challenged with new substrates. Further, mixed cultures are sensitive to "shock loads" of novel substrates that can severely disturb a biological treatment system. Introduction of chlorinated analogs into systems used to

---

J. C. Spain, C. A. Pettigrew, and B. E. Haigler • U.S. Air Force Engineering and Services Center, Tyndall AFB, Florida, 32403.

treat nonhalogenated aromatic compounds has resulted in failure of biotreatment systems (Knackmuss, 1981).

PATHWAYS IN PURE CULTURES

We have studied a strain of *Pseudomonas* selected for its ability to grow on 1,4-dichlorobenzene (Spain and Nishino, 1987). The isolate, strain JS1, and its non-encapsulated derivatives, strains JS6 and JS150, also degrade a wide range of other substituted aromatic compounds and may have advantages for use in treatment systems. It responds rapidly to changes in substrates and readily degrades mixtures of aromatic compounds under laboratory conditions.

The pseudomonad initiates the attack on dichlorobenzene by means of a dioxygenase enzyme which catalyzes the insertion of two hydroxyl groups in the ring (Fig. 1). The dioxygenase enzyme system appears to be similar to toluene dioxygenase which has been extensively characterized by David Gibson and his associates (Gibson and Subramanian, 1984). It exhibits a very broad substrate range and can oxidize a variety of compounds. A modified ortho ring-fission pathway similar to that used for 2,4-dichlorophenoxy acetate (Evans et al, 1971), chlorobenzoates (Dorn and Knackmuss, 1978), and chlorobenzene (Reineke and Knackmuss, 1984) catalyzes the oxidation of the substituted catechols formed in the initial reaction. The modified ortho pathway is also very nonspecific and allows the degradation of a variety of substituted catechols. In addition to the pathway described above, oxygenases for initial attack on the aromatic rings of naphthalene, salicylate, benzoate, phenol, and benzene can be induced (C.A. Pettigrew, B.E. Haigler, and J.C. Spain unpublished). In addition to the modified ortho pathway, the strain also carries the genes for both the classical ortho (Ornston and Stanier, 1966) and the meta (Dagley and Gibson, 1965) ring-fission pathways for the substituted catechols produced by the initial reactions. The presence of the genes for a wide array of pathways allows the organism to use an extensive list of aromatic compounds as the sole source of carbon and energy (Table 1). Preliminary experiments (Table 2) indicate that the meta pathway is induced during growth on all of the substrates except benzoate. In contrast, the modified ortho pathway appears to be more closely regulated and is only induced by halogenated substrates. These experiments were done with cells grown on single substrates and little is known about regulation of the pathways in the presence of mixtures of substrates.

SELECTION OF NOVEL STRAINS

The presence of the genes for the enzymes in the pathways described above provides the basis for facile selection of new strains able to grow on novel substrates. It might be possible to combine the various reactions into new pathways, change the regulation of existing pathways, or change structural genes to alter enzyme specificity. For example, the wild-type strain cannot grow on *p*-chlorotoluene as the sole source of carbon. The meta ring-fission pathway is in-

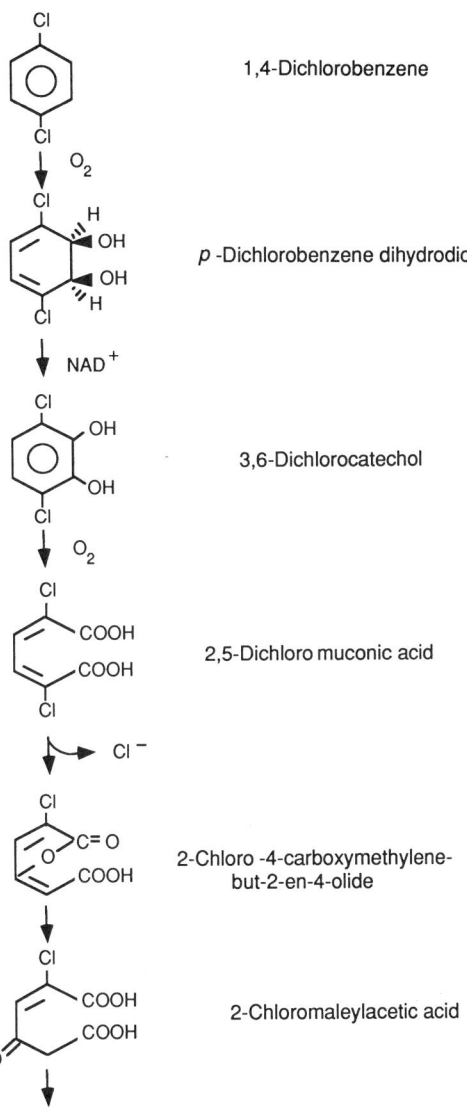

Figure 1. Pathway for biodegradation of 1,4-dichlorobenzene by *Pseudomonas* sp. Strain JS6 (Spain and Nishino, 1987).

duced by *p*-chlorotoluene and it causes the accumulation of toxic levels of the dead-end metabolite, chloromethylcatechol. However, *p*-dichlorobenzene-grown cells can utilize *p*-chlorotoluene by means of the initial dioxygenase-catalyzed reaction and subsequent reactions of the modified ortho pathway. Spontaneous mutants able to grow on *p*-chlorotoluene were readily isolated from agar plates when the wild-type strain was grown with low levels of *p*-chlorotoluene (Haigler and Spain, 1989). The induction pattern in the mutants was altered so that the modified

Table 1. Disposition of Various Substrates (Pettigrew, C.A., B.E. Haigler, and J.C. Spain, unpublished results)

Growth substrates of wild type	
Chlorobenzene	Toluene
Bromobenzene	Benzene
Iodobenzene	Ethylbenzene
1,4-Dichlorobenzene	4-Hydroxybenzoate
1,4-Bromochlorobenzene	Naphthalene
1,4-Dibromobenzene	Salicylate
Hydroquinone	

Growth substrates of derivative strains	
3-Chlorobenzoate	4-Chlorotoluene
4-Chlorobenzoate	4-Bromotoluene
m-Toluate	4-Iodotoluene

Mineralized	
1,2-Dichlorobenzene	o-Cresol
1,3-Dichlorobenzene	m-Cresol
4-Chlorophenol	p-Cresol
2-Chlorophenol	1,4-Xylene
3-Chlorophenol	2,5-Xylenol
2,5-Dichlorophenol	

Partially metabolized	
1,2,4-Trichlorobenzene	Indan
Trichloroethylene	Indene
Nitrobenzene	Indole
Fluorobenzene	

Table 2. Ring Fission Pathways Induced after Growth on Various Substrates (Pettigrew, C.A., B.E. Haigler, and J.C. Spain, unpublished results)

	Pathway		
Growth substrate	Meta	Ortho	"Modified" Ortho
Toluene	+	–	–
Benzene	+	+	–
Chlorobenzene	+	–	+
Benzoate	–	+	–
Naphthalene	+	–	–
Salicylate	+	–	–

ortho pathway was induced in response to the presence of *p*-chlorotoluene. Similar simple selection procedures have resulted in the isolation of mutant strains able to grow on several other compounds (Table 1). It is likely that the pseudomonad can use a similar strategy in natural habitats to respond quickly to the presence of these and other xenobiotic compounds. Thus the inherent plasticity of the gene pool can give the organism a considerable flexibility to take advantage of novel substrates. In a rapidly changing environment, such as a hazardous waste site, the advantage of this flexibility could offset the disadvantage of the energy required for duplication of the extensive genetic information.

COMETABOLISM

"Cometabolism" or "cooxidation" describe a process where substrates are transformed by nonspecific enzymes to products that accumulate. The reactions result in partial metabolism and the organism derives no carbon or energy from the process (Alexander, 1979). Examples include the cooxidation of trichloroethylene by methylotrophs (Wilson and Wilson, 1985), and the cooxidation of a variety of organic compounds by toluene dioxygenase in either *Pseudomonas putida*, or in *E. coli* (Gibson et al, 1990). *Pseudomonas* sp. strain JS6, also produces a dioxygenase that oxidizes a wide range of substituted aromatic compounds including alkyl and halogen substituted benzenes, phenols, and trichloroethylene (Pettigrew, Haigler, and Spain, unpublished). With some substrates that cannot support growth, only the initial dioxygenase reactions are carried out and transformation products accumulate (Table 1). For example, toluene-grown cells convert *p*-chlorotoluene to 1-chloro 4-methylcatechol and chlorophenols to the corresponding chlorocatechols (Fig. 2). Other substrates such as the isomeric cresols are completely degraded and the organism derives carbon and energy from the process even though cells cannot grow on cresols in the absence of toluene or phenol. Growth of strain JS6 on chlorobenzene and the concomitant induction of the modified ortho pathway allows the complete degradation of chlorophenols and *p*-chlorotoluene (Fig. 3). Thus, it is clear that the nature of the inducer determines the extent of biodegradation in this system. A number of compounds that cannot serve as the sole source of carbon can be completely degraded and serve as sources of carbon and energy when the appropriate pathways are induced. This finding confuses the concept of cometabolism. It becomes very difficult to maintain an arbitrary distinction between substrates that serve as the sole carbon source, those that serve as a carbon source in induced cells, and those that are transformed but provide no carbon to the cell. Indeed, growth on multiple substrates at low concentrations is likely to be the norm in natural systems where it would not be necessary for each substrate to act as an inducer. Growth of pure cultures on single substrates appears to be a laboratory artifact which has been very useful for studies of metabolic pathways and their regulation. Perhaps terms such as "cometabolism", "cooxidation", "transformation", and "complete metabolism" should be avoided and replaced by the term "metabolism."

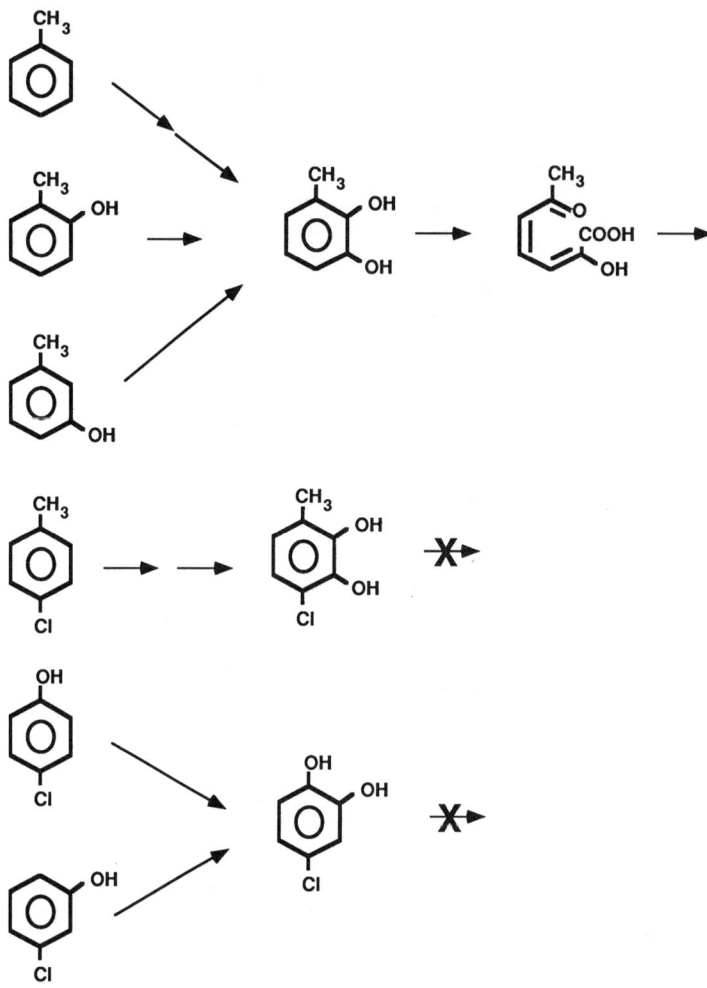

Figure 2. Metabolism of substituted aromatic compounds by toluene-grown cells containing the meta ring-fission pathway.

BIODEGRADATION OF MIXTURES

The preliminary work described above established that the *Pseudomonas* isolate could grow on a wide range of substrates but gave no insight on the degradation of mixtures. It has been well established (Taeger et al. 1988) that the meta ring-fission pathway for degradation of alkyl substituted aromatic compounds is blocked by chloroaromatic compounds. Recently, however, strains have been constructed with the ability to degrade methyl benzoate by the modified ortho pathway (Rojo et al, 1987). Such strains readily grow on mixtures of chloro- and methylbenzoates. Subsequently, spontaneous mutants of *Alcaligenes eutrophus* JMP134, and a *Pseudomonas* strain isolated from soil were also shown to degrade mixtures of 3-

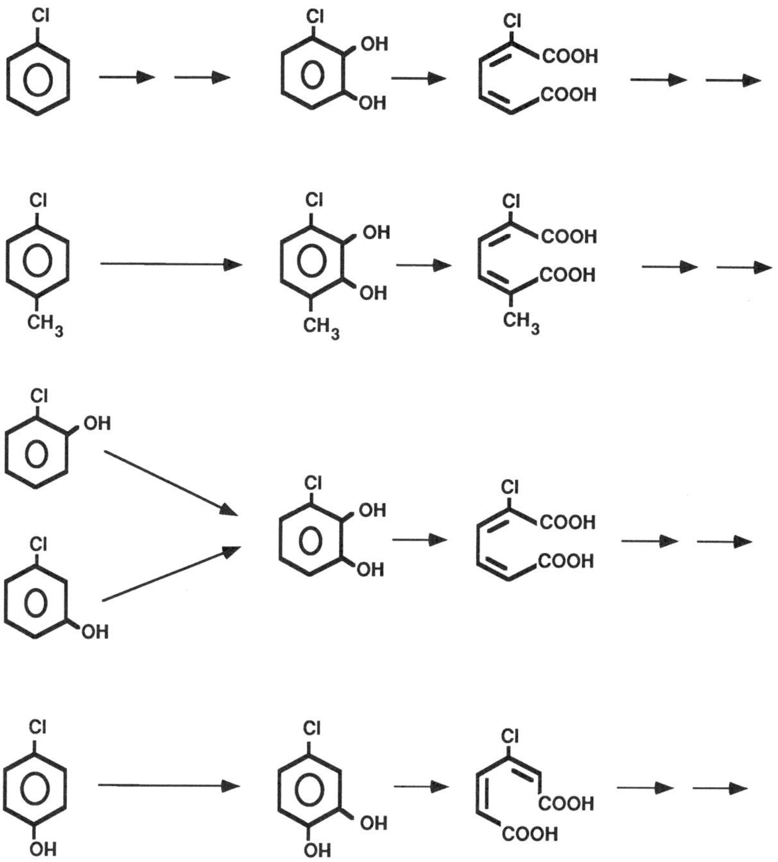

Figure 3. Metabolism of substituted aromatic compounds by chlorobenzene-grown cells containing the modified ortho ring-fission pathway.

chlorobenzoate and 3-methylbenzoate via the modified ortho pathway (Taeger et al, 1988). These studies suggested that *Pseudomonas* sp. strain JS6 might be able to grow on mixtures of toluene and chlorobenzene. In preliminary batch culture experiments toluene-grown cells converted chlorobenzene to chlorocatechol which accumulated to toxic levels. In contrast, chlorobenzene-grown cells could degrade toluene with transient accumulation of only traces of 2-methyllactone and 3-methylcatechol. When cells were grown on toluene in a chemostat then switched to a 1:1 mixture of toluene and chlorobenzene, growth ceased and 3-chloro-, and 3-methylcatechol accumulated in the medium (Fig. 4) (Pettigrew, Haigler, and Spain in press). After three hours the metabolites disappeared and growth resumed. Enzyme assays revealed that the enzymes of the modified ortho pathway were fully induced and that only low levels of the catechol 2,3-dioxygenase, indicative of the meta pathway, could be detected. The patterns of enzyme activity suggested that both toluene and chlorobenzene were degraded via the modified ortho pathway.

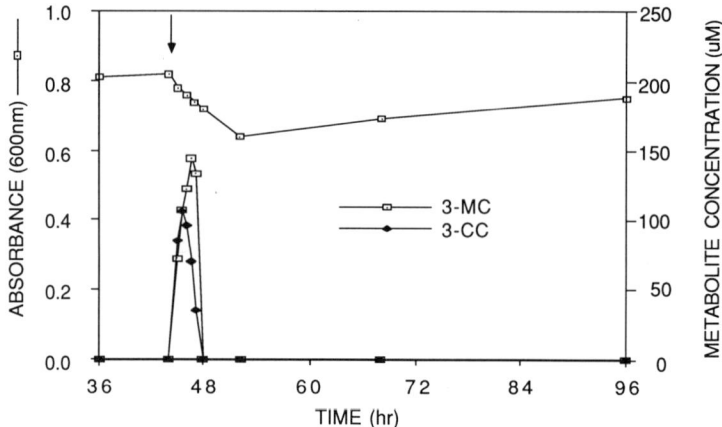

Figure 4. Growth of Strain JS6 in a chemostat. Cells were grown to steady state with toluene as the sole source of carbon. After 44 h, the substrate feed was switched to a mixture containing equal amounts of toluene and chlorobenzene (Pettigrew, Haigler, and Spain, in press). Top curve: Absorbance.

To our knowledge this was the first observation of toluene degradation by an ortho ring-fission pathway. However, the small amount of 2,3-oxygenase activity did not allow rigorous determination of which pathway accounted for toluene metabolism. Therefore, we designed an experiment to test whether toluene could be degraded in the complete absence of the meta fission pathway. A spontaneous mutant of strain JS6, designated strain JS62, lacks a functional catechol 2,3-dioxygenase (Pettigrew, Haigler, and Spain in press). In the chemostat the mutant strain grew better than the wild type on chlorobenzene (Fig. 5), probably because some of the chlorocatechol is misrouted into the unproductive meta pathway in the wild type. When the substrate feed was switched to toluene the wild-type strain grew faster because the meta pathway is more efficient for alkyl-substituted compounds. The mutant strain grew on toluene, and enzyme assays revealed fully induced levels of catechol 1,2-dioxygenase but no detectable catechol 2,3-dioxygenase (Table 3). The above experiments suggested that the broad substrate range of the initial dioxygenase and the modified ortho ring-fission pathway would allow growth on mixtures of other substituted aromatic compounds. Wild-type cells grown under conditions that did not lead to induction of the modified ortho pathway exhibited typical inhibition and misrouting when presented with mixtures of substituted aromatic compounds. We have tested a variety of more complex mixtures and discovered that as long as chlorobenzene is included to induce the modified ortho pathway high rates of biodegradation can be sustained. For example, up to 124 mg of a mixture containing chlorobenzene, toluene, benzene, *p*-chlorotoluene, naphthalene, and *p*-dichlorobenzene can be mineralized per hour in a 1.3 liter chemostat (Pettigrew, Haigler, and Spain unpublished). Removal efficiencies are 97 % in the chemostat and when the effluent is passed over a small trickling filter the concentrations are reduced to below the detection limits. We are currently conducting

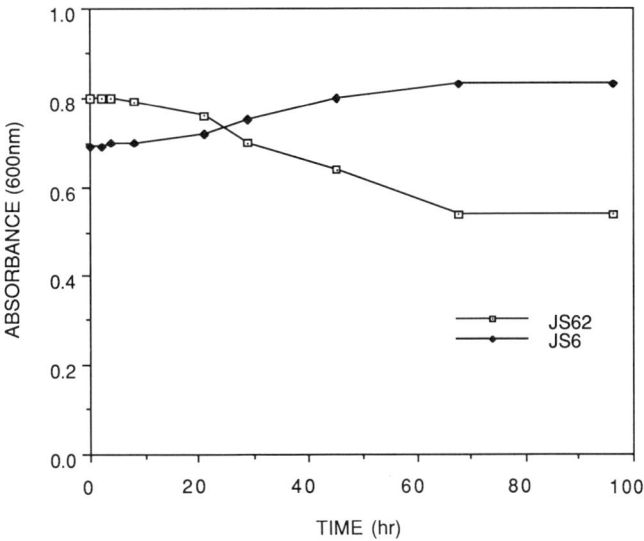

Figure 5. Chemostat cultivation of Strains JS6 and JS62 grown to steady state on chlorobenzene then switched to toluene at T=0 (Pettigrew, Haigler, and Spain, in press).

Table 3. Enzyme Activities in Cell Extracts of JS62 and JS6 (Pettigrew, C.A., B.E. Haigler, and J.C. Spain, in press)

Enzyme assayed and assay substrate	Specific activity (µmol/min per mg protein) at steady state[1]			
	Toluene		Chlorobenzene	
	JS62	JS6	JS62	JS6
Catechol 1,2-dioxygenase				
Catechol	0.039	0.003	0.067	0.042
3-Methylcatechol	0.094	ND	ND	ND
3-Chlorocatechol	0.044	<0.001	0.074	0.048
Catechol 2,3-dioxygenase				
Catechol	<0.001	0.142	<0.001	0.016
3-Methylcatechol	<0.001	0.349	<0.001	0.034

[1]Data correspond to Fig. 5; ND, not done.

experiments with immobilized cells to test whether similar results can be obtained with dilute waste streams.

CONCLUSIONS

Our results indicate that pure cultures can degrade a wide range of substituted aromatic compounds when provided with the appropriate inducer. Alkyl- and chloro substituted compounds can be degraded simultaneously after very brief acclimation periods. Induction of the modified ortho ring fission pathway appears to require the presence of a chloroaromatic substrate, but once the pathway is induced it can accommodate alkyl substituents as well. More information is needed on the regulation of the ring-fission pathways and on biodegradation of multicomponent mixtures.

REFERENCES

Alexander, M., 1979, Role of cometabolism, in: *Microbial Degradation of Pollutants in Marine Environments: Proceedings of the Workshop*, (A.W. Bourquin and P.H. Pritchard, eds.) U.S. Environmental Protection Agency, EPA-600/9-79-012, pp.67-75.

Bartha, R. and Bossert, I., 1984, The treatment and disposal of petroleum wastes, in: *Petroleum Microbiology*, (R.M. Atlas, ed.), Macmillan Publishing Company, New York, New York, pp. 553-577.

Dagley, S., and Gibson, D.T., 1965, The bacterial degradation of catechol, *Biochem. J.* 85:466-474.

Dorn, E., and Knackmuss, H.-J., 1978, Chemical structure and biodegradability of halogenated aromatic compounds. Two catechol 1,2-dioxygenases from a 3-chlorobenzoate-grown pseudomonad, *Biochem. J.* 174:73-84.

Evans, W.C., Smith, B.S.W., Fernley, H.N.,and Davies, J.I., 1971, Bacterial metabolism of 2,4-dichlorophenoxyacetate, *Biochem. J.* 122:543-551.

Frick, T.D., Crawford, R.L., Martinson, M., Chresand, T., and Bateson, G., 1988, Microbiological cleanup of groundwater contaminated by pentachlorophenol, in: *Environmental Biotechnology: Reducing Risks from Environmental Chemicals Through Biotechnology*, (G.S. Omenn, ed.), Plenum Press, New York, New York, pp.173-192.

Gibson, D.T.,and Subramanian, V.,1984, Microbial degradation of hydrocarbons, in: *Microbial Degradation of Organic Compounds*, (D.T. Gibson, ed.) Marcel Dekker, Inc., New York, pp. 181-252.

Gibson, D.T., Zylstra, G.J., and Chauhan, S., 1990, Biotransformations catalyzed by toluene dioxygenase from *Pseudomonas putida* F1, in : *Pseudomonas : Biotransformations, Pathogenesis, and Evolving Biotechnology*, (S. Silver, A.M. Chakrabarty, B. Iglewski, and S. Kaplan, eds.), American Society for Microbiology, Washington, D.C., pp. 121-132.

Haigler, B.E., and Spain, J.C., 1989, Degradation of *p*-chlorotoluene by a mutant of *Pseudomonas* sp. Strain JS6, *Appl. Environ. Microbiol.* 55:372-379.

Knackmuss, H.-J., 1981, Degradation of halogenated and sulfonated hydrocarbons, in: *Microbial Degradation of Xenobiotics and Recalcitrant Compounds*, (T. Leisinger, R. Hutter, A.M. Cook, and J. Nuesch, eds.), Academic Press, London, pp. 190-212.

Morgan, P., and Watkinson, R.J., 1989, Microbiological methods for the cleanup of soil and ground water contaminated with halogenated organic compounds, *FEMS Microbiol. Rev.* 63:277-300.

Ornston, L.N., and Stanier, R.Y., 1966, The conversion of catechol and protocatechuate to B-keto adipate by *Pseudomonas putida*-biochemistry, *J. Biol. Chem.* 241:3776-3786.

Reineke, W., and Knackmuss, H.-J., 1984, Microbial metabolism of haloaromatics: isolation and properties of a chlorobenzene-degrading bacterium, *Appl. Environ. Microbiol.* 47:395-402.

Rojo, F., Pieper, D.H., Engesser, K.-H., Knackmuss, H.-J., and Timmis, K.T., 1987, Assemblage of ortho cleavage route for simultaneous degradation of chloro- and methylaromatics, *Science* 238:1395-1398.

Spain, J.C., and Nishino, S.F., 1987, Degradation of 1,4-dichlorobenzene by a *Pseudomonas* sp., *Appl. Environ. Microbiol.* 53:1010-1019.

Taeger, K., Knackmuss, H.-J., and Schmidt, E., 1988, Biodegradability of mixtures of chloro- and methylsubstituted aromatics: Simultaneous degradation of 3-chlorobenzoate and 3-methylbenzoate, *Appl. Microbiol. Biotechnol.* 28:603-608.

Thomas, J.M., Ward, C.H., 1989, In situ biorestoration of organic contaminants in the subsurface, *Environ. Sci. Technol.* 23:760-766.

Wilson, J.T., and Wilson, B.H., 1985, Biotransformation of trichloroethylene in soil, Appl. Environ. Microbiol. 49:242-243.

The Field Implementation of Bioremediation: An EPA Perspective

Fran V. Kremer and Walter W. Kovalick, Jr.

BACKGROUND

As we approach the treatment of hazardous wastes, which are increasingly more diverse with respect to the contaminants and the contaminated matrices, we will be more reliant on innovative technologies for improved treatment efficiencies and lower costs. To meet these objectives, the U.S. Environmental Protection Agency's (EPA) Administrator, William Reilly, has sought to develop an agenda for the 1990's to identify strategies for increasing the use of bioremediation for the treatment of hazardous wastes. To develop this agenda, assistance has been received from biotreatment companies, site cleanup contractors, industry, academia, environmental organizations, and other Federal agencies, in addition to the various offices within EPA.

One of the initial recommendations from this consortium was the need to expand our field experience using this technology. Even though bioremediation is a viable technology to treat some hazardous wastes, it has not been fully utilized for the many different types of wastes and site conditions requiring remediation. It was recommended that the Agency serve as a focal point in fostering field tests, demonstrations, and evaluations of bioremediation, using good test protocols and documentation of results.

Based on this recommendation, the Office of Solid Waste and Emergency Response (OSWER) and the Office of Research and Development (ORD) have instituted a Bioremediation Field Program. This program provides assistance to the Regions and the states in conducting field tests and carrying out evaluations of site cleanups using bioremediation. Sites considered in this program include Superfund sites, RCRA corrective action sites and Underground Storage Tank (UST) sites. The program is designed to:

Fran V. Kremer • U.S. Environmental Protection Agency, Cincinnati, Ohio 45268. Walter W. Kovalick, Jr. • Technology Innovation Office, Office of Solid Waste and Emergency Response, U.S. Environmental Protection Agency, Washington, D.C. 20460.

1. more fully assess and document performance of full-scale field applications of bioremediation,
2. provide technical assistance at various stages of site remediation, from site characterization to full-scale implementation, and
3. regularly provide information on bioremediation projects being undertaken nationally.

Each of these aspects of the Program is described below.

FULL-SCALE FIELD EVALUATIONS

As solid full-scale performance data is needed to assess the capabilities of this technology, evaluations of field operations are being undertaken. Sites considered for evaluation have field biological units for treatment of wastes in situ or ex situ i.e. treatment of soils or groundwater in place or treatment in a reactor or land treatment facility. Based on nominations of sites received from the Regions and states, four sites have been selected for field evaluation.

Creosote Site

Operations at a wood treating facility in Montana has resulted in contamination in a waste pit area, a tank farm area, and a former butt dip. Biological treatment is being utilized in three ways: in situ treatment of the groundwater and associated soils, treatment of the groundwater in a reactor, and a land treatment facility for soils. The Record of Decision requires a cleanup level in the upper aquifer of 40 ppt total carcinogenic PAHs.

The in situ remediation of the upper aquifer includes the application of oxygen and nutrients to enhance biodegradation in the aquifer. Plans for expansion of the in situ treatment include the installation of additional injection wells. The injection well system will consist of three separate well systems. A source area injection system will be located at the head of the plume to enhance degradation within and downgradient of the source area. An intermediate system in the tank farm source area will enhance degradation in the middle of the plume. The third injection system is a downgradient system to enhance degradation in the plume off site.

Construction was completed in April of 1990 on a fixed film bioreactor to treat groundwater which is being pumped as a containment measure. The influent is pretreated in an oil separator; oxygen and nutrients are added to the heated reactor. The indigenous microorganisms on site serve as the source of inoculum for the bioreactor. The reactor has been providing up to 99% removal of polyaromatic hydrocarbons. Reactor operating parameters are being further evaluated to improve treatment efficiency.

For the land treatment facility, 45,000 yd^3 were excavated from source areas. Rocks greater than one inch were removed. A pilot study was undertaken in a one-acre lined facility. Effluent from the bioreactor was added to the soils as a source of biomass and nutrients. Based on results of the pilot study, the land treatment facility is being scaled up.

Trichloroethylene Site

A facility manufacturing brakes had a degreasing operation which resulted in the contamination of the groundwater with trichloroethylene (TCE). Secondary products, cis- and trans-dichloroethylene (DCE) and vinyl chloride (VC), are present downgradient with the groundwater flowing toward a lake. The site itself is located on a sand dune.

Work is being undertaken to treat the TCE and it's degradation products in situ. The biorestoration system design being considered for this site is essentially operated as a pump and treat system, with the treatment being carried out in the subsurface. Upon extraction of the groundwater, methane and oxygen are added to the groundwater, at the surface, in alternating pulses. This technique is expected to avoid potential clogging problems. The contaminated groundwater, with the amendments, is then reinjected into the contaminated zone. This then results in the subsurface biodegradation of the recirculated contaminants and the contaminants which have remained in place. No nutrient addition is required as the natural nutrient availability is adequate. New injection techniques will be evaluated to enhance the transport of the amendments within the aquifer. This is one of the first full-scale in situ projects conducted for the treatment of TCE and it's by-products.

Ethylene Glycol Site

Remedial action is underway at an ethylene glycol site to treat contaminated groundwater. An inner tier treatment system consists of above ground treatment operations using equalization/pH adjustment, physical-chemical precipitation for metals removal, biological treatment in a sequencing batch reactor (SBR), air stripping and carbon filtration. The treated groundwater is then discharged to polishing ponds, with final discharge through an National Pollutant Discharge Elimination System (NPDES) outfall.

The SBR consists of a batch biological activated sludge system which operates in a sequence of cycles. The cycle is comprised of a fill, treatment, settle, decant, and idle stages. This unit is presently operated on a partial fill cycle, i.e. for a full cycle, one-third of the total daily volume is introduced three times during the day. Aeration continues throughout the fill cycle until the subsequent settling period. At the end of the third partial fill cycle, the SBR continues to aerate until the settling and decant periods. Nutrients are added at the beginning of each cycle.

Underground Storage Tank Site

A utility in Colorado has a service center where a leak in an oil catch basin occurred. This has resulted in contamination primarily in the soil below the facility but also in the ground-water. The bottom of the basin was approximately 9 feet below grade and was in use for 29 years. Significant quantities of adsorbed and dissolved hydrocarbons are present in the subsurface around the waste oil pit. The highest concentrations of adsorbed hydrocarbons were upgradient from the pit. These were composed primarily of the heavier, less soluble hydrocarbons which were detected at 12,000 ppm total oil and grease. Dissolved hydrocarbons have been measured at 80 ppm total oil and grease and 1.8 ppm total dissolved gasoline hydrocarbons.

The site is underlain at a depth of about 35 to 38 feet by interbedded claystone and sandstone bedrock. Overlying the bedrock are alluvial sand and gravels deposited in the floodplain. These dense deposits average 25 to 30 feet thick. Above the sand and gravel rests a layer of very sandy and silty clay, four to ten feet thick.

To date, nutrient and hydrogen peroxide addition systems, an infiltration gallery and infusion well points have been installed. A water and soil boring sampling program has been initiated to evaluate the degradation of adsorbed hydrocarbons and to ensure the enriched water is reaching the contaminated areas.

TECHNICAL ASSISTANCE

Technical assistance is available to the Regions and the States on treatability and field pilot studies. This is to ensure adequate site characterizations, proper design of treatability studies, and interpretation of results. In some cases, the EPA may conduct the treatability work. This assistance is available through the EPA's Technical Support Centers. Presently, assistance is being provided on a number of creosote sites and chemical facilities.

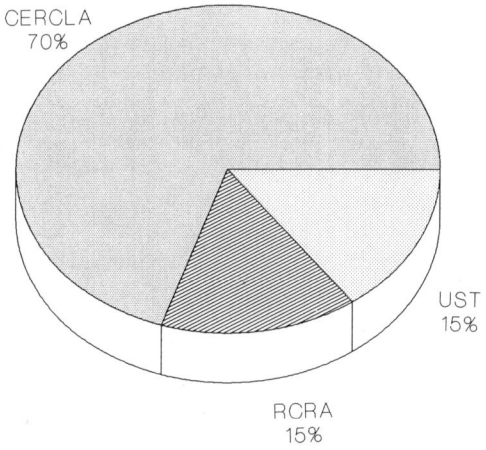

Figure 1. EPA program distribution of the sites planning or using bioremediation.

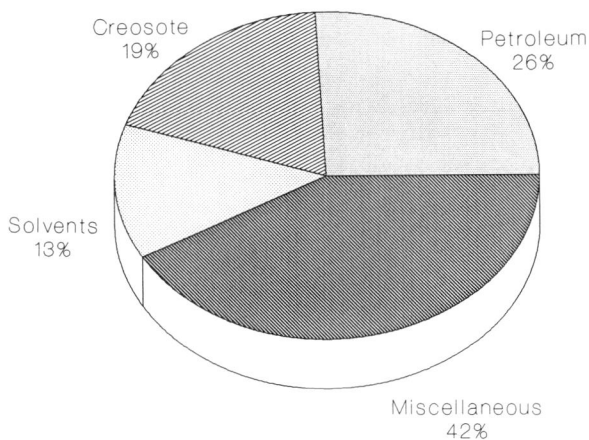

Figure 2. Major waste types being biologically remediated.

DEVELOPMENT OF TREATABILITY DATABASE

Data is being compiled on laboratory-, pilot-, and full-scale projects in order that EPA will have a central repository of treatment information. Treatability data is being collected from the Regions, States, other Federal agencies and the private sector. This data will be available through the Risk Reduction Engineering Laboratory's Treatability Database Program (513-569-7503) and also through the Alternative Treatment Technology Information Center (ATTIC) (301-816-9153). The Treatability Database Program provides specific information on the treatability of specific chemicals for a variety of technologies, including bioremediation. ATTIC is an on-line information retrieval network that provides current information on innovative treatment methods for hazardous wastes. Work is also underway on the development of a bioremediation expert system which will contain treatability data and design information on biosystems. The Treatability Database Program and ATTIC are presently available; the bioremediation expert system will be available in the Fall of 1991.

To date, over 130 sites have been identified across the country where bioremediation is being planned, treatability studies are being conducted, is under full-scale operation, or has been completed. These sites include CERCLA, RCRA and UST sites. As depicted in Figure 1, most of these sites are Superfund sites. In reviewing the types of wastes being remediated (Figure 2), petroleum, creosote and solvent wastes comprise over half of the waste types being biologically treated.

An Historical Perspective: Does Good Science or Good Press Generate Demand?

T. G. Zitrides

As one in the bioremediation industry, I'd answer the title's rhetorical question with the word "both." Bioremediation, if we disregard inflated claims, is good science. And lately it has been receiving a good press, which hasn't hurt our cause either. Neither the scientific community nor the press, however, has been the primary influence on demand for bioremediation. That role has been and continues to be played by the regulator. Our business is regulation driven; it is only to the extent that public or scientific opinion influences regulations that it can be said to have a major effect. When the EPA's William Reilly touts the potential of bioremediation, and it is widely reported in the press, then we have scored a major point.

Today I would like to discuss the public perception of our tools—microorganisms—and how it has affected, both positively and negatively, the development of bioremediation. At the same time I would like to show how attitudes in the scientific and industrial community have diverged from public perception, and what their influence has been. Finally, I would like to touch upon what we, as an industry, can do to influence public opinion and to obtain greater acceptance by the regulatory community.

Pasteur is a good starting point. He loved microorganisms, and never tired of pointing out the good that they do, from raising bread to fermenting grape juice. He knew that the higher forms of life could not survive without them. And, what is his legacy? "Wash your hands before you eat." "Don't touch that, it is full of germs." Most people today, when they think of bacteria at all, have visions of the black plague, rabies, typhoid, cholera and smallpox. Paul deKruif's book *Microbe Hunters* has a lot longer press run than Roseberry's *Life of Man*, in which the author pointed out that while there are a few bad actors among the bacteria most are beneficial, indeed necessary, to the proper functioning of the body, and that we are carrying around billions of the creatures, no matter how often we wash our hands. It is just very hard to accept, especially when among those billions there

---

T. G. Zitrides • BioScience Management, Inc. Bethlehem, Pennsylvania 18017.

are indeed some disease-causing opportunists, just waiting for a break in the body's defenses.

After Pasteur, the multitude of break-throughs in identification of disease vectors reinforced public phobias about microorganisms. Bacteria replaced evil spirits as the cause of human ills. Real neuroses developed. There's a character in the wonderful newspaper play, "Front Page," who goes around with a spray can of Lysol, crying "Germs abound!" The other reporters ridicule and play jokes on him, but the character is funny because it's an exaggeration of an attitude we have all been expose to from grade school on.

That attitude has had real effects. For example, one of the few microbes of which the public in general is aware is coliforms. Coliforms are bad. We don't want them in our drinking water or in treated sewage effluent. There are many varieties of microorganisms showing similarities to coliforms, however, that are useful in biodegradation, as well as being harmless to humans. Our industry has run into obstacles using them because of public and regulatory reaction.

On the other hand, the scientific and engineering communities in general, while accepting the role of bacteria in disease, have remained truer to Pasteur's original version of predominantly beneficial effects. This attitude has also had unforeseen results. During the 19th century and at least half of the 20th, it was assumed that microorganisms in the ecosystem would automatically degrade and recycle whatever compounds we threw at them in whatever quantity. This complacency led to abuses that we are still trying to cope with. It is enshrined in classical biology in the form of the ubiquity principle, the hypothesis that all bacteria exist everywhere, and that when a food source becomes available a species will show up eventually that is capable of digesting it. Industrial wastewater treatment plant operators who have tried to start up a plant using a municipal sludge seed to degrade solvents can define the length of time denoted by the word "eventually." The ubiquity principle has also led to some unintentional humor, as for example when Alaskan authorities, faced with the EPA experiments in bioremediation after the Valdez spill, resisted the use of "non-indigenous" naturally-occurring bacteria.

As the above example points out, the ubiquity principle was unknown to the general public, which included politicians, who recently drafted federal legislation regulating the use of naturally-occurring microbial cultures through registration and peruse public disclosure requirements.

In relations with the mass media, the bioremediation industry is thus faced not only with the normal amount of skepticism about a new approach, but also with ingrained phobias and prejudices.

All of us involved in bioremediation have experienced the following dialogue, in one form or another, whenever the application or stimulation of bacteria is described to a layman: "Well, it sounds like a good idea, but I'm not so sure about releasing microorganisms into the environment. . ."

"They are in the environment already. All we have done is to beef up their numbers and their capacity to digest the oil."

"Okay. Suppose they do eat up all the oil, what do they start on then?"

You can explain the relationship between food sources, reaction conditions, and growth rates until you are blue in the face, but you can tell that your questioner is still unconvinced. The problem is that we are dealing with complex questions of microbial ecology that do not lend themselves to one-liner answers.

In this regard, our partners in the laboratory have not always advanced the cause of field application. There is some confusion about the types of microorganisms being applied in the field. The press is accustomed to news releases from, for example, genetic engineers working on manipulations that may offer significant possibilities for the destruction of PCB's and other refractory compounds in extreme environments.

Such press releases are what Russell Baker calls "new news," as compared to field application or stimulation of naturally-occurring, adapted bacteria, which has been going on for more than twenty years and is widely reported. Such "new news" reinforces the public perception the bioremediation is still a laboratory curiosity, and that toxic-degrading microbes are weird and unnatural species that it would be better not to unleash upon the fragile, germ-free environment.

Now there is a word—"unleash." We in the bioremediation industry are going to have to be very careful in relations with the press to use favorable wording. Bacteria, microbes and germs are bad. Cultures are good, microorganisms are neutral. Rather than "unleash genetically altered bacteria", you "apply beneficial cultures in a controlled environment" to break down toxic wastes.

This caution is made only half in fun. The prevailing attitudes toward microorganisms have indeed held back the application of safe and viable technology. More than a decade ago, when there was little hard data about the effects of bioaugmentation, a major oil refinery decided to test the effectiveness of selected hydrocarbon-degrading strains in its wastewater treatment plant. The plant was ideal for the test because it consisted of two identical treatment trains each treating the same inflow. One side of the plant served as a control with an indigenous biomass. The other side was augmented with naturally occurring cultures selected and adapted to degrade refinery hydrocarbons. The results were conclusive. The augmented train showed improvements in every measured characteristic . . . and, when the adapted cultures were applied to the untreated train, its performance improved to that of the treated train.

Case studies, in the real world, are one of the best ways to demonstrate the effectiveness of any product, microbial cultures included. But the oil company decided to keep the results of the test confidential. The oil company was not going to admit either to having a performance problem in the first place or to unleashing microbes upon the environment in the second place.

This reluctance to talk about successful bioremediation projects still occurs and may be delaying more widespread public acceptance, although in view of the complex regulations and legal liabilities surrounding cleanup sites, I can sympathize with this viewpoint.

The press itself has some characteristics that may have affected public acceptance of bioremediation. Here I would like to distinguish between the general news media and the trade press. Trade journals, in recent years, have generally done an excellent job in reporting on bioremediation. They may follow fashions, like the news media, but their attention span is longer, and they do follow up. Their readers are looking for practical information, and they have an economic interest in finding out what finally happened.

Let's look more closely at attention span. News, to a newspaper or a broadcaster, is only news for a day. Unless there are continuing developments, the time drops out of sight in 24 hours. The reporter gets assigned to other breaking news, and unless he or she has a special interest, that's the end of the topic. The use of microbial cultures in the Mega-Borg spill is a case in point. This novel approach to an on-water spill was applied, various experts commented and speculated, and poof, the story vanished. If you were following it in the daily newspaper, you never discovered whether the project was successful or not.

The corollary to attention span is lack of follow-up. A few newspapers have made a half-hearted effort to report on the final outcome of breaking news, but their hearts weren't in it. The column usually appears on page 23 under the heading "Whatever Happened To?" Newspapers also have short memories. The stories about bioremedial applications presented as news could have been, and probably were, written twenty years ago, but each time, the project was treated like a breakthrough, without historical perspective. This tends to make biological solutions appear untried. I'm not blaming newspapers for any of this. It's simply the way the craft of journalism works, and we must find better ways to work with it.

Another characteristic of newspapers, more prevalent today than in the past, is their attention to fairness — getting the opposing view. As in most emerging technological endeavors, controversy abounds in bioremediation and a newspaper or magazine is always able to find an expert to downplay the results of an application, suggest better alternatives, or speculate about possible repercussions. I guess the general result of this tendency is good, but in matters of science and technology it can confuse a reader who isn't sure what weight to place upon conflicting opinions.

The newspapers' use of experts, to be fair, is due in many cases to the lack of capacity on staff to evaluate the significance of technological developments. Even the best science writers can't be knowledgeable about every field, and bioremediation is a relatively new one. As it develops, we can hope that it will develop its own experts. I was very impressed with the technical background of European science writers I met at a recent environmental exhibit in Germany.

There are other encouraging signs. Biotechnology continues to be in the news, and bioremediation, as it's used more often, will no longer be such a novelty. Major general magazines, such as the *National Geographic* have not only commented favorably upon bioremediation in Alaska, but have done so without speculating. The interest of William Reilly at the EPA in the form of public pronouncements and internal follow-up has helped bioremediation's credibility immensely and is raising its status from a novelty to an accepted technology.

The biotechnology era is also eroding the prejudice against microorganisms and advancing scientific understanding and education in ecology is stressing the inter-relatedness of all things, from elephants to microbes. It is traditional to have a blessing of the animals on the Saint's Day of Francis of Assisi. This month in New York, among household pets and animals from the zoo, a petri dish of single-celled microorganisms received a blessing as well.

A great deal remains to be done in the way of education about microorganisms in general and their role in ecosystems in particular. This effort must begin in grade school, before the "wash your hands" syndrome is firmly established and then continue at all levels. Kids who have been exposed to mutant Ninja turtles should be amenable to a comic book or game in which bacteria are the good guys, gobbling up waste products and converting them to water and carbon dioxide. The more complex aspects can come later.

Biotechnology is a highly complex field. Those who make decisions regarding bioremediation must have the knowledge and the data to enable them to decide rationally and without prejudice. At present even regulators who are favorably disposed to biological solutions are hampered. There is a widespread lack of understanding of basic principles of biodegradation and microbial ecology. There is inadequate hard data upon which to make decisions. And, we don't have a good standardized methodology to evaluate the hard data once we get it. Among others, the Applied BioTreatment Association by publishing a compendium of bioremediation case studies, by providing educational seminars, and by facilitating the exchange of information between the industry and the regulatory community is setting an example of what needs to be done. But, we can and must do more in all of these areas. Bioremediation is both good news and good science, but neither is adequate without the confidence of the regulatory community who in the end make the final decision on which remedial technology to use. We will gain and hold this confidence through better education and awareness and consistent documented performance.

Ways to Identify and Obtain Rights to Technology from Federal Facilities

John C. Corey, Gerald J. Hooker, and Cindy K. Tew

INTRODUCTION

Bioremediation of hazardous waste materials is in its infancy. The applications associated with this pollution abatement technique will continue to increase rapidly as we learn to optimize physical and chemical conditions for the degradation process. In the future, biodegradation will be possible but not necessarily economical for most compounds. Factors influencing future biodegradation approaches will include the time and area available for the process, regulations, and cost of competing technologies. This paper emphasizes recent opportunities for industry to work in conjunction with the federal government that should reduce the time and the cost of biotechnology development for the private sector.

The U.S. Federal Government has a vested interest in seeing that the research conducted at the over 700 federal laboratories in the U.S. is used efficiently to make the world a safer, healthier, and a better place to live. At the present time approximately 1/6 (100,000 individuals) of the U.S. scientists and engineers work for the Federal Government. The federal R&D budget is approximately $20 billion. In DOE alone, the national laboratories are home to 22,000 scientists and engineers who perform scientific research and develop new technology in such fields as basic science, applied energy, environmental restoration and waste management. Annually, they perform over $5 billion of cutting edge R&D.

LEGISLATIVE BACKGROUND

To assure that the developments in the government laboratories have wide distribution in the private sector, Congress has passed a series of Technology Transfer Acts including the Bayh-Dole Act of 1980, the Stevenson-Wydler Act of 1980,

---

John C. Corey, Gerald J. Hooker, and Cindy K. Tew • Westinghouse Savannah River Company, Aiken, South Carolina 29802.

the Technology Transfer Act of 1986, and the National Competitiveness Technology Transfer Act of 1989. Each of these acts further strengthens the legislative intent to move technology from the government to the private sector more efficiently than before. Some of the key steps have included mandating the sharing of royalties with the government inventors from inventions licensed to the private sector, the establishment of technology transfer offices to facilitate the interface between industry and the government laboratories, the permission to conduct co-operative research and development agreements between the government laboratories and the private sector on subjects of common interest, the ability to negotiate intellectual property agreements prior to the beginning of the cooperative research and development agreement, permission to perform work for others, the encouraging of exchange of technical personnel for specific periods of time between industry and government laboratories, the making available to industry of one-of-a-kind scientific user facilities at government laboratories to assist American industry, and the encouragement of patent licensing by the government laboratories.

FEDERAL LABORATORY CONSORTIUM

The need for more effective use of Federally developed technology led to the establishment of the Federal Laboratory Consortium for Technology Transfer (FLC). The FLC is a service organization that provides a basic link between individual laboratory members and potential users of government developments.

The FLC Statement of Goals and Objectives stipulates that: "It is the policy of the FLC to identify and mobilize the necessary resources to provide the environment, the organization, and the necessary technology transfer mechanisms required to facilitate the fullest possible utilization of Federally sponsored research and development results by both public and private sector potential users."

To accomplish its goals and objectives, the FLC operates as an association of Federal research and development centers and laboratories that work together to maximize the transfer of technology. The Stevenson-Wydler Technology Transfer Innovation Act of 1980 mandated the creation of Offices of Research and Technology Applications (ORTA's) to support the technology transfer process. Thus, ORTA became the generic name of the technology transfer offices located within federal laboratories. The responsibilities of these offices are to receive and respond to technical requests and to implement those technology transfers that best fit the laboratory's mission. The FLC provides a network of ORTA personnel to meet with other technology transfer personnel to exchange information and experiences.

More than 300 of the largest Federal government research and development laboratories and centers representing 12 Federal agencies are presently members of the FLC. These agencies are:

- Department of Agriculture
- Department of Commerce
- Department of Defense
- Department of Energy

- Department of Health and Human Services
- Department of the Interior
- Department of Justice
- Department of Transportation
- Environmental Protection Agency
- National Aeronautics and Space Administration
- National Science Foundation
- Tennessee Valley Authority

The FLC provides a forum for sharing technology transfer related experiences and processes. The strength of the network is the ability to put the potential user in contact with a Federal laboratory person with expertise in a specific area of interest. Once the contact is found, the arrangements for transfer take place between the user and the laboratory.

The FLC holds two national meetings annually, providing a forum for formal and informal exchanges of information among member laboratory representatives, Federal laboratory representatives, and representatives from state and local government, industry and the Congress. General sessions, special sessions and workshops are held to discuss new directions, policies and methodologies relevant to technology transfer practitioners.

To access the FLC you may write or telephone:

FLC Administrators Office
P.O. Box 545
224 W. Washington St., Suite 3
Sequim, WA 98382-0545
Phone 206-683-1005

OFFICE OF RESEARCH AND TECHNOLOGY APPLICATIONS (ORTA'S)

Many of the Federal Laboratories have an ORTA. The office is normally staffed with a person with a broad perspective on the laboratory's capabilities and functions. Frequently, this office also houses the intellectual property section of the laboratory.

The offices are key to initiating strategic alliances between industry and laboratories. These alliances will facilitate the next generation of Technology.

The legislation encourages a number of activities to assist industry to leverage it's resources. These activities include technical personnel exchange, cooperative research and development agreements, scientific user facilities, patent licensing, consulting by laboratory scientists and R&D work for others that requires special capabilities and unique facilities at national laboratories.

TECHNICAL PERSONNEL EXCHANGE

Under this program, industry scientists and engineers work at one of the laboratories for a significant period of time, e.g. six months, in areas of technology transfer of interest to the industrial company. This exchange is one of the best ways to encourage dissemination of information. Unfortunately, our country has a culture that tends to reduce the enormous value that this program offers. Frequently this move by the industrial scientists is not advantageous to an American's career. In other cultures, this activity is often a required stepping stone to increased corporate responsibility.

COOPERATIVE RESEARCH AND DEVELOPMENT AGREEMENTS (CRADAS)

These agreements enable industry to work more effectively with government departments to develop processes of value to industry and the government. The various government departments have model Cooperative Research and Development Agreements for use. Accompanying this they have Joint Work Statements and Cooperative Research and Development Agreement Guidelines that provide instructions with respect to the specific terms and conditions. The key points in the new CRADAs are that it is not a funds out vehicle (no funds from the government to the partner). It is a funds in or no funds exchanged document. Proprietary information can be withheld from dissemination to the public for a fixed period of time. The parties to the CRADAs are responsible for compliance with the export control laws. The parties to the agreement have the first option to elect to retain title to any invention made by their respective employees but DOE retains a non-exclusive, non-transferable, irrevocable, paid-up, license to practice or to have practiced for or on behalf of the United States every invention under the CRADA.

SCIENTIFIC USER FACILITIES

Another opportunity for industry and universities to interact with government laboratories is by doing research at the unique scientific user facilities located at the laboratories. In DOE alone, there are about 200 of these User Facilities that are available for scientific research. Use of these facilities should accelerate the development of technology and reduce the R&D expenditures for industry.

PATENT LICENSING

Government laboratories are anxious to conduct patent licensing of technologies developed at government expense. Frequently the technologies will need maturation thereby increasing the probability that further patents belonging to the licensor will be developed. The value of the technology will be further enhanced by this intellectual property. The Department of Commerce does much of the licensing for government laboratories. Technologies developed by the Department of Energy are available through Robert Marchick (telephone 202-586-2802).

CONSULTING

Consulting guidelines are in place with each laboratory having its own parameters within which its employees can consult in their areas of technical expertise.

R&D WORK FOR OTHERS

The special capabilities and unique scientific facilities at the national laboratories are available to U.S. private sector firms, universities or other agencies for R&D as long as the facility does not compete with the private sector. Any inventions that result from this R&D are automatically titled to the third party in advance through class patent waivers.

AREAS OF OPPORTUNITY

Each of DOE's laboratories has something to offer. I would like to share with you examples from two of the lesser known laboratories to tweak your interest. The first of these is the Idaho National Engineering Laboratory, Idaho Falls, Idaho and the second is the Savannah River Laboratory near Aiken, South Carolina. Needless to say, the larger national laboratories such as Oak Ridge, Los Alamos, Argonne, Livermore National Laboratory, Sandia, Brookhaven, etc. have even more to offer in the area of biotechnology.

IDAHO NATIONAL ENGINEERING LABORATORY

This laboratory has a center for bioprocessing and biotechnology under LaMar Johnson. The center provides multidisciplinary approaches devoted to research and scale-up activity. The foci are on developing innovative bioprocesses of energy resources and other natural resources for the purpose of improving the usefulness and value of those relatively low value-high volume resources and development of innovative biological processing methods for raw and intermediate materials. Selected project titles demonstrate the breadth of research activities underway. These include bioremediation and bioreclamation, development of genetic engineering capability for biomining bacteria, selenium detoxification, isolation and genetic characterization of coal desulfurizing bacteria, biological fouling and corrosion, continuous bioprocessing of organic solvents, and in situ biodegradation of chlorinated solvents. Industrial partners would be welcome in each of these areas.

SAVANNAH RIVER LABORATORY

As part of DOE's Subsurface Science Program, a comprehensive program to examine the occurrence of microorganisms at various depths in the earth's surface is being conducted. The initial research was conducted at the Savannah River Site. This research was motivated by the need to restore contaminated soils and ground water at DOE sites. As a spinoff, important new discoveries concerning the presence, abundance and diversity of microbial communities at depth have been made. The research program had a number of university and national laboratory participants from Brookhaven National Laboratory, Cornell University, Florida State University, Oak Ridge National laboratory, Pacific Northwest Laboratory, Pennsylvania State University, University of Oklahoma, and the University of Tennessee. At the present time DOE's culture collection at Florida State University contains over 4,000 strains of microorganisms which were found at depth. The culture collection is available to industry for testing for useful applications by contacting the authors of this manuscript.

The research was initiated by DOE's Office of Energy Research at the Savannah River Site in 1986. The research has expanded in cooperation with the Idaho National Engineering Laboratory, The Hanford Site and the Nevada Test Site. Drilling took place at Idaho this summer into the volcanic layers present at INEL. Drilling is currently taking place at Hanford in the Columbia River sediments. The subsurface microbial populations are distinctly different at all three locations. The number of isolates and the subsurface microbial population are highest at Savannah River. It is anticipated that the suitability of in-situ remediation techniques will vary widely also in these locales. The program has also developed advanced aseptic drilling and sampling methods that have wide application to evaluating and restoring DOE's contaminated sites.

CONCLUSIONS

The government laboratories are extremely enthused about the Technology Transfer Acts that encourage the transfer of technology to the private sector. The ORTAs have very simple goals: to be the best that we can be, to be professional, to be helpful, to be available, and to develop long-lasting relationships that will allow us to continue to serve industry in an effective manner.

An Overview of Current Attitudes on the Use of Biotreatment for Cleanup

William J. Lacy

INTRODUCTION

One key indicator of the current attitudes on the use of biotreatment for environmental cleanup is to assess the EPA's Office of Research and Development current programs and its five year plan.

By all accounts it falls between very promising and excellent. Some examples of the ongoing work among other projects include:

1. *In Situ* Biorestoration of a gasline spill
2. Use of white Rot Fungus in a Rotating Bio-Contactor
3. Development of Bench-Scale Treatment system for biological removal of Trichloroethylene (TCE).
4. Biodegradation of PCB-contaminated Soil using recombinant DNA bacteria.
5. Chem-Bio Treatment test to destroy drum stored process wastes.
6. Determining the Genotoxicity of microbial and mammalian metabolites of biosystem pollution control pollutants.

As you heard from Dr. John Skinner Deputy Assistant Administration of EPA ORD. The principal purpose of these programs, therefore, is to make acceptable to the user community and the general public improved and novel biological treatment systems whose aim is the reduction or elimination of the risk associated with hazardous waste and other forms of environmental pollution, and to accomplish this in both an effective and least expensive manner.

Spearheaded by EPA Administrators William Reilly, top agency officials recently met with executives from industries involved in biological treatment of hazardous waste to begin eliminating scientific and legal barriers to the broader use

---

William J. Lacy • Lacy & Company, Alexandria, Virginia 22309.

of biotechnology as a solution for toxic waste cleanup.[1] At a meeting near Washington DC in late February, Reilly stressed the need for a breakthrough in costs of cleanup technology, in order to avoid a much longer time frame for cleanup that has been anticipated. EPA is looking for the help of industry in developing wider use of microbes at NPL sites and hazardous-waste facilities undergoing RCRA corrective actions.

"We have enormously complicated and expensive problems" Reilly said. "The costs of cleaning up those sites with technology we have is prohibitive... We need a breakthrough in the technology of cleanup. Work with us to give us some reasonable sense of what we may expect in five to ten years." Delegates from biotechnology industries responded to the call with requests for better information from the agency, a more positive regulatory stance, and support for demonstration projects. One example of regulatory roadblocks was a 1988 regulation (never formally proposed) that would have required companies to register naturally-occurring microorganisms used commercially. When the industry responded that the rule would mean eliminate the use of biotechnology for cleanup, the Office of Toxic Substances withdrew the regulation.

Clarification of definitions and the extent of federal control also will be needed to move ahead, according to meeting reports. There are differing opinions as to whether bioremediation is an "emerging technology," with some participants advocating that it has become a proven remediation method and others claiming there is still a lack of data on the effectiveness of the treatment. Some industry representatives assert that there is substantial evidence for biological treatment being a "best demonstrated available technology" for hazardous wastes under RCRA. Such a designation would enable the biotreatment of wastes coming under land-disposal bans and also would enable the treatment to be used for Superfund cleanups. Others point out that there are risks, and that the industry is not completely certain of what organisms will do in nature. "We often haven't learned from our failures," stated one meeting participant.

For its part, EPA must decide on the extent of the federal control over the use of microbes, and on the definition of "Commercial purpose" and "release to the environment." Currently, regulations dating back to 1986 require EPA approval prior to any commercial use of naturally-occurring or man-made organism that can cause disease in humans, animals, or plants. John Skinner, the agency's Assistant Administrator for Research & Development, indicated that the agency would move forward with providing more information and conducting more demonstration projects, but he avoided announcing any specific actions.

Our environment is contaminated with toxic chemicals, complex mixtures of unknown chemicals, and pollutants of all kinds. The need for remedial action has resulted in the creation of environmental legislation and regulations requiring that action be taken to identify and reduce or eliminate these hazards in a cost effective and environmentally responsible manner. These regulations are the result of a number of legislative mandates: the Clean Water Act (protect and restore aquifers);

---

[1] Hazardous Material Control Research Institute "Focus" April, 1990.

the Resource Conservation and Recovery Act (destroy or detoxify hazardous wastes); the Comprehensive Environmental Response, Compensation, and Liability Act (restore Superfund sites and decontaminate polluted soils); the Toxic Substance Control Act (environmental safety evaluation procedure); and the Federal Insecticide, Fungicide, and Rodenticide Act (wastes pesticide disposal).

The Comprehensive Environmental Response, Compensation, and Liability Act (CERCLA), commonly known as Superfund, is the most important legislation relating to the objective of this program. It was enacted in 1980 to provide broad authority to response directly to release of hazardous substance that endanger public health or the environment.

The Superfund Amendments and Reauthorization Act of 1986 (SARA) strengthens the authorities for remedial actions and enforcement procedures. It also provides new authorities for research and development, training, health assessments, community right-to-know, and public participation!! Section 209 of SARA specifically addresses the need for a comprehensive Federal program promoting a variety of research developments and demonstrations. New and improved data bases, risk assessment methods, and control technologies are the products of this program.

While human health and ecological risk assessment define existing or expected risks, it is the implementation of control technologies that actually mitigates or eliminates the public health and ecological problems. Superfund sites contain an enormous number of hazardous substances and complex chemical mixtures. The goal of control technology is to provide permanent treatment technology to clean up Superfund sites including any polluted groundwater aquifers. The challenge in this engineering program is to provide cost-effective control technologies that can treat these complex mixtures and achieve high risk reduction efficiencies for pollutants that are toxic at very low concentrations.

The principle purpose of the EPA program, therefore, is to make acceptable to the user community and the public at large new biological treatment systems that reduce or eliminate the risk associated with hazardous wastes and other forms of pollution. Chemical and physical ways have been developed to rid the environment of these pollutants. Unfortunately, these technologies do not restore the site to its original state. There is a need for relatively inexpensive but effective methods of completely destroying the contaminants in place or on site without further damage to the environment.

WHAT IS BIOSYSTEM TREATMENT?

Biosystem treatment is the organized and controlled use of microorganisms and their products for cleaning up toxic and hazardous wastes. (The term "biotechnology", as currently used by EPA, refers to the use of microorganisms that have been genetically engineered to accomplish biodegradation.) The microorganisms used in the biosystem treatment process are either selected from natural populations exhibiting a desired capability or genetically manipulated by man to carry out specific metabolic functions at an enhanced level.

Biological degradation offers the potential for effective, safe, and inexpensive cleanup of hazardous waste contaminants at many Superfund sites. Current treatment processes either move the chemical from one area to another or attempt to confine their further spread. Both techniques leave the waste essentially intact. Biological treatment can degrade the chemicals into benign products that are cycled within the biosphere. Application of this technology to the cleanup of Superfund sites is becoming possible because of increased information on the biology of the degrading microorganisms and their numerous potential treatment applications. The objective of the EPA initiative is to take advantage of this newly inexpensive and effective biological approach towards cleaning up Superfund sites and other hazardous and toxic wastes sites.

There are several benefits that biological treatment offers to the RCRA and Superfund programs:

1. Lower initial investment in the development of the treatment system.
2. Cost-effective reactor design and operational requirements.
3. The promise of minimum, site disruption and hazard avoidance associated with contaminated materials transport.
4. The ability to exploit and broad versatility of microorganisms towards mixed wastes and to accommodate the constraints of the site.
5. The ability to tailor treatment process towards specific compounds or groups of compounds at specific sites.
6. The potential to eliminate excavation and transportation costs.
7. Minimization of air emissions associated with treatment.
8. Likely decreased use of scarce natural resources related to energy requirements.

WHAT IS THE HISTORY OF BIOSYSTEM TREATMENT AT SUPERFUND SITES?

Microbial treatment has already been successfully used both in the United States and in other countries for on-site and in situ treatment of organic contamination of soils at hazardous wastes sites. Use of enzymes such as hydrolases for detoxification of organophosphate pesticides in soils has also been demonstrated. Research and actual cleanup of soils and aquifers contaminated by hydrocarbons, phenol, cyanides, and chlorinated solvents such as trichloroethylene have taken place.

Biodegradation of hazardous wastes, or bioremediation, is an emerging waste treatment technology that uses naturally occurring microorganisms that have been acclimated to thrive on a variety of organic compounds. Many of these compounds, such as volatile organic compounds (VOCs), phenolics, polynuclear aromatics (PNSs), and polychlorinated biphenyls (PCBs), can be found in the most toxic of hazardous wastes. When added to wastewaters, soils, or sludges, these select "bugs" go to work ingesting the contaminated materials. Through the bug's digestion, the

VOCs, phenolics, PNAs and PCBs are then converted to carbon dioxide and water, or bioaccumulate in the microorganism's cytoplasm.

The EPA has already ranked several remediation options available to a client with respect to the specific chlorinated organics found in a site with contaminated pond sludge—most notably, benzene, toluene, naphthalene, phenols and PCBs. Of the possible remediation options, including washing and stabilization, solvent extraction, and incineration, EPA ultimately recommended bioremediation with biological residue stabilization because it best satisfied the Agency's performance criteria.

For the purpose of this site, the EPA defined a variety of wastes to be remediated: all material containing over 25 ppm of PCBs and dike surface solids, including oily soil on the inside dike slope and grossly contaminated soil and sludge deposits visible in the dike.

One of the most important treatment goals was to reduce the level of PCBs in the pond residue to less than 50 ppm. If that performance level were reached, the client could return the stabilized residue to the pond to be capped. Failure to reduce the PCBs to less than 50 ppm would force the client to return the stabilized residue to a landfill in the pond area that met stringent approval guidelines defined in the Resource Conservation and Recovery Act.

The report *Tracking Superfund*[2] recently released by the Hazardous Waste Treatment Council and five environmental group, finds fault with EPA's application of bioremediation during cleanup of Superfund sites in 1989, claiming that the agency's decisions were made with inadequate information. "Few of EPA's FY-1988 cleanup decisions based the selection of bioremediation on treatability studies that would have defined site-specific factors critical to the success or failure of the approach," according to the report, which covers well over 100 records-of-decision during that fiscal year.

During FY-1988, EPA chose bioremediation in 6 of its 54 records of decision—a rate of about 11 percent. The report cites the French Limited site (Crosby, TX)—contaminated with petrochemical wastes—as one example of the inappropriate use of bioremediation. Bioremediation has been considered as a promising, less expensive alternative to incineration. However, the report contends that EPA must pay more attention to selection factors when making its decisions. For example, the agency needs to discover if organic chemicals are truly biodegradable. Some chemicals, including dioxins, dibenzofurans, and highly chlorinated PCBs, are not readily biodegradable; if they are found in concentrations exceeding health-based levels, bioremediation should not be used as the only cleanup technology, the report states.

Likewise, bioremediation is not effective for waste that contains large quantities of heavy metals (e.g., copper, zinc, cadmium) that are toxic to bacteria. In each of these cases, the report notes, bioremediation may be suitable when used with other technologies in a treatment train approach.

---

[2]Available from Hazardous Waste Treatment Council, Suite 310, 1440 New York Avenue, N.W. Washington, D.C. 20005.

The Connecticut-based, Business Communications Company recently released an assessment of the environmental management market, which includes hazardous waste, city wastewater, city solid waste processing, and sites converting waste to energy.[3] Over the next half-decade, the market is expected to rise from $823 million to an excess of 1 billion. The report targets hazardous waste bioremediation as the most rapidly advancing aspect of the market, with a projected leap from a $34 million industry (1989) to a $153 million one (1995). Research and development of technologies to better handle hazardous waste are estimated to get a funding increase from $80 million (1989) to $200 Million (1995).

The company also named as important marketing areas the microbially-based processing of underground storage tank leaks, oil spills, and discharge from paper factories and microbial treatments applicable to the four waste treatment sections named above. The reclamation of oil field wastes is seen as the largest portion of the market, but with little change anticipated in the current $400 million figure.

The projections based on the current market for the biological treatment of hazardous wastes are estimated to be from $1–2 billion per year, primarily applied to industrial sites for the treatment of industrial wastewater. The reasonable market for product sales in 1990 is estimated to be $50–$100 million, mostly on-site, though there will be more off-site applications of pretreatment technologies plus treatment in select cases where specific waste is centrally pooled or mixing with other wastes is required.

The development of a viable biological technology over the next decade will require that many legislative/regulatory factors be taken into consideration as well as existing industrial barriers. The Resource Conservation and Recovery Act may inhibit the development and commercialization of some biotechnologies in their current form by discouraging new on-site demonstrations through EPA's complex permitting process. Currently, it encourages the upgrading of existing installations that could assist biotechnologies; mainly it provides more incentive to provide the waste to an off-site disposer who usually is reluctant to use biotechnology due to the variable nature of the wastes received at facilities (biological technologies require a fairly consistent quantity and quality of waste and there are no sure ways to assure the cost-effective supply off-site).

One of the significant institutional barriers to utilization of biotechnology is in professional attitudes which are usually based on education. Usually chemical, civil, mechanical, and electrical engineers have only a limited practical experience or education in biology, and on the other hand, biologists have little or none at all in engineering. Each professional needs to gain from the new techniques, but they must work together and pool their talents to be successful. Another institutional barrier is constituted by the fact that hazardous wastes are regulated under a solid waste program for a solid waste industry which has traditionally used two things, transport and landfill, as its infrastructure.

---

[3]Business Communications Co., 25 Van Zant Street, Norwalk, CT. 06855. Environmental Management through Biotechnology: Microorganism & Enzymes for Waste Treatment. (No. C-110).

When one evaluates the influence of various regulatory, economic, and technical factors on the development of biotechnology as applied to hazardous waste, technical achievements in both the public and private sector prevail. These have been stimulated by a strong economy. Though regulations might appear to be a market factor, the actual liability the waste poses to industry is considered sufficient motivation to force use of cost-effective biotechnology even if there will be only a regulatory enforcement program.

In the Washington Post of 24, September 1990, Jack Anderson and Dale Van Atta had an interesting and cogent column entitled "Bureaucracy Stifles Biotechnology" in which they pointed out the great delays caused by the U.S. Patent and Trademark Office.

They pointed out that America has the chance to excel in biotechnology that could lead to new cancer treatments, better medical devices, and environmental and agricultural breakthroughs. But, according to their column, because of the snail's pace of American bureaucracy, the Japanese are poised to surpass the United States in this field also.

What Are the Critical Issues Necessary to Win the Confidence of State Regulators? Views of a Project Manager on Bioremediation Sites

Frank R. Peduto

A cooperative relationship between the regulator and the contractor is critical to the successful completion of any project where environmental issues are at stake. Frequently, however, this relationship becomes adversarial and when this has a negative impact on the work to be done, there is more to lose than just a few bruised egos. This paper will explore this relationship in the context of a bioremediation site and examine what the necessary ingredients are that can foster a cooperative relationship.

In order to establish some credibility with you on this issue, I believe it is necessary that you have an appreciation of how the Oil Spill Program operates in New York State.

Article 12 of the Navigation Law of New York, established in 1978, gives the Department of Environmental Conservation the authority to enforce an oil spill cleanup program. The main tenets of the law are:

1. Any spill, leak, discharge, etc. of petroleum to the waters of New York State is illegal.
2. The owner/operator, regardless of fault, is responsible for the cleanup.
3. In the event a responsible party is uncooperative or can not be identified, the Department of Environmental Conservation (DEC) is responsible for cleanup of the spill.
4. In the case of an uncooperative responsible party, reimbursement for the total cleanup cost, in addition to possible fines, would be sought upon completion of the cleanup.
5. A fee of one cent to four cents per barrel stored by the major oil storage facilities will be used to support the program.

Frank R. Peduto • Bureau of Spill Prevention and Response, New York State Department of Environmental Conservation, Albany, New York 12233.

In addition to unidentified and uncooperative spillers there are many who simply do not have the necessary resources to remediate a spill for which they were responsible. As a result, the oil spill program assumes responsibility for hundreds of spill cleanups. As a regulatory agency, we do not have the physical resources to conduct cleanup operations ourselves. However, under the Navigation Law the Department has the authority to maintain a system of standby contractors to perform the actual work. These contractors and their prices are pre-approved on a time and materials basis. Under this scenario the DEC assumes responsibility for the cleanup and identification of the remediation process, hires a contractor and evaluates the result. As regulators in the New York State Spill Program, we are experienced players in each of the roles that can be played out on any remediation site. We have been a responsible party, a consultant, a cleanup contractor and of course a regulator. Having functioned in each of these roles, I believe I have a unique perspective from which I can discuss a typical bioremediation project and identify what issues need to be addressed to ensure a cooperative relationship between all parties and ultimately a successful site remediation.

Before we discuss the elements of a cooperative relationship it is necessary to define the players involved. There are three major participants in the operation of a bioremediation project. These are the responsible party (RP), contractor/consultant and the regulator.

The responsible party (RP) is the person who has been identified as being legally responsible for the spill. They will ultimately be held accountable for cleaning the site and is the party who will pay the bill. This is frequently the first person to meet the state regulator.

The contractor/consultant is the person hired by the RP to engineer and clean up the spill. Although these are frequently two distinct parties I have combined them for a purpose. During the negotiations with the regulator there should be one voice speaking on behalf of the RP. While the RP always reserves the right to involve himself in all discussions, the consultant should be the RP's project manager through which all information will flow. This is necessary because few firms have complete turn-key spill cleanup operations. This is especially true of specialty services such as bioremediation. There will inevitably be a host of sub-contractors from the bioremediation specialist to the backhoe operator. All these people and all their services will have to be coordinated. Neither the regulator nor the RP wants to talk to all of these people on a continuous basis. Outside of some preliminary meeting with the bioremediation contractor, the only person the regulator should need to talk to is the project manager.

The regulator represents the state or federal agency under whose jurisdiction the incident occurred. This individual is responsible for assessing the environmental impact of the contamination and must be familiar with all the regulatory requirements applicable to the operation. This individual will function in the same capacity as the contractor's project manager. He or she is the eyes and ears of the agency.

The title of this presentation mentions winning the confidence of state regulators. There are many different ways to win someone's confidence, but there is one sure way to lose it, and that is not being completely truthful as to when and how the spill took place. This is necessary not only to establish credibility but information like this also helps the project manager determine the best plan to attack the cleanup.

Commitment for the RP usually means money. They should understand the magnitude of the issue at stake and commit the necessary resources. They should demand the same level of commitment from their contractor.

The contractor/consultant should know what they are capable of achieving. If you are an experienced project manager and after you have conducted your site investigation and feasibility study, you should be able to fairly accurately predict what level of remediation you can achieve and how long it will take. There should be little problem doing this on an ex-situ site where all soil is aboveground and the entrainment of oxygen and nutrients is under your control. If you don't believe you can achieve a regulatory standard with the resources you have at your disposal, you should discuss this with the regulator before the project begins. Negotiate before you start and not at the end.

In situ projects present a much more difficult problem. It will require more extensive investigation. Even under these conditions, however, barring any unforeseen circumstances, you should still be able to predict what cleanup levels can be achieved. A good investigation and feasibility study will demonstrate to the regulator how well you understand the task at hand.

Be prepared to commit the proper resources. This starts with the project manager and continues with other necessary support staff. If at all possible, keep the same project manager throughout the length of the project. One way to have your credibility questioned is to have a revolving door policy with your manager. If it is necessary to replace him/her, do so with as little impact as possible to the regulator or the RP. The new manager should be well briefed on the project. Don't make the regulator bring your manager up to speed. Notify the regulator as soon as the change is made.

Provide feedback to both the responsible party and the regulator on a regular basis. This could manifest itself in the form of a written status report or just a periodic phone call. Bioremediation projects are characterized by long periods of physical inactivity as the bugs go about their business. Don't allow long periods to go by without some contact.

Be candid when reporting to the regulator. If problems have come up make him/her aware of them, especially if these problems will cause significant delays.

Follow established procedures. Follow the plan as agreed to by the regulator. If there is a need to deviate from the plan, get approval first.

Know what is expected of you. Know what cleanup criteria you are expected to achieve and how you are expected to demonstrate that it has been achieved.

Demonstrating whether or not you have achieved a performance standard will usually manifest itself in the form of laboratory analysis. Design a sampling plan and submit it to the regulator for approval. Do this early in the life of the project. How many samples will have to be taken? What kind of samples (soil, water, grab or composite, split spoon, etc.)? What analytical method is to be used? Who will collect the samples? Is the laboratory capable of doing what you want?

Sampling costs a lot of money. Don't put yourself in a position of having to redo samples because you failed to satisfy the regulator. Remember, there usually are no strict regulations for sampling. They are usually developed on a site by site basis.

The regulator must have a good working knowledge of the laws and regulatory issues associated with the project. They should be able to define what the acceptable cleanup limits will be. If there are no specific limits, then the methodology which will be used to define final cleanup must be identified. The regulator should never leave the "How Clean is Clean Enough?" question unanswered. It is necessary in order to develop a successful plan, predict cost and to estimate a time of completion. It is also the regulator's responsibility to define what will be acceptable as proof of completion. If there are standardized sampling protocols identified, they should be made available. If not, then the proof of completion should be discussed and agreed upon by both parties in advance of the start-up of the operation.

One of the most important jobs of the regulator is to get the project manager through the bureaucracy and red tape as painlessly as possible. Unlike the private sector where one individual can speak for all, regulatory agencies are not designed for "one stop shopping". If your remediation project incorporates several technologies you may find yourself seeking several permits. No one person does all these things in any municipal agency. However, you can direct the project manager to the right people to accomplish what has to be done.

In situ bioremediation is a perfect example. It is frequently designed in concert with other technologies. Soil vapor extraction is commonly used on gasoline spills to remove the volatile organics. It also functions as a source of oxygen supply for bacterial growth. If vapor extraction is used, it may require an air permit.

When groundwater is being treated some of the groundwater is reinjected and mixed with nutrients or other chemicals. That portion which is not reinjected may require treatment through an air stripper or carbon column. Regardless of the method, discharging waters will likely require a discharge permit.

Proper guidance by the regulator can save a lot of time for the project manager and more importantly, will help the project begin with a minimum of delay.

Once bioremediation is introduced as a remediation technology for a site, the contractor should be prepared to present evidence to support the claim that the process will work at this particular site. Your experience or that of your bioremediation subcontractor should be made available. Successful as well as not so successful projects (with explanations) can be presented. If the site is to be treated in-situ then evidence of treatability will probably be gathered in a feasibility study.

Feasibility studies can range from the very extensive at an in-situ site to a considerably smaller study at ex-situ sites.

Every bioremediation project requires a complete understanding of three essential characteristics: chemistry, microbiology, and hydrogeology. Collecting the necessary data to define these parameters will only come from a detailed examination of the site characteristics, which will investigate the soil and groundwater in detail.

Some of the soil data that should be collected and presented are:

a. physical parameters such as grain size, porosity, moisture content, density and possibly pump test data to determine permeability

b. soil type
   c. chemical composition (inorganic chemicals—iron, manganese, etc.
   d. contaminant concentration
   e. nutrient concentrations
   f. bacterial counts

Similarly, groundwater sampling should produce comparable data such as chemical composition, contaminant concentration, nutrient concentration, bacterial counts, and dissolved oxygen level.

The contractor should demonstrate his concern for impacting the environment during the treatment phase by identifying these impacts and discussing how they will be mitigated.

What will happen to any discharge water and how will it be treated? What air emissions are expected and how will they be addressed?

Other aesthetic issues should also be addressed such as odors and site appearance.

Above ground (ex-situ) treatment sites may not require as extensive a study. Groundwater will not need to be addressed and if aeration is mechanically introduced by tilling, the hydrogeology of the site is of lesser significance. While an extensive study may not be required, there is a risk of treating these sites almost too casually. Contractors who suggest a minimum preliminary evaluation and recommend sprinkling some fertilizer over the contaminated soil should be looked at skeptically.

At no time is the relationship between the contractor, the regulator and the responsible party more vulnerable than when the project is at its expected termination. If cleanup of the site is approaching projected completion time and cleanup goals have not been achieved, there must be some accountability. The regulator need only look to the RP. The RP will look to the contractor. The contractor has no one else to look to but himself/herself.

Is there a documented reason why results were not achieved? If progress reports were being provided along the way, the situation should not be a surprise at least. In any event, if there were no extenuating circumstances, is the contractor prepared to assume responsibility? The New York State Department of Environmental Conservation's Oil Spill Response Bureau is attempting to draft a bioremediation specification based on two primary elements; a performance based contract bid on a lump sum basis. The primary objective is to have a preestablished cleanup level with an agreed upon price. While it is expected that bids will be higher to allow for contingencies, this will be offset by the guaranteed performance criteria within an acceptable time frame. If the performance criteria are not achieved, cleanup will continue at no additional cost. The contractor has an incentive to do the best job in the shortest time to realize the greatest profit. The responsible party will know the extent of his/her liability. The regulator is assured that the site will be cleaned up in the shortest time frame and if the treatment period is extended, it will be at no additional cost to the responsible party.

This technique will provide for a level of accountability for all parties involved. While there has been concern expressed over a guaranteed result it is not believed to be impractical. If you are settling a service you should have faith in that service's ability to perform to expectations. We can no longer keep pouring money into projects with no end in site. This approach will also deter those individuals who have a limited understanding of the technology from making a quick profit.

Combining all the elements which have been discussed and reducing it to the written word results in a product known as a project specification. A specification is a detailed plan of something to be done. More specifically, a project's specifications are a set of instructions detailing what must be done, how it will be done, and proof that it was done. Other issues that must be included are contractor payment, equipment requirements, reports and a guarantee of a deliverable product.

In summary, identifying all of these issues in a clear and concise manner agreed upon by all interested parties will provide for the greatest opportunity to achieve all the goals of a successful site remediation.

Federal Regulations: How They Impact Research and Commercialization of Biological Treatment

Sue Markland Day

INTRODUCTION

Biology-based waste treatment is a creature of this nation's environmental protection laws. Without the Superfund; water, solid and hazardous waste management programs; and the requirement for remediation of leaking underground storage tanks, little research would be underway and little commercial demand would exist for bioremediation.

It is a fact that governmental policies control the commercialization of waste treatment methodologies. The choice of specific treatment technologies is, to a large part, controlled by laws, regulations, and policies. Such governmental documents may encourage the uses of treatment or may place constraints on a technology. Often these two conflicting purposes are contained in the same statute.

The recent rediscovery of bioremediation provides an excellent study in governmental impact on market creation and application control. This chapter will begin by exploring the statutory mandates defining a niche for biology-based waste treatment; then characterizing major environmental regulations as they uniquely apply to environmental biotechnology; and conclude with the factors which will contribute to the future regulation of this waste treatment methodology.

MARKET CREATION

Created by environmental regulations, the waste management and remediation market for the United States has been estimated to top $23 billion.[1] That figure is growing. For example, the USEPA estimated in July, 1990, that its proposed corrective action rule, covering the 4,679 treatment, storage and disposal facilities permitted under the Resource Conservation and Recovery Act (RCRA), may result

---

Sue Markland Day • The Center for Environmental Biotechnology and The Waste Management Research and Education Institute, The University of Tennessee, Knoxville, Tennessee 37996.

in as many as 64,000 solid waste units requiring remediation.[2] The Regulatory Impact Analysis for that proposed rule estimated a total present value cost for the corrective action program to range between $7 and $42 billion for non-federal facilities and between $3 and $ 18 billion for federal facilities. Former USEPA Deputy Assistant Administrator, Al Alm, has said that some estimates go as high as 3% of the U.S. Gross National Product will be dedicated to environmental restoration by the end of this century.

In addition to making permanent treatment a national priority (Table 1), a new tactic in environmental legislation is to incorporate statutory language which directs the waste generator to certify that efforts are underway to decrease waste production at its source. For example, the Clean Water Act's pretreatment program's July 24, 1990 final rule states, " . . . the Industrial User shall certify that it has a program in place to reduce the volume and toxicity of hazardous wastes generated to the degree it has determined to be economically practical." Language in PL98-616, the Hazardous and Solid Waste Amendments (HSWA) to RCRA emphasized waste minimization by requiring waste generators (1) to certify on their hazardous waste manifests and annual permit reports that they have programs in place to reduce the volume or quantity and toxicity of their hazardous wastes as much as economically practical and (2) to describe on their biennial reports the efforts they have undertaken during the year to reduce the volume and toxicity of their hazardous waste and to compare these efforts to previous years.

Although these Congressional goals for permanent treatment and waste reduction can apply to all permanent waste treatment methodologies, biology-based approaches offer destruction of selected hazardous wastes, without toxic residues.[3] For those pollutants which can be completely converted to carbon dioxide, water and cell mass, bioremediation may be the regulator's management method of choice. Moreover, in the future, bioengineered micro-organisms may expand the capabilities of the naturally occurring organisms so that more recalcitrant contaminants can be degraded.

APPLICATION CONTROL

Because the advantages and limitations of bioremediation have not yet been formulated, and because many of today's environmental rules were conceptualized before the Environmental Protection Agency realized that a biotreatment industry existed, the current federal regulatory structure is more complicated for bioremediation than for other competitive treatment methodologies. Eight environmental laws and regulations spell out the operational and performance standards which must be followed when biology-based waste treatment is utilized.

The federal government regulates both the software (the microbial products) and the hardware (the reaction units and the methodologies). At present, naturally occurring microorganisms, usually in the form of cocktails or consortia, are bioremediation's software. To date, no genetically engineered organisms designed for waste mineralization have been utilized in the field. For that reason, because

Table 1. Key Environmental Regulations as They Apply to Bioremediation

Regulation	Coverage	How it works
Federal Water Pollution Control Act, as amended by PL100-4	Market creation	The discharge of toxic pollutants in toxic amounts is prohibited by national policy.
Clean Water Act Pretreatment Standards	Market creation	Requires industrial users to pretreat pollutants before discharging them to POTWs—includes waste minimization requirements, input controls, and notification requirements.
Resource Conservation & Recovery Act (RCRA) as amended by HSWA PL98-616	Market creation	Makes treatment of waste to minimize the present and future threat to human health and the environment national policy.
RCRA Toxicity Characteristic Rule	Defines regulatory scope	Adds 25 organic chemicals to Subtitle C regulatory universe and sets regulatory levels for organic chemicals.
RCRA Mixture Rule	Defines regulatory scope	Any compound, including soils and microorganisms, is defined as hazardous if it has been in contact with a listed hazardous waste.
RCRA Land Disposal Restrictions	Specifies handling of process wastes, contaminated soils and groundwater	The rule is *technology-based*, restricting treatment system design, setting cleanup levels, and specifying treatment technologies for some waste streams.
RCRA Soil and Debris Standards (under development)	Clarifies LDRs for contaminated soils and debris (also called old wastes)	Intended to set cleanup standards for "old" waste which reflect soil absorption/desorption properties of RCRA regulated wastes.
RCRA Corrective Action (proposed rule, July 27, 1990)	Solid waste management units located at RCRA permitted facilities	*Health-based* action levels for hazardous wastes are specified as triggers for remedial actions. Setting cleanup standards for each management unit is delegated to the Regional Administrator.
Underground Storage Tank Remediation (RCRA Subtitle I)	Market creation	Applies to both products and wastes; no national remedy standards—site by site decisions by the implementing agency, usually the State.
Permits	Delineates treatment process	Specifies cleanup levels and design or performance standards.
CERCLA, as amended by SARA PL99-499	Market creation; presently there is a strong agency push to use bioremediation at Superfund sites	A remedy shall be selected which utilizes permanent solutions and alternative treatment technologies.

(continued)

Table 1. *(continued)*

Regulation	Coverage	How it works
CERCLA National Contingency Plan	Selected abandoned contaminated sites; remedial action sites not covered by CERCLA default to the State government programs.	To expedite cleanups, remedial actions at NPL sites do not require a RCRA permit; cleanup standards are determined site by site but must reflect any legally applicable or relevant and appropriate requirements, standards, criteria or limitations under Federal or more stringent State laws. Remedial action decisions may take *cost* into account.
CERCLA—Subpart H of the National Contingency Plan	Biological additives used in navigable waters	Microbial products must be listed on the NCP Product Schedule.
Toxic Substances Control Act	Genetically engineered organisms	Regulatory criteria is "unreasonable risk." • Requires notification of US EPA and data, 90 days before production or use of a "new" chemical. • Current rules do not cover naturally occurring organisms or activities undertaken by small businesses and universities.
PCB remediation under Toxic Substances Control Act	Specifies handling of PCBs	Residue levels must be equivalent to those obtained by incineration for approval of an alternative treatment technology; for PCB spills, soil replacement, not onsite treatment, is specified.
Federal Insecticide, Fungicide, and Rodenticide Act	Microbial products used as pesticides	A pesticide is defined as "any substance or mixture of substances intended for preventing, destroying, repelling, or mitigating any pest. . . . Competition with indigenous organisms could fall under this definition.
Infectious Waste (RCRA Subtitle H)	Addresses disposal of spent cell mass	A federal pilot program requires manifesting of any microbial sludges if transported offsite; some state programs may require sterilization of cells before disposal.
National Environmental Protecction Act (NEPA)	Actions by federal agencies or activities which use federal funds	Requires an assessment of the environmental impact of, or any alternatives to, a federal action.

this book chapter has been written for immediate application, the hardware component will be emphasized first.

Scope

The description of the rule's coverage, in other words, the rule's scope is the first critical element of environmental regulation. In the case of a waste treatment technology, two such parameters are important: the wastes which are regulated and the operational/performance standards established for a particular treatment methodology.

Regulated Universe. For bioremediation, the key wastes to be managed under the Clean Water Act are the promulgated priority pollutants. Under the Resource Conservation and Recovery Act, managed wastes are either listed by name or by characteristic. Petroleum sludges and creosote, two major bioremediation applications, are scheduled to be listed as RCRA hazardous wastes. On March 29, 1990, a new procedure for defining the toxicity characteristic became final.[4] The Toxicity Characteristic Rule replaces the Extraction Procedure (EP) leach test with the TCLP, adds 25 organic chemicals to the Subtitle C regulatory universe, and sets trigger levels for these organic chemicals. The majority of wastes which are treated biologically may be considered organic, therefore, the expansion of the RCRA regulatory universe by the TC rule to include pollutants such as benzene, carbon tetrachloride, and trichloroethylene, disproportionally impacts biotreatment.*

Provisions of the Clean Air Act encourage the use of low sulfur coal as a energy feedstock. The inclusion of sulfur as an unwanted coal contaminant establishes that element as a potential biological treatment target. The Clean Air Act also focuses on air toxics as significant wastes to be managed. Such compounds as benzene, carbon tetrachloride, phenols, and toluene, which have been demonstrated as biodegradable, are included under this regulatory program as hazardous air pollutants.

The regulation of residues resulting from waste treatment can also impact bioremediation. For example, under a federal pilot program covering infectious waste, transportation of spent microorganism sludges off site requires manifesting. In some states, these sludges are defined as infectious and must be sterilized before land disposal. Ocean dumping regulations have limited the disposal of cell sludges created by Puerto Rico's pharmaceutical industry and have the potential to restrict the management of spent cell masses from bioreactors used in waste treatment. The RCRA Subtitle C mixture rule regulates cell mass if the microorganisms are used to degrade RCRA regulated hazardous waste. Under RCRA any compound, including soil and microorganisms, is defined as hazardous if it has been in contact with a waste listed as hazardous under RCRA. The waste plus cell mass, even if none of the toxic original compound remains, must be delisted (a time consuming rulemaking process) before the EPA officially recognizes that the problem has been solved.

Of particular interest to the biotreatment industry is the concentration of residual contaminant allowed at the site or after the treatment process. This too is a scope issue, because, as in many waste treatment technologies, today's biological

---

*Although the inclusion of a waste stream under federal regulations creates an incentive for treatment, it also introduces complicated record keeping and may restrict the design and operation of a particular treatment method. This creates a quandary which is particularly evident to the bioremediation industry. Currently firms view the land disposal restrictions as a major impediment to landbased biological treatment and for that reason, are not enthusiastic about capturing more wastes under the RCRA hazardous waste rules.

treatment may not degrade the target pollutants to background levels. If a regulation establishes residual concentration levels, the rule may limit the treatment choices.

Actions Regulated. One of the structures regulated under RCRA is a tank that is used for waste treatment activities; therefore, a RCRA TSDF permit and/or Clean Water or Clean Air discharge permits for bioreactors, such as mobile treatment units, soil slurry reactors and pump/treatment reaction vessels is required. Only work on federal superfund sites is exempted from this permit requirement; however, requirements similar to those in permits may be included in compliance order specifications for federal superfund sites. Though a facility permit may not be required for in situ bioremediation projects, public notice and cleanup standards specified by the RCRA, UST or CERCLA programs may be in order. If the biological treatment effort does not fall under environmental regulations, if the activity is paid for by federal funds compliance with the National Environmental Policy Act is required.

Procedures and Process Control: *Key Requirements in Environmental Regulations*

Clean Water Act. Under section 1004 of RCRA hazardous waste mixed with domestic sewage is not regulated under RCRA, but instead under the Clean Water Act. To address the hazardous waste streams feeding into the Publicly Owned Treatment Works (POTWs) standards are being promulgated, focusing on 126 toxic pollutants that are listed in Appendix A to 4OCFR part 423. These toxic pollutants include benzene, acetone, and other biologically degradable compounds.

Published July 24, 1990, the pretreatment rule attempts to put teeth into the enforcement of the standards by the POTWs.[5] The rule also establishes self reporting for industrial dischargers; eliminates discharges of pollutants into sewers throughout the publicly managed and owned waste collection system; sets toxicity testing requirements for POTW effluents; and requires that all industrial users certify waste minimization programs. The rule does not ban the discharge of solvents to POTWs. It does commit to issue four new technology-based pretreatment standards and to revise these existing standards. In general, regulations which specify technologies inhibit the introduction of more sophisticated approaches and institutionalize the status quo. Therefore, the pretreatment program has two different impacts on the use of biotreatment. By improving enforcement, the rules support the expanded use of bioremediation before the wastes are discharged; however, cleanup standards which rely on technology designation do not provide much incentive for new, improved waste treatment methodologies to be developed.

Resource Conservation and Recovery Act. *Permits.* Permits provide the rules under which a particular treatment regime can be executed. RCRA Subtitle C permits are required for the treatment of hazardous waste. Table 2 defines the different types of facilities; which must have RCRA land disposal permits. One contaminant which does not fall under the RCRA program but is still one of interest to the

Table 2. Hazardous Waste Management Units Requiring Permits

Land-Based Management Units

A **surface impoundment** is a depression or diked area (e.g., pit, pond, or lagoon) used for storage, treatment, or disposal. The impoundment is open on the surface and designed to hold an accumulation of waste in liquid or semi-solid form. 40 CFR Part 264 Subpart K requires new impoundments to have double liners and a leachate collection system (minimum technological requirements).

Waste piles, used for treatment or storage of a non-containerized accumulation of solid, non-flowing hazardous waste, are regulated under 40 CFR Part 264 Subpart L. Waste piles used for disposal must comply with Subpart N requirements.

Land treatment is the process of using soils and microorganisms as a medium to biologically treat hazardous waste. Subpart M of 40 CFR Part 264 covers this treatment unit.

40 CFR Part 264 Subpart N contains the requirements for **landfills.** As with surface impoundments, hazardous waste landfills must install double liners and leachate collection systems. Disposal of liquids is restricted.

Non-Land-Based Waste Management Units

Tanks are stationary devices designed to contain an accumulation of hazardous waste and constructed primarily of non-earthen materials. 40 CFR Part 264 Subpart J of RCRA hazardous waste management standards specifies secondary containment and release detection systems for tanks.

---

biotreatment industry is polychlorinated biphenyl. To treat PCBs, a Toxic Substances Control Act (TSCA) permit is required.

Totally enclosed treatment facilities, waste water treatment units associated with a NPDES permitted discharge, and elementary neutralization units do not require RCRA treatment, storage or disposal facility (TSDF) permits. Some bioreactors may fall under this exclusion, but others may require a tank permit which specifies tank design, not the treatment process itself. Individual RCRA permits are required for each nonCERCLA location where onsite treatment is underway, no matter whether the treatment unit is permanent or mobile. In situ treatment is handled under corrective action orders.

The RCRA permit, and other comparable documents such as the CERCLA Record of Decision (ROD) and the corrective action compliance order, are important to biotreatment for three reasons. First, cleanup standards and design and/or performance of the treatment technology can be specified. The latter will be discussed under the Land Disposal Restrictions section. The former sets the remediation targets, determining the cleanup goals and, potentially, limiting the treatment choices.

A comparison of cleanup targets for some compounds which have been commercially bioremediated found in the land disposal restrictions and the proposed corrective action rule demonstrates the difference between the risk- and technology-based remediation targets. For example, USEPA has established the BDAT performance standards for phenol at 2.7 mg/kg (technology based) and the corrective action level for the same pollutant at 3,000 mg/kg (health based).

For those biological procedures which can degrade selected contaminants to non-detection levels, the biotreatment market would profit from the lowest acceptable residual concentration. But for those pollutants which can be biologically degraded to a residual level which is non-threatening to human health and the environment, but can not be completely eliminated, as is possible with incineration, then technology-based standards would restrict bioremediation markets.

Of the three permit elements (Table 2), the third, the waste stream or site characterization, may be the most critical. Without specialized information on the existing microorganisms and their degradative capabilities, the microtoxicity of the target wastes, the climate conditions (moisture, temperature), and the bioavailability of the contaminants, a decision on whether bioremediation is the appropriate choice can not be made. Not all federal rules and guidances, often used by State and Regional professionals when reviewing permits, have been updated to include characterization data specific for biology-based treatment. For example, the July, 1990 proposed RCRA corrective action rule only specifies physical and chemical analysis, not biological, for the remedial investigations.

Moreover, no standard EPA protocols currently exist for this data gathering or for bioremediation treatability studies. Bioremediation treatability protocols are pivotal in the demonstration of the successful detoxification or mineralization of pollutants by biological processes. There is a need for uncomplicated, inexpensive standardized protocols.

Few RCRA TSDFs permits have been issued by the EPA for biological treatment. The lack of examples of federal or state permits is a problem in the short term. Representative permits are useful to the permit applicant because they illustrate the type of information required by the regulatory agency and the favored application format.

Another type of RCRA permit which should be of interest to the biotreatment industry is the Research Development and Demonstration (RD&D) permit. In 1984, the Resource Conservation and Recovery Act was amended to give the USEPA new permit authority: the RD&D permits for innovative and experimental hazardous waste treatment technologies or processes. The purpose of the RD&D permit was to aid the development of safe alternatives to land disposal of hazardous wastes and allow applicants to conduct experimental testing of new hazardous waste treatment technologies or processes by modifying or waiving RCRA permit applications and procedural requirements. Congress directed the Agency to limit the quantities of hazardous waste addressed and to require financial responsibility assurances for RD&D permits. As of January, 1990, the Office of Solid Waste's Permits branch is not aware of an RD&D permit issued for a biological process.

Land Disposal Restrictions. On June 1, 1990, the final rule in a series of regulations to implement Section 3004 (m)(l) of RCRA was published.[6] The Agency, under the Hazardous and Solid Waste Amendments (HSWA), was required to set "levels or methods of treatment, if any, which substantially diminish the toxicity of the waste or substantially reduce the likelihood of migration of hazardous constituents from the waste so that short-term and long-term threats to human health and the environment are minimized."[7] The LDRs prohibit the continued land disposal of hazardous waste unless treatment standards have been met or disposal is in a unit from which there will be "no migration." According to the Agency, "for the purposes of the LDR program, the statute specifically defines land disposal to include, but not limited to, any placement of hazardous waste in a landfill, surface impoundment, waste pile, injection well, land treatment facility, salt dome or salt bed formation, or underground mine or cave." Descriptions of the different land-based hazardous waste management units can be found in Table 2.

It is the term "placement" which impacts biological treatment. As described in other chapters of this book, biology-based treatment systems have historically included landfarming, the application of a mixture of wastes plus organisms directly on the soil without liners or other barriers to migration. Moreover, in situ bioremediation often relies on recirculation of contaminated ground water to optimize soil contaminant degradation, to contain the wastes within a management area, and to simultaneously treat the polluted ground water. Under the LDRs, these two operational designs are not acceptable.

Because of these restrictions, the use of variances may become more important. For example, a "no migration" variance is a possible avenue to the use of a landfill or waste pile arrangement for biological treatment. "A reasonable degree of certainty, that there will be no migration of hazardous constituents from the disposal unit or injection zone for as long as the wastes remain hazardous,"[8] must be demonstrated for this variance. Two USEPA documents may be useful when making a case for this exemption: the Office of Solid Waste, July, 1986 "Permit Guidance Manual on Hazardous Waste Land Treatment Demonstrations" and the RCRA air toxics regulations currently under development.

A second hurdle created by the land disposal restrictions is the establishment of treatment levels which are technology based. These residual contaminant levels are either promulgated as a concentration or as a particular treatment technology. The LDR standards were developed by asking how wastes from specific industrial processes, or end-of-the-pipe wastes, should be managed. The waste residue concentrations were not determined through risk analysis, but instead, by the performance of available technologies.

This creates three problems: old technologies, such as incineration, have a competitive edge; the level of cleanup is not health-based and does not provide an opportunity for cost analysis; and the Best Demonstrated Available Technologies (BDAT) standards do not work well for contaminated soils and debris. Any biological treatment unit or system, including tanks serving as soil and water bioreactors, has to produce effluents and soils, if they are to be land disposed, with final contamination levels often established by a competitor technology (incineration).

Corrective Action Rule. Congress created this program component as a mechanism to compel cleanup of contaminated lands/groundwater by the TSDF permit holders. It was published in July, 1990 as a proposed rule.[9] The proposed corrective action (CA) rule does not specify a cleanup level, as is done in the Land Disposal Restrictions regulations; instead, the extent of the remedial action is to be negotiated and is risk-based. The choice of treatment technology, therefore, is similar to the process used for Superfund sites. However, there is one major difference—corrective actions must observe the land disposal requirements. Because of the timing of the two rules' promulgations, this proposed CA rule recognizes the potential problem of LDR placement when an onsite treatment methodology is to be used. Its solution is the establishment of a corrective action management unit (a CAMU) in which contaminated media can be moved during the cleanup operation. However, until this concept becomes final, a variance for selected bioremediation operations will be necessary.

The proposed CA rule does not only attempt to solve problems: it also may create a new one for the biotreatment industry. The proposed rule removes underground storage tank (UST) cleanups at facilities with a TSDF permit from the UST program and places these sites under RCRA.

Underground Storage Tanks. USEPA issued final technical performance standards and associated regulations implementing the UST program on September 23, 1988.[10] This program covers underground tanks storing regulated substances, including petroleum products (e.g. gasoline and crude oil) and Superfund-defined hazardous substances. Tanks storing hazardous wastes are regulated under Subtitle C of RCRA. Similar to Superfund and the corrective action program, the federal UST program does not have promulgated cleanup standards, relying on site-by-site decisions by the implementing agency, usually the State. The regulatory red tape for this program appears to be considerably less than that for RCRA hazardous waste management. Today, the remediation of soils and groundwater contaminated by USTs is the single largest application of non-wastewater environmental biotechnology. It is the absence of national remedy standards and the ease of petroleum product biodegradation which can be credited for this.

Comprehensive Environmental Response, Compensation and Recovery Act or Superfund. The 1986 Superfund Amendments and Reauthorization Act (SARA) eliminated permit requirements for onsite Superfund remedial actions. This amendment also established the federal cleanup program as a mechanism to test out new treatment technologies. Such a mandate is important to the applications of environmental biotechnology. For example, there is a potentially large demand for heavy metal removal at Superfund sites. This could be accomplished by biosequestering or bioleaching already in use by the mining industry.

As mentioned previously, the key to qualifying bioremediation as a potential remedial action choice is a site characterization designed to gather critical information on the "friendliness" of the site to microbial action. Superfund guidances still remain in the state and federal field offices which lack instructions on biology related data gathering.

Unlike RCRA, CERCLA is a statute which allows the cost of decontamination to be considered when choosing a treatment technology. In the long run, if the bioremediation industry can substantiate cost savings, this unique element of Superfund will prove advantageous.

The disadvantage of the CERCLA program is a mound of administrative red tape, including a chain of custody procedures, public notification requirements, and potential contractor liability. No prescreening guidances for bioremediation contractors have been finalized by the USEPA; although, such federal guidelines have been written for incinerator contractors.

Toxic Substances Control Act. RCRA does not govern the removal and treatment of polychlorinated biphenyls (PCBs). PCBs are regulated under TSCA, 40 CFR Part 761. Although there are limited exceptions, all handling and disposal of this compound require a permit.

Progress in the research and development of this waste's mineralization could be hindered by the lack of TSCA program experience within the hazardous waste treatment industry. To compound the problem of limited familiarity, the federal TSCA program is not delegated to the States. Unlike the RCRA program, the TSCA program requires communication with USEPA's headquarters or its regional offices.

A brief compare and contrast of the RCRA and PCB programs reveals several differences. First, the RCRA program has a permit exemption for treatability studies; the PCB program does not. Second, a research permit is required of individuals who do not own the contaminated site (are not the generators doing site remediation work) but are either interested in research to develop new PCB treatment technologies, such as bioremediation, or in contracting for PCB treatability or pilot studies. These permits are issued by USEPA headquarters. Under RCRA RD&D permits are usually issued at the regional or state level. For those seeking to clean up a site, a PCB disposal permit is required. Such permits are routinely issued for incineration, chemical waste landfills, or high efficiency boilers. Any alternative method requires a variance from either the State or USEPA's Regional Office. The applicant for a variance must demonstrate that the treatment technology will result in residue levels equivalent to those obtainable through incineration.

Because the PCB disposal permit is technology specified, penetration of this treatment market may only be possible after a regulatory change. The criteria for PCB remediation would have to be changed from a specified treatment method to a risk-based residue concentration.

National Environmental Policy Act. This law requires any federally funded activity to be analyzed for its environmental impact. For example, if bioremediation is planned for a site on federal property, at least a short form Environmental Assessment document must be submitted to USEPA. If the project is considered major, the Council of Environmental Quality will also be involved. To date, few federal cleanups have relied on bioremediation but considerations unique to biology-based treatment, such as potential byproducts, dispersion of non-indigenous organisms, or disruption of carbon or nitrogen cycles, deserve attention in these analyses.

PRODUCT CONTROL

Toxic Substances Control Act

Information on chemical risks from those who manufacture and process chemicals may be gathered by the Environmental Protection Agency under this Act. TSCA's regulatory trigger is a chemical's unreasonable risk at any stage in its lifecycle: the manufacturing, processing, distribution in commerce, use or disposal. Similar to the Patent Office, USEPA has defined micro-organisms as chemicals for the purpose of this regulation. Therefore, the Toxic Substances Control Act regulates the planned introduction of non-indigenous microorganisms to the environment.

TSCA, a notification statute, requires the manufacturers or importers of new chemicals to notify USEPA 90 days in advance (a premanufacturing notice or PMN), except for those categories expressly excluded from coverage by the statute. If a chemical is not listed on the TSCA inventory of existing chemicals, it is considered "new." Even though a chemical may already be listed on the inventory, the Agency may decide that additional uses are significant and require a 90 day notification (significant new use notice or SNUN) before allowing the market for a product to be expanded.

Since publication of the Coordinated Framework for Regulation of Biotechnology in 1985, USEPA has been drafting TSCA regulations to cover biotechnology products and processes (which are not covered by USDA or FDA and are not regulated by EPA under the Federal Insecticide, Fungicide and Rodenticide Act).

Proposed amendments to the TSCA regulations are now scheduled for 1991. In the winter of 1988, a draft rule was published in the Federal Register which proposed to regulate non-engineered microorganisms. In a rare demonstration of unity, the national environmental groups joined with the biotreatment industry to protest the inclusion of naturally occurring organisms under TSCA. The major argument against the expanded universe was one of limited Agency manpower to review the notifications. Because the Agency has received a rare public-private mandate, the coverage of non-genetically engineered microorganisms by the TSCA rule does not look likely at this time.

However, it is this federal act which can be credited as the critical "stopper" to the transition of genetically engineered microorganisms from the laboratory reaction vessels and petri dishes to the field. Other chapters in this book discuss the field applications of microorganisms, emphasizing the need for better proof that the waste is being biodegraded or better site characterization in order to predict bioavailability, but the question of how to gain public and regulator approval for the use of a bioengineered organism has not been addressed. That gap in the discussion exists because no company, federal laboratory, or university has attempted to use a genetically engineered organism outside the laboratory.

It is the fear of the unknown that is the stopper. But scientific or engineering uncertainty is not the cause. It is the certainty that even the most benign deliberate release will require years of work to satisfy the USEPA and to pave the way for testing of more sophisticated environmental products or processes. Tables 3 and 4 compare the existing TSCA data requirements with the draft TSCA rule.

National Contingency Plan (CERCLA)

Subpart H of the NCP regulates the use of any biological additive in the navigable waters of the United States, including territories, and adjoining shorelines, the waters of the contiguous zone and the high seas. To be used, an individual microorganism or a cocktail of microbes had to be listed on the NCP Product Sched-

Table 3. Information Required under Current TSCA Regulation

1. Identity of the chemical
2. Its molecular structure
3. Proposed categories of use
4. An estimate of the amount to be manufactured, imported, or processed
5. The byproducts resulting from the manufacture, processing, use and disposal of the chemical
6. Estimates of exposure
7. Any test data related to the health and environmental effects of the chemical *plus*
8. Any data generated by a test rule

Table 4. Information Required by Draft TSCA Biotechnology Rule

1. Identity of the organism
2. List of published health and environmental data on the microbe plus its unmodified parent
3. Summary of unpublished health and environmental data
4. An estimate of the amount of the organism to be produced annually (including expected viability)
5. Information on the sites where the organism is manufactured, processed, used, and disposed of
6. Worker exposure information
7. Estimated amount of the organism to be released annually
8. Detailed description of transport parameters
9. Description of intended use including all engineering controls and personal protective equipment [biological controls are *not* acceptable]
10. Information on locations, conditions, and purpose of intended release; anticipated survival, multiplication, and dissemination; identification of target organisms; a description of host range; involvement in biogeochemical or biological recycling processes; etc.
11. Description of intended disposal (e.g., efficacy of inactivation)
12. Exposure potential and dispersal routes
13. Other information that the submitter believes necessary for risk assessment

ule. The procedure is found in 40 CFR 300.86. However, the March 8, 1990 amendments to the NCP allows the on scene coordinator to authorize the use of biological additives, including products not listed on the NCP Product List, when the use of the product is necessary to prevent or substantially reduce a hazard to human life.[11]

Federal Insecticide, Fungicide, and Rodenticide Act

It is possible that a biological product which is advertized as competing with indigenous or pre-existing organisms in order to improve a process' performance might classify as a pesticide. In 40 CFR Subpart A 152.3 (s), a pesticide is defined as "any substance or mixture of substances intended for preventing, destroying, repelling, or mitigating any pest. . ." An organism (except man) is defined as a pest under circumstances that make it deleterious to man or the environment. USEPA registration is mandated if a firm claims pesticidal actions for its microbial product.

Plant Pest Act

The Department of Agriculture regulates the transport of microorganisms across state lines if that organism is a plant pest. The burden of proof is on the sender/receiver to demonstrate that his particular microbe is not a plant pathogen. Microorganisms are considered possible suspects if a member of the same genus is a plant pathogen. Since many of the bioremediation workhorses are soil microorganisms, a relative might fit this criteria. Organisms which are exempt from this requirement are periodically listed in the Federal Register.

CONCLUSION

As the biotreatment industry becomes more at ease with the federal and state regulators and as the regulators gain experience in biological treatment, many of today's problems with inadvertent regulatory glitches, which discourage bioremediation, will be corrected. Because direct changes in regulations take time, it is the opinion of the author that regulatory interpretations and updated guidances could be utilized by USEPA to bridge the gap. This mechanism has been used by other more mature treatment vendors and users to rectify implementation problems.

The universities' environmental biotechnology research and their arrangements for transferring discoveries to the private sector will be a significant factor in tomorrow's bioremediation regulatory climate.

Current commercial biological treatment relies on a limited number of reaction protocols and on naturally occurring microbes. University work is more complex, with genetically engineered organisms created for both monitoring activity and for degrading recalcitrant compounds. The second generation of bioremediation is posed at the graduate schools for field application.

But, a comfort level does not presently exist between the university scientists and engineers and the regulators. This was demonstrated in the fall of 1989 when university investigators and TSCA bureaucrats met for two days in Washington, D.C. to carve out a system of regulation for university research. According to Ronald Evans, Biotechnology Program Leader for TSCA, no consensus resulted from that gathering and over a year later no approach has been decided.

Today's federal and state regulators have little experience working with the academic community. In my opinion, the academic investigators' inexperience with regulation may result in handing over discoveries to the private sector for development, including dealing with the regulators, prematurely. This nation's expertise in ecological impacts, in the biochemical pathways utilized by microorganisms to degrade wastes, in the techniques necessary for new organism development, and in the engineering and soil science principles of bioavailability is concentrated in the universities. In other words, the skill required to generate the information of interest to the regulators is not yet in the private sector.

That leads into a second pivotal factor in tomorrow's environmental biotechnology regulation. Today too few multi-disciplinary, technically grounded professionals are available to the federal and state regulatory agencies. This same shortage is impacting the generators, such as the Departments of Defense and Energy, who need such expertise to manage remedial action projects, and the biotreatment firms themselves. Universities must have the organizational structure and the monies necessary for equipment, laboratory space and graduate scholarships in order to educate the future environmental biotechnology workforce.

In spite of an anticipated rapid transition from basic research to in-the-field applications and an expanding university graduate pool, the future of bioremediation will not be bright without public confidence. For that reason, the continued debate over the use of technology standards versus health or risk-based standards as treatment objectives should be moved from the regulatory offices into the public arena. Moreover this debate should be focused on specifics, such as the "not in my backyard" syndrome which encourages onsite treatment, the potential transfer of pollutants among different media, and the advantages/ disadvantaged of different treatment methodologies. The risk dialogue should also apply to the safety issue questions associated with the planned release of genetically-altered microorganisms into the environment.

It is the author's opinion that the citizens and the academic communities are best suited to organize such a public debate. I leave those groups in this audience with that challenge.

REFERENCES

1. Henley, Mike, "Europe Poses Multi-billion Enviro Market," *WasteTech News*, Volume 2, Number 24, August 13, 1990, p. 1.
2. "Corrective Action for Solid Waste Management Units at Hazardous Waste Management Facilities: Proposed Rule, Part II, Environmental Protection Agency (40 CFR Parts 264, 265, 270, and 271)," *Federal Register*, Vol. 55, No. 145 (July 27, 1990).
3. *New Developments in Biotechnology: U.S. Investment in Biotechnology*, Office of Technology Assessment, July, 1988.
4. "Hazardous Waste Management System: Identification and Listing of Hazardous Waste; Toxicity Characteristics Revisions, Environmental Protection Agency (40 CFR Parts 261, 264, 265, 268, 271, and 302)," *Federal Register*, Vol. 55, No. 61 (March 29, 1990).
5. "General Pretreatment and National Pollutant Discharge Elimination System Regulations; Final Rule (40 CFR Parts 122 and 403)," *Federal Register*, Vol. 55, No. 142 (July 24, 1990).
6. "Land Disposal Restrictions for Third Scheduled Wastes; Rule, Part II Environmental Protection Agency, Part II (40 CFR Part 148 et al.)," *Federal Register*, Vol. 55, No. 106 (June 1, 1990).

Disclaimer: This chapter should not be viewed as providing legal advice. It is meant to serve only as an overview of key regulatory requirement. Each application of a treatment technology is unique and requires detailed discussions with the appropriate local, state and/or federal regulator.

7. PL98-616, The Hazardous and Solid Waste Amendments to the Resource Conservation and Recovery Act.
8. 40 CFR Part 268.6 "Petitions to Allow Land Disposal of a Waste Prohibited under Subpart C of Part 268."
9. "Corrective Action for Solid Waste Management Units at Hazardous Waste Management Facilities: Proposed Rule, Part II, Environmental Protection Agency (40 CFR Parts 264, 265, 270, and 271)," *Federal Register*, Vol. 55, No. 145 (July 27, 1990).
10. "EPA Technical Standards and Corrective Action Requirements for Owners and Operators of Underground Storage Tanks (40 CFR 280); *Federal Register*, Vol. 53, p. 37194 (September 23, 1988).
11. National Oil and Hazardous Substance Pollution Contingency Plan," Environmental Protection Agency (40 CFR Part 300), *Federal Register* (March 8, 1990).

Polluted Heterogeneous Environments: Macro-scale Fluxes, Micro-scale Mechanisms, and Molecular Scale Control

Geoffrey Hamer and Armin Heitzer

INTRODUCTION

Most large-scale processing ventures are initiated by ill-defined statements of need rather than by the discovery of either a new product or a novel process route. One of the best examples that has been contrary to this general pattern during the past decade has been the dramatic development of biotechnology, where one has seen both products and processes emerging as a result of research push rather than market pull. In contrast to this, environmental biotechnology is developing in response to a perceived need; the clean-up of indiscriminate pollution that has occurred from the advent of the industrial revolution until relatively recent times. With increasing public awareness and sensitivity to the crises that either have or will develop with respect to enviromental safety, health and quality, it is only now that politico-economic policies are being implemented that allow the development of technological responses to such crises. Newly evolving bioremediation and biorestoration technologies for soil and ground water clean-up, respectively, represent only one response scenario. Their future will depend on both their efficacy and their economics. Unless they continue to offer clear advantages on both counts, they will rapidly become obsolete. The maintenance of their relative attractiveness depends on the effectiveness and availability of underpinning fundamental research and its exploitation in solving practical problems.

In both terrestrial and aquatic environments, Protista, i.e., bacteria, fungi, cyanobacteria and algae, act as major mediators of pollutant fluxes by changing driving forces between sources and sinks and by changing both the chemical and physical status of environments subjected to macro-scale pollutant fluxes. The fundamental problems in assessing levels of mediation in any polluted environment stem from an inability to either characterize or quantify the communities of protists

---

Geoffrey Hamer and Armin Heitzer • Institute of Aquatic Sciences and Water Pollution Control, Swiss Federal Institute of Technology–Zürich, CH-8600 Dübendorf, Switzerland.

present, on the one hand, and to differentiate between perturbed and non-perturbed states within such communities, on the other hand.

More than a century ago, Koch (1881) announced a procedure that permitted for the first time the isolation of pure cultures of single species of bacteria, i.e., monocultures. The subsequent refinement of the procedure allowed microbiologists to study microbes at the defined species level and for decades thereafter, most investigations were carried out with monocultures growing on single substrates, giving rise to our basic fundamental knowledge of the biochemistry, genetics and growth physiology of microbes (Bull and Quayle, 1982). However, increasing knowledge has been accompanied by an increasing realization that monocultures and single substrate systems are remote from the realities encountered in Nature. As a result, the past two decades have seen the initiation of concerted attempts to add back to experimental systems the very degrees of complexity that so bedevilled the attempts of 19th century microbiologists to isolate pure cultures.

The primary concept of the present symposium, i.e., moving from the flask to the field, is an enigma, particularly when one considers how little of the research that has posed as either environmental microbiology or environmental biotechnology is of any application in the real problems confronting the practitioners of bioremediation. The concept of homogeneity in multi-phase systems, such as microbial cultures, is, by definition, a paradox, and it is only the size of individual microbes that has encouraged such generalizations. In fact, most culture systems, including chemostats, exhibit distinct macro-scale heterogeneities such as wall growth (Topiwala and Hamer, 1971).

Perhaps the three most important constraints in the consideration of any microbial system is that no microbial system is ever in equilibrium (steady state), but changes with respect to time; that gradients exist in all microbial systems, thereby creating conditions at reaction sites that are entirely different from conditions measured by either probes inserted into or samples removed from systems; that when the physiological states of microbes differ, their dynamic response to identical perturbations will also differ (Hamer, 1990).

Until the enunciation of the "Superfund Strategies" some five years ago (Hirschhorn, 1988) the major emphasis of pollution control for environmental quality maintenance for the previous 25 years had been efforts to reduce the effects of polluting discharges, particularly gaseous and liquid effluent streams, by reducing bulk pollutant concentrations in such streams. This was in marked contrast to earlier philosophies where, on the assumption of a virtually unlimited capacity of receiving environments to accept pollutants, pollution management was, almost exclusively, based on dispersal and dilution in receiving environments rather than on treatment. Essentially, apparent amelioration by dilution was sought in order to avoid incurring costs for treatment. As far as dispersal and dilution policies were concerned, waste solids and sludges were largely ignored, as were residues from repetitive and frequently excessive applications of toxic chemicals to land.

For decades, highly concentrated solid and semi-solid wastes have been subject to disposal by landfilling. Erroneously, dumping was frequently thought to imply immobilization. However, this ignored well established processes for pollutant transport and transfer between various environmental compartments. Waste dumping (disposal) and gross environmental abuse are by no means the prerogatives of primary industries concerned with natural resource exploitation, where frequently activities are undertaken with the collusion of Governments, but have been very widely performed by most secondary and tertiary industries, as well as by households and municipalities. In recent years, a critical point with respect to pollutant accumulation and adverse effects from pollutant transport is either being approached or has been reached, such that it is no longer possible for the frequently surreptitious adverse effects of pollutant dumping and/or release to remain unnoticed and unattended.

Biodegradation is a well known, but incompletely understood phenomenon. It occurs in virtually all aquatic and terrestrial environments, but degrees and levels are only rarely fully quantifiable. For much of the present century, laboratory test methods have been available for assessing both biodegradation and biodegradability, usually with particular reference to aquatic environments. However, such tests are at best only qualitative indicators of what might be expected to happen with respect to biodegradative activity in aerobic wastewater biotreatment plants, rivers, lakes and estuaries and as far as the latter three environments are concerned, it is predominantly the biodegradative activity occurring in the water column, rather than that occurring in associated sediments and soils, that are assessed. The result of failing to ascertain pollutant biodegradation and mobilization in complex matrices such as soils and sediments is evidenced by the hazardous state of many such environments because of accidental spillages, continuous or intermittent leakages, landfilling and excessive application to land, in the specific case of agricultural biocides.

In the present chapter, the objective is to evaluate those limited aspects of microbiological information that enable more effective scientific underpinning of practical bioremediation and biorestoration operations. Whilst clearly, very many important gaps in knowledge exist, there is no doubt whatsoever that failure to exploit what is already known remains a frequent, costly and humiliating occurrence.

Subsurface Heterogeneity

When considering subsurface environments, both macroscopic and microscopic heterogeneities are clearly evident. Typical soil microenvironments comprize three principal inorganic particulates: sand, silt and clay; a physically- and chemically-diverse organic fraction; an aqueous phase containing unevenly distributed organic and inorganic matter; a gaseous phase of markedly different composition from the overlying atmosphere; and, of course, a diversity of Protista (Burns, 1980).

The soil clays are generally considered to be the most influential inorganic particles affecting biological activity, because of their extensive surface areas and their ion exchange properties, which result in major sorptive capacity. Sand and silt are comparatively inert in comparison to clays. Although they tend to retain neither water nor films of humic matter, they mediate both gas and water diffusion, as well as aggregate formation. Soil organic matter comprizes several fractions, biotic debris, biochemically well defined breakdown products and an extremely varied range of colloidal and polymeric humic substances. Heavily polluted soils additionally contain mixtures of organic and/or inorganic chemicals with which they have been polluted and their partial breakdown products. The physical properties of the microbes present in soil are determined by their surface charge, whilst their metabolic activity is determined by their physiological state and pertaining environmental conditions.

It has been shown that three principal types of colloids, clay, humic matter and microbes, exist in typical soils. Colloids are particles with diameters of less than 10 µm. Because of their ability to sorb and stabilize both organic and inorganic matter, and under appropriate conditions, become mobile, they have become implicated in accelerated subsurface pollutant transport (McCarthy and Zachara, 1989). However, transport of essentially unmodified pollutants, although of very considerable importance, must be considered to be subordinate to the biotransformation of pollutants as far as bioremediation and biorestoration are concerned.

Distribution of Microbes

As has already been mentioned, soils are extremely complex microbial habitats, being dominated by particulate matter exhibiting extensive variability with respect to type, structure and surface properties. According to Marshall (1980), the habitats available to the soil microbes are governed by the structural organization of the various particulate fractions present. Fluctuating wet and dry conditions in soils influence the availability of organic matter, the degree of weathering, the leaching and redeposition of soluble materials and the translocation of smaller particulate fractions, including soil colloids.

A soil microhabitat can be described as a restricted but variable volume of soil wherein specific reactions directly affect the soil microbes and, simultaneously, are directly affected by the soil microbes, i.e., the microhabitat is represented by the sum of the physical, chemical and biological variables with which the microbes interact, rather than being represented by a particular site.

It seems probable that when microbes are attached to particulate matter, one of two mechanisms is involved; reversible sorption, where microbes are attracted to surfaces of like charge under conditions where van der Waals attraction exceeds electrostatic diffuse double-layer repulsion and permanent adhesion, where the microbes are anchored to the surface by means of polymer bridging. In addition, in wet soils, microbes are also found suspended in the bulk aqueous phase, either with or without associated colloidal matter, and physically entrapped within various aggregates.

Water Availability

All microbes require water for growth, but in spite of this microbial growth can occur in environments that exhibit a wide range of water availabilities. In any habitat, water availability does not necessarily depend on the total water content, but is a complex function of both sorption and solution factors. The most common ways in which water availability is expressed are either in terms of water activity or water potential. When water availability is affected by sorption, the effect is said to be matric and when affected by solute interactions, is said to be osmotic. In simple systems where changes result solely because of changing solute concentrations, it can be treated most conveniently by considering water activity. However, in complex systems, such as soils, where changes in temperature, solute concentrations and interactions between the volumetric water content and the geometrical and physicochemical properties of solid matrices are involved, water potential is considered to be a more convenient basis. (Griffin and Luard, 1979).

Historically, most microbiologists have approached the question of water availability from the viewpoints of either osmotolerance and osmophilicity or halotolerance and halophilicity; essentially in consideration either of artificial environments used for the preservation of biological matter where water availability has been depressed by the inclusion of high concentrations of organic or inorganic solutes or hyper-saline natural environments such as salt lakes in hot arid regions where the sodium chloride concentration has reached disproportionately high levels because of long term imbalance between evaporation and the combination of wet precipitation and run-off. Frequently, failure to differentiate purely osmotic effects from those effects resulting from elevated sodium chloride concentrations has occurred.

For soil environments, water potential is clearly the most appropriate basis for describing water availability because it incorporates all the variations in matric potential derived from liquid-solid and liquid-gas interfaces involving capillarity, repulsion between charged colloidal particles and sorption by surfaces, membranes and macromolecules. The water content of soil and the matric potential both depend on the pore size distribution and the relationship between them is best represented by the water sorption isotherm for the system (Griffin, 1981). However, because of different rate controlling mechanisms during the drying and wetting of soils, water sorption isotherms for soils exhibit hysteresis. Water potential is a selective factor with respect to microbes in soils. For movement in a soil matrix unicellular microbes require a continuum of water-filled pores of requisite diameter to provide appropriate pathways, whereas most filamentous fungi have the ability to bridge air-filled pores and to penetrate semi-solid matter. This is clearly evidenced by greater domination of the latter with decreasing water potential. In general, bacterial activity is reduced at -100 kPa and becomes negligible at -500 kPa, whilst a range of xerotolerant and xerophilic fungi have growth optima below -5 MPa (Griffin, 1981). This situation has clear implications for soil systems where it is proposed to enhance natural microbial activity in order to accelerate pollutant biodegradation.

Soil Atmosphere

The gas phase present in soils is very different from the atmosphere above the soil surface. A significant fraction of many of the trace gases found in the atmosphere is derived from biogenic, particularly microbial, activity in the subsurface, in natural water bodies and in sediments. Such gases include methane, hydrogen, carbon dioxide, carbon monoxide, nitrous oxide, hydrogen sulphide and ammonia. Such gases are all important intermediates in or mineralization products of the biogeochemical cycles that are responsible for global elemental cycling and, depending on conditions, are either substrates for or products of microbial metabolism. The concentrations of such gases in soil atmospheres depends on a wide range of factors such as water content, oxygen availability, temperature, sorption, pH, carbonaceous matter availability, etc., as well as on the microflora for process mediation. Unlike other substrates and nutrients, apparently zero concentrations of gaseous substrates and nutrients does not imply zero availability, because the diffusion mediated transport of gases in either gas phases or when dissolved in liquids is comparatively rapid so that although microbial activity occurs under limiting conditions, substrate or nutrient starvation is averted.

All too frequently, the various biogeochemical cycles are considered in isolation with many important interactions between cycles being ignored. Although, this stems from the failure of laboratory investigations to engage in the study of microbial consortia, it also stems from artificial boundaries that have been constructed between chemoorganotrophic metabolism, on the one hand, and chemolithotrophic metabolism, on the other hand, together with a failure to recognize the metabolic flexibility of facultative chemolithotrophic bacteria, e.g., *Thiobacillus* and *Alcaligenes* spp., which exhibit mixotrophy (Kuenen and Bos, 1989), depending on the pertaining environmental conditions, particularly alternative substrate availabilities. Mixotrophic bacteria have clear selective and survival advantages under some complex mixed culture/mixed substrate conditions with respect to either obligate chemoorganotrophic or chemolithotrophic bacteria, and this is most probably an important feature governing the preponderance of *Alcaligenes* spp. in activated sludge wastewater treatment processes.

MICROBIAL COMMUNITIES, CONSORTIA AND MOIETIES

Bull and Slater (1982) have pointed out that in virtually no open, unprotected environments have conditions either selected for or preserved monospecies populations of microbes and that on the contrary, most such environments are characterized by a diversity of microbial species. Further, they noted that changes in environmental conditions induce successional changes in population composition and that the self-regulation of community composition is made possible by homeostatic mechanisms based on interactions between community members, with negative feedback as a crucial element.

A wide range of potential interactions in binary cultures have been identified by Bungay and Bungay (1968) and throughout the interim the proposed basis has, very largely, dominated analysis of mixed culture interactions. However, such an approach assumes that the major interactive structure in complex natural environments actually involves individual microbial species. On the contrary, it seems much more probable that interactions of the type specified by Bungay and Bungay (1968) occur not between individual species, but between individual microbial moieties. A microbial moiety can be described as an essentially stable mixed culture comprising, under all but the most extreme conditions, more than 90 percent of a dominant primary substrate utilizing strain growing together with an associated strain that exhibits specific substrate dependency on a by-product that is overproduced by the dominant strain and with incidental strains, usually less than four in number, that are essentially indifferent to the metabolism of the dominant strain, but grow on lysis products which, if permitted to accumulate, would inhibit the growth of the dominant strain. The interactive structure of such moieties clearly depends on protocooperation.

Probably the best elucidated example of protocooperation is that of methane-utilizing moieties, which are of course, of direct interest in subsurface environments. Although methanotrophic bacteria were first isolated in the early years of the century (Söhngen, 1905), considerable difficulties were experienced in growing such bacteria in pure culture until some 20 years ago and, even then the quest for rapid growth that was clearly indicated from *in situ* ecological observations still eluded laboratory researchers. However, Wilkinson *et al.* (1974) sought to resolve this apparent difficulty, by investigating the growth characteristics of an essentially stable mixed methanotrophic culture. The findings clearly indicated a marked dependence of the dominant obligate, type II methanotrophic strain on the activities of an associated facultative methylotrophic *Hyphomicrobium* sp., which removed trace, but inhibitory, concentrations of the by-product methanol, thereby permitting rapid growth of the dominant methanotroph. Ill-defined lysis products that were also formed were removed by the activities of two incidental chemoorganotophic strains, a *Flavobacterium* sp. and an *Acinetobacter* sp. Further, the associated *Hyphomicrobium* sp. did not compete with the obligate methanotroph for oxygen at dissolved oxygen concentration corresponding to less than 30 percent of saturation with air at atmospheric pressure because of its ability to utilize nitrate as an alternative electron acceptor. (Wilkinson, 1972).

Perhaps even more pertinent to the subsurface environment is the comprehensive role that methanotrophic moieties, of the type described above, can play in the nitrogen cycle. First, the incidental strains are able to deaminate complex nitrogenous compounds, such as proteinaceous lysis products, into ammonia; second, obligate methylotrophs, particularly *Methylococcus* spp.which are type I methanotrophs, are able to oxidize ammonia not only to nitrite, but also to nitrate; third, facultative methylotrophs such as *Hyphomicrobium* spp.are able to denitrify nitrate to form dinitrogen (Hamer, 1986). Additionally, methanotrophs have also been shown to be able to fix dinitrogen. Laboratory observations of dinitrogen fixation by methanotrophic bacteria were first made by Davis *et al.* (1964) using mass balance techniques, but because of interference with the acetylene reduction assay for nitrogenase activity by methane monooxygenase, nitrogen fixation by both type I and II methanotrophs was not accepted until failures with respect to the nitrogenase assay technique were explained (de Bont and Mulder, 1974; de Bont, 1976).

The non-specific nature of the methane monoxygenase also gave rise to the discovery that methanotrophs possessed extensive co-oxidative capacities for compounds other than methane (de Bont and Mulder, 1974), such that some 10 years ago, Higgins *et al.* (1980) speculated on the importance of methanotrophs in pollutant biodegradation within the biosphere, a concept which unfortunately is still in need of extensive research before it can become fully proven.

Reverting briefly to the question of the nitrogen cycle, a few additional points are worthy of mention. First, denitrification can no longer be considered as a process which occurs exclusively in anoxic environments, but must be considered to occur under both oxic and anoxic conditions (Robertson and Kuenen, 1984; 1990). Second, the hitherto neglected role of chemoorganotrophic nitrifiers that also denitrify under aerobic conditions (Robertson et al., 1989) should be reevaluated with respect to the assumed, exclusive, role of chemolithotrophic nitrifiers as the dominant strains for nitrification (Bazin et al, 1976). Third, the effects of both oxygen (Davies et al., 1989) and competition for electrons (Omlin et al., 1990) on nitrous oxide formation during denitrification needs further elucidation, particularly as far as soil environments are concerned, where both oxygen and electron donor availability can vary markedly.

PHYSIOLOGICAL STATE AND STRESS

On a physiological basis, microbes can be classified as viable and active, non-viable and active, dormant and dead (Mason et al., 1986), but what is really important is not the classification *per se*, but transitions from one physiological state to another and within physiological states, particularly within the active and potentially active states. Dead microbial cells can be defined as cells that are totally devoid of metabolic activity but still possessing an intact cell wall. They are incapable of changing their physiological state and are only of interest because of their abiotic capacity to act as sorbants and as a possible source of substrates and nutrients for the "cryptic" growth of other microbes. In general, dead cells result from extremely severe stresses such as high concentrations of toxic chemical agents or prolonged exposure to markedly elevated temperatures. Milder stresses, of which starvation is frequently cited as the most common (Poindexter, 1981; Dawes, 1984), result in either changes in the physiological situation (Konstantinov and Yoshida, 1989) within a specific physiological state or, in the case of dormancy, a change of state.

The quantitative evaluation of microbial activity has been seriously retarded by the erroneous assumption that microbes can be regarded as particles of constant elemental composition and properties, which are invarient with respect to the environment, changes therein, time, previous history, etc. In other words, the concepts of phenotypic variability and fluctuating macromolecular composition of microbial cells, which very largely determine culture dynamics, have frequently been disregarded, so that knowledge concerning both the dynamic response of cultures and the manifestations of unsteady state growth is meager. Recently, Wanner and Egli (1990) have highlighted the enormous intrinsic flexibility of microbes in their response to changes in the availability of essential nutrients.

Some of the most illuminating investigations on phenotypic variability were published more than 30 years ago. Schaechter et al., (1958) showed the dramatic effect of specific growth rate on cell size and the RNA, DNA and protein contents per cell for *Salmonella typhimurium* growing exponentially in batch culture where the specific growth rate was controlled by medium composition. Additionally, it was reported (Kjeldgaard et al., 1958) that when microbial cells were transferred from a medium permitting a high specific growth rate to one permitting only a much reduced specific growth rate, a period of equilibration was required before

Polluted Heterogeneous Environments

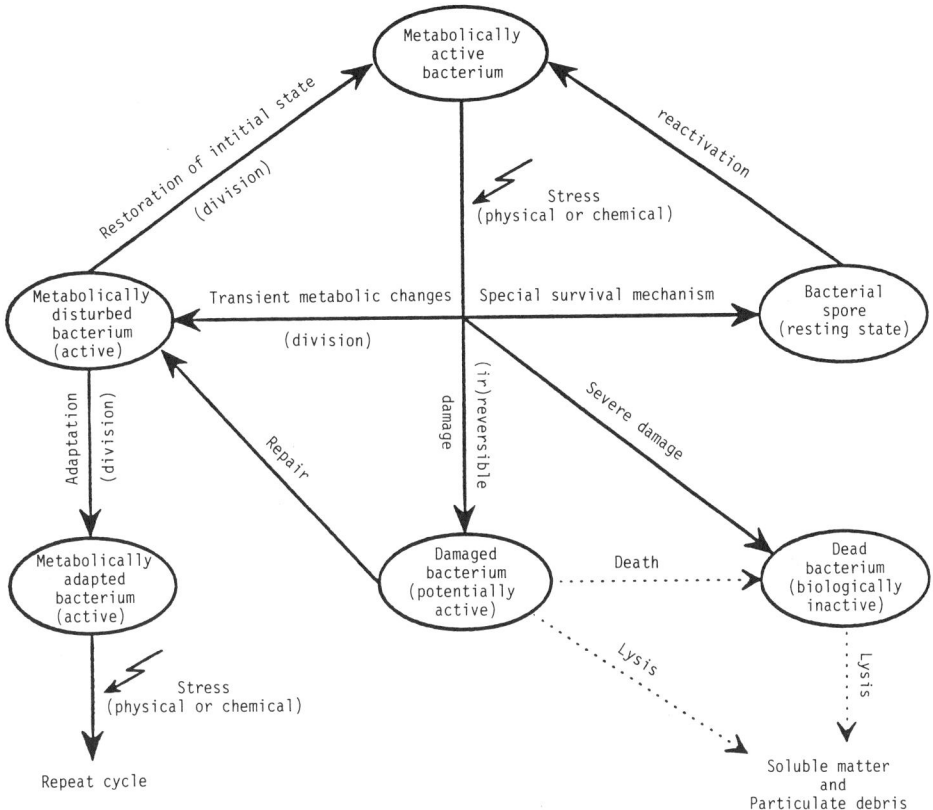

Figure 1. The generalized stress cycle.

cell size and composition corresponded to those typical for the new growth medium and growth proceeded at the maximum specific growth rate attainable in the new growth medium. Simultaneously, Herbert (1958) reported essentially similar findings for *Aerobacter aerogenes (Klebsiella pneumoniae)* growing in chemostat culture at various imposed specific growth (dilution) rates, but unfortunately, no results concerning transient state behaviour during changes in growth rate were provided. In these batch and chemostat culture studies the imposed changes in either growth medium composition or in dilution rate would now be considered as subjection of the culture to mild stress.

The concept of a generalized stress cycle has recently been proposed by Heitzer (1990) and this is shown in an extended form in Figure 1. It is postulated that irrespective of the nature of the stress imposed on a growing bacterium, it is intensity (magnitude and exposure time) that will govern subsequent metabolic effects.

Most bacteria have evolved mechanisms that allow rapid growth under favourable conditions, on the one hand, but also permit survival under conditions that are unfavourable for growth, on the other hand. Clearly, these two requirements are in conflict (Neidhardt, 1987). The proteins necessary for balanced growth are usually present in requisite quantities, whilst those required for protection, repair or growth under either extreme or adverse conditions are usually only present in very low concentrations and synthesized when needed.

The most extensively studied response to a potentially adverse environmental change is the seemingly universal heat shock response (Lindquist and Craig, 1988), which occurs when either cells or organisms are subjected to a temperature increase in either their optimum or their superoptimum temperature range for growth. The heat shock response involves rapid transient synthesis of a specific set of proteins and in the case of *Escherichia coli*, the most extensively studied bacterium, 18 heat shock proteins are now known (Neidhardt et al., 1984; Raina and Georgopoulos, 1990).

In addition to temperature increases, a number of other changes or agents induce a stress response. Of particular relevance in the context of either bioremediation or restoration are induction by substrate and nutrient starvation (Matin et al., 1989), anaerobiosis (Spector et al., 1986), hydrogen peroxide challenge (Jenkins et al., 1988), osmotic (salt) effects (Hecker et al;., 1989), ultraviolet light (Walker, 1987), cold (Jones et al., 1987), acid (pH) (Heyde and Portalier, 1990) and cadmium chloride (Neidhardt and VanBogelen, 1987). Although little is known concerning the regulatory details of most of these other mediated inductions of stress protein synthesis, under some inducing conditions, only subsets of proteins are produced (Lindquist, 1986), supporting the view that bacteria have evolved a wide variety of multigene systems in order to respond in a coordinated manner to adverse changes in their growth environment.

The response of a growing bacterium to perturbations resulting from external stress is first, to cease the current mode of growth and metabolism; second, to protect the cell against any adverse consequences due to the initial response; third, to develop strategies that can deal with the changed conditions; fourth; to implement such strategies. In such a sequence the specific effects involve both gene expression and enzyme activity. The timing of the individual stages is variable.

In the generalized stress cycle (Figure 1), the several overall results of stress on a bacterium are presented. In the special case of a spore forming species, sporulation and subsequent reactivation from the dormant state will occur after stress removal. In the case of very severe stress, a dead, totally biologically inactive, bacterium can result, but for the most part death occurs simultaneously with lysis, such that soluble matter and fine particulate debris become available for the "cryptic" growth of other bacteria that are comparatively unaffected by the stress. With decreasing severity of stress either an extensively damaged, potentially active, bacterium, which can die, lyse or undergo repair or a metabolically disturbed, active, bacterium, which is able to divide will be produced. A metabolically disturbed bacterium can divide and depending on conditions, either undergo restoration to the initial metabolic state upon stress removal or adapt to the stress and enter a repeat cycle. Essentially, it is the stress response, irrespective of its mode of induction, that gives rise to metabolically disturbed bacteria. One must suspect that in virtually all natural environments, either polluted or otherwise, it is metabolically disturbed bacteria that mediate the reactions involved in various biogeochemical cycles. However, it should also be noted that, as far as the heat shock response is concerned, bacterial cultures that have entered the stationary phase exhibit a markedly reduced response compared to exponentially growing cultures when exposed to the same heat stress, thereby further complicating interpretation of stress phenomena in complex environments.

SUBSTRATE/NUTRIENT LIMITED GROWTH

In all but the most heavily polluted environments, the microbes present have to cope with both the simultaneous presence and low concentrations of a multiplicity of homologous substrates and nutrients, i.e., compounds that satisfy the same physiological function (Egli and Schmidt, 1990), such that in Nature, it is most unlikely that a single compound will be limiting the growth of either any specific microbial species or any specific moiety at any moment in time. Whilst in recent years considerable interest has developed in dual carbon energy substrate limited growth (Egli *et al.*, 1986) and in dual electron acceptor mediated growth (Robertson and Kuenen, 1990), the question of simultaneous growth limitation by multiple non-homologous substrate/nutrient mixtures has escaped widespread interest, in spite of the fact that in soil environments such a state will be encountered, particularly as far as gaseous macro-substrates and -nutrients that mediate the growth of microbial moieties are concerned. The fundamental problem with gaseous macro-substrates and -nutrients is that, prior to utilization by microbes, they must first be dissolved in water. Whilst some such substrates and nutrients are very soluble in water, e.g., ammonia, the majority, e.g., oxygen, methane, carbon monoxide, are only slightly soluble in water, such that under equilibrium conditions dissolved concentrations of such gases will be low, even when the partial pressure of the gas in the gas phase is high. Of course, any microbial utilization of gaseous substrates or nutrients from solution will further reduce their concentrations in systems where gas transfer is either impeded or where the gas must first be desorbed from a surface.

One of the few analyses of microbial growth under dual non-homologous substrate/nutrient limitation concerns the growth of obligately aerobic methanotrophic bacteria, where methane and oxygen can become simultaneously growth limiting (Harrison, 1972; Hamer *et al.*, 1975). Essentially, the model developed indicates that, at very low growth rates, growth is limited by a nutrient other than either oxygen or methane, but as growth rate increases, the possibilities of dual methane/oxygen limitation increase depending on the equilibrium saturation concentration of each gas, the rates of transfer of the two gases and the rates of utilization of the two gases. Whilst the solubilities of methane and oxygen are similar, their availabilities depend on the partial pressure of the gases in the gas phase and their respective transfer rates and utilization depends on the methane-based biomass yield coefficient, which is, of course, a variable depending on growth conditions and perturbations therein. In fact, the utilization of ammonia as the nitrogen source results in higher methane-based biomass yield coefficients than does the utilization of nitrate which, in turn, results in higher methane-based biomass yield coefficients, than when methanotrophs are fixing dinitrogen.

Another most important aspect of substrate and nutrient availabilities is that the physiological potential of microbes depends on their abilities to synthesize enzymes (Hamer et al., 1985). Enzyme (protein) synthesis is a function of the availabilitiy of the nitrogen source. This has obvious implications when the biodegradation of a multiplicity of carbonaceous pollutants occurs in either nitrogenous nutrient deficient or nitrogenous nutrient variable environments.

When medium carbon:nitrogen ratios are below ca. 7, distinct carbon limitation, as evidenced by the presence of dissolved ammonium-nitrogen, occurs in chemostat cultures of a methylotrophic *Hyphomicrobium* sp., growing at a dilution rate of 0.054 h^{-1}, and with this same bacterium, distinct nitrogen limitation occurs under similar growth conditions at carbon:nitrogen ratios above ca. 12.5 (Gräzer-Lampart et al., 1986). Between these two ratios there is dual limitation, as evidenced by the absence of carbon substrate (methanol) and ammonium-nitrogen in the growth medium. However, what is much more important is that both the overall protein concentration and specific enzyme concentrations were markedly reduced as the carbon:nitrogen ratio exceeds 7. Further, Egli and Schmidt (1990) have demonstrated that the onset of dual limitation is a function of the dilution rate in chemostat cultures.

A final important point concerning the question of substrate and nutrient limited growth is how appropriate are Monod type saturation kinetics at very low substrate or nutrient concentrations? Shehata and Marr (1971) have investigated the effect of substrate concentration on growth and found that saturation type kinetics did not predict behaviour over the entire range of nutrient concentrations used, i.e., the use of parameters estimated under very low nutrient concentrations resulted in an over estimation of the specific growth rate constant at high nutrient concentrations. In a similar vein, Koch and Wang (1982) have examined the question of how close to the theoretical diffusion limit bacterial substrate uptake systems operate, whilst Rutgers et al., (1987) have provided preliminary evidence in support of describing bacterial growth on the basis of concepts originating from non-equilibrium thermodynamics proposed by Westerhoff et al. (1982).

MICROBIAL GROWTH ON SURFACES

As far as attached growth is concerned, the only predominant features that have been discussed in detail until recently are nutrient and substrate diffusion gradients. Because most investigations have concerned aerobic wastewater treatment systems, the nutrient that is inevitably subject to study is oxygen. Whilst attempts have been made to measure oxygen gradients in biofilms with microelectrodes, probing always results in structural alterations. Most theoretical analyses of oxygen diffusion and transfer in biofilms have been remarkably naive, such that the fact that facultative anaerobic bacteria play significant roles in non-aseptic biofilms has escaped recognition. Further, only recently have the complementary roles of chemoorganotrophs and chemolitotrophs in biofilms been subject to consideration (Wanner and Gujer, 1986). Such problems apart, the comparative physiology of attached and discretely suspended microbes has hardly merited discussion. In an attempt to review knowledge on this important area, Fletcher (1984) was only able to suggest that in attached cells, some elastic deformation in the cell envelope, that could affect either substrate uptake mechanisms or respiratory efficiency, might occur. The dilemma was to identify any experimentally verified mechanisms which are responsible for changes in physiological activity which occur when bacteria are associated with surfaces.

Since attached microbial cells were first studied, there have been repeated claims that attachment results in activation of cellular metabolism. However, the fundamental hypothesis in need of proof is whether cellular function and metabolism are affected directly and to any significant extent by the fact that a microbial cell is in direct contact with surface. Suggestions such as permeability changes and of improved nutrient availability are common, whilst factors such as surface tension, osmotic pressure and water activity may have subtle consequences on cell function.

Until recently, a lack of availability of appropriate techniques for the critical examination of attached microbe activity has caused the hypothesis that states that the consequences of attachment affect cell physiology and kinetic behaviour, beyond those factors pertaining simply to the microenvironment, to remain unproven. However, the introduction of novel non-invasive *in vivo* analytical techniques such as nuclear magnetic resonance based on both ^{31}P and ^{13}C spectra, when coupled with fluorescence, DNA, RNA, protein and more conventional microbial system parameter measurements will provide the comprehensive analytical approach needed for comparing attached microbial cells with cells of the same strain grown in suspension culture under various conditions.

GENERAL DISCUSSION AND CONCLUSIONS

In the previous sections of this chapter a diversity of microbiological considerations pertinent to bioremediation and biorestoration practice have been introduced. To date, empirical approaches have largely dominated bioremediation and biorestoration practice. For their sustained economic attractiveness, an understanding of some of the microbiological considerations discussed will be essential, not only desirable. The future will require a transformation from the speculation and conjecture of the past to a firm factual and quantitative basis that describes the microbiology of real environments. This will require imaginative conceptual thinking followed by validation of concepts. Systems will need to be mathematically modelled, using as much structure and segregation as possible, so that critical variables can be identified. Whilst for the identified microbial resource some potential mechanisms are known, estimations concerning the magnitude and effects of the unidentified microbial resource must not be neglected. The obsession for studying easily enriched and isolated microbes which are of little environmental significance must be curbed, if the total microbial resource is to be exploited.

Clearly, the techniques of molecular biology will have a major role to play in the development of our understanding concerning the microbial mediation of pollutant pathways within polluted environments. However, such techniques can only be used for the generation of real information in conjunction with microbial physiology, microbial ecology, microbial biochemistry and system modelling, employing, as the basis for microbial biomass description, macromolecular composition and physiological state, rather than elemental composition which is physiologically meaningless.

Specific research areas in need of stimulation include a much improved understanding of:

1. Gaseous substrate and nutrient utilization and the processes mediated, directly and indirectly, by gaseous substrate-utilizers;
2. Microbial moiety, consortia and community structure and activity;
3. The total microbial resource, i.e., the 80 percent that awaits isolation;
4. The way in which the chemistry and physics of complex environments impact on microbiology;
5. The availability of sorbed pollutants to growing microbes;
6. The kinetics of microbial growth at very low concentrations of multiple homologous substrates;
7. Whether bioaugmentation can be made to work effectively and reproducibly in polluted environments;
8. Whether the additional biodegradation potential offered by filamentous fungi is realizable for bioremediation and biorestoration;
9. Whether specific actions adopted for *in situ* bioremediation and biorestoration impair biodegradative activity, e.g., hydrogen peroxide injection as a means of oxygenation;
10. The problems introduced in biodegradation when polluting chemicals are "off-specification" as a result of either manufacturing process malfunction or a catastrophe during production or storage.

However, it is obvious that this list is by no means complete.

REFERENCES

Bazin, M.J., Saunders, P.T., and Prosser, J.I., 1976, Models of microbial interactions in soil, *Crit. Rev. Microbiol.* 4:463.

Bull, A.T., and Quayle, J.R., 1982, New dimensions in microbiology: an introduction, *Phil. Trans. R. Soc. Lond. B* 197:447.

Bull, A.T., and Slater, J.H., 1982, Microbial interactions and community structure, in: *Microbial Interactions and Communities*, (A.T. Bull and J.H. Slater, eds.), Vol. I, Academic Press, London, pp. 13-44.

Bungay, H.R., and Bungay, M.L., 1968, Microbial interactions in continuous culture, *Adv. Appl. Microbiol.* 10 : 269.

Burns, R.G., 1980, Microbial adhesion to soil surfaces: consequences for growth and enzyme activities in: *Microbial Adhesion to Surfaces*, (R.C.W. Berkeley, J.M. Lynch, J. Melling, P.R. Rutter and B. Vincent, eds.), SCI/Ellis Horwood, Chichester, pp 249-262.

de Bont, J.A.M., 1976, Nitrogen fixation by methane-utilizing bacteria, *Antonie van Leeuwenhoek* 42 : 245.

de Bont, J.A.M., and Mulder, E.G., 1974, Nitrogen fixation and co-oxidation of ethylene by a methane-utilizing bacterium, *J. Gen. Microbiol.* 83 : 113.

Davies, K.J.P., Lloyd, D., and Boddy, L.,1989, The effect of oxygen on denitrification in *Paracoccus denitrificans* and *Pseudomonas aeruginosa*, *J. Gen. Microbiol.* 135 : 2445.

Davis, J.B., Coty, V.F., and Stanley, J.P., 1964, Atmospheric nitrogen fixation by methane-oxidizing bacteria, *J. Bacteriol.* 88 : 468.

Dawes, E.A., 1984, Stress of unbalanced growth and starvation in microorganisms, in: *The Revival of Injured Microbes,* (M.H.E. Andrew and A.D. Russel, eds.), Academic Press, London, pp. 19-43.

Egli, T., and Schmidt, C.R., 1990, Dual-nutrient-limited growth of microbes, with special reference to carbon and nitrogen substrates, in: *Mixed and Multiple Substrates and Feedstocks,* (G. Hamer, T.. Egli and M. Snozzi, eds.), Hartung-Gorre-Verlag, Konstanz, pp. 45-53.

Egli, T., Bosshard, C., and Hamer, G., 1986, Simultaneous utilization of methanol-glucose nixtures by *Hansenula polymorpha* in chemostat: influence of dilution rate and mixture composition on utilization pattern, *Biotechnol. Bioengng.* 28 : 1735.

Fletcher, M., 1984, Comparative physiology of attached and free living bacteria, in: *Microbial Adhesion and Aggregation,* (K.C. Marshall, ed.), Springer-Verlag, Berlin, pp. 223-232.

Gräzer-Lampart, S.D., Egli, T., and Hamer, G., 1986, Growth of *Hyphomicrobium* ZV620 in the chemostat: regulation of NH_4+-assimilating enzymes and cellular composition, *J. Gen. Microbiol.* 132 : 3337.

Griffin, D.M., 1981, Water and microbial stress, *Adv. Microb. Ecol.* 5 : 91.

Griffin, D.M., and Luard, E.J., 1979, Water stress and microbial ecology, in: *Strategies of Microbial Life in Extreme Environments,* (M. Shilo, ed.), Verlag Chemie, Weinheim, pp. 49-63.

Hamer, G., 1986, Transformation of nitrogen compounds in wastewater treatment systems, in: *Perspectives in Microbial Ecology,* (M. Megusar and M. Gantar, eds.), Slovene Soc. Microbiol., Ljubljana, pp. 74-79.

Hamer, G., 1990, Immobilized microbes: interfaces, gradients and physiology, in: *Physiology of Immobilized Cells,* (J.A.M. de Bont, J. Visser, B. Mattiasson and J. Tromper, eds.), Elsevier, Amsterdam, pp. 15-24.

Hamer, G., Egli, T., and Mechsner, K., 1985, Biological treatment of industrial wastewater: a microbiological basis for process performance, *J. Appl. Bacteriol. Symp. Suppl.* 59 (14):127.

Hamer, G., Harrison, D.E.F., Harwood, J.H., and Topiwala, H.H., 1975, SCP production from methane, in: *Single-Cell Protein II,* (S.R. Tannenbaum and D.I.C. Wang, eds.), MIT Press, Cambridge, pp. 357-369.

Harrison, D.E.F., 1972, Physiological effects of dissolved oxygen tension and redox potential on growing populations of microorganisms, *J. Appl. Chem. Biotechnol.* 22 : 417.

Hecker, M., Völker, U., and Heim, C., 1989, RelA-independent (p) ppGpp accumulation and heat shock protein induction after salt stress in *Bacillus subtilis, FEMS Microbiol. Lett.* 58 : 125.

Heitzer, A., 1990, Kinetic and physiological aspects of bacterial growth at superoptimum temperatures, *Doctoral Thesis 9217, ETH Zch,* pp. 1 - 108.

Herbert, D., 1958, Some principles of continuous culture, in: *Recent Progress in Microbiology,* (G. Tunevall, ed.), Blackwell, Oxford, pp. 381 - 396.

Heyde, M., and Portalier, R., 1990, Acid shock proteins of *Escherichia coli, FEMS Microbiol. Lett.*69 : 19.

Higgins, I.J. Best, D.J., and Hammond, R.C., 1980, New findings in methane-utilizing bacteria highlight their importance in the biosphere and their commercial potential, *Nature* 286 : 561.

Hirschhorn, J.S., 1988, Superfund strategies and technologies: a role for biotechnology, in: *Environmental Biotechnology,* (G. S. Omenn, ed.), Plenum Press, New York, pp. 419-429.

Jenkins, D.E., Schultz, J.E., and Matin, A., 1988, Starvation-induced cross protection against heat or H_2O_2 challenge in *Escherichia coli, J. Bacteriol.* 170 : 3910.

Jones, P.G., Van Bogelen, R.A., and Neidhardt, F.C., 1987, Induction of proteins in response to low temperature in *Escherichia coli, J. Bacteriol.* 169 : 2092.

Kjeldgaard, N. O., Maalo, O., and Schaechter, M., 1958, The transition between different physiological states during balanced growth of *Salmonella typhimurium, J. Gen. Microbiol* 19 : 607.

Koch, A.L. and Wang, C.H., 1982, How close to the theoretical diffusion limit do bacterial uptake systems function? *Arch. Microbiol.* 131 : 36.

Koch, R., 1881, Zur Untersuchung von pathogenen Organismen, *Mitt. Kaiserl. Gesundht Berl.* 1 : 1.

Konstantinov, K., and Yoshida, T., 1989, Physiological state control of fermentation processes, *Biotechnol. Bioengng.* 33 : 1145.

Kuenen, J.G., and Bos, P., 1989, Habitats and ecological niches of chemolitho(auto)trophic bacteria, in: *Autotrophic Bacteria,* (H.G. Schlegel and B. Bowien, eds.), Science Tech Publ., Madison/Springer Verlag, Berlin, pp. 53-80.

Lindquist, S., 1986, The heat-shock response, *Annu. Rev. Biochem.* 55 : 1151.

Lindquist, S., and Craig, E.A., 1988, The heat shock proteins, *Annu. Rev. Genet.* 22 : 631.

Marshall, K.C., 1980, Adsorption of microorganisms to soils and sediments, in: *Adsorption of Microorganisms to Surfaces,* (G. Bitton and K.C. Marshall, eds.), Wiley, New York, pp. 317-329.

Mason, C.A., Hamer, G. and Bryers, J.D., 1986, The death and lysis of microorganisms in environmental processes, *FEMS Microbiol. Rev.* 39 : 373.

Matin, A., Auger, E.A., Blum, P.H., and Schultz, J.E., 1989, Genetic basis of starvation survival in non-differentiating bacteria, *Annu. Rev. Microbiol* 43 : 293.

McCarthy, J.F., and Zachara, J.M., 1989, Subsurface transport of contaminants, *Environ. Sci. Technol.* 23 : 496.

Neidhardt, F.C., 1987, Multigene systems and regulons, in: *Escherichia coli and Salmonella typhimurium Cellular and Molecular Biology,* Vol. 2, (F.C. Neidhardt, ed.), Am. Soc. Microbiol., Washington, pp. 1313-1317.

Neidhardt, F.C., and Van Bogelen, R.A., 1987, Heat shock response, in: *Escherichia coli and Salmonella typhimurium Cellular and Molecular Biology,* Vol. 2 (F.C. Neidhardt, ed.) Am. Soc. Microbiol. Washington pp. 1334-1345.

Neidhardt, F.C., Van Bogelen, R.A., and Vaughn, V., 1984, The genetics and regulation of heat shock proteins, *Annu. Rev. Genet.* 18 : 295.

Omlin, D., Snozzi, M., and Hamer G., 1990, Transients between oxic and anoxic growth in continuous cultures of *Paracoccus denitrificans,* Presented at *6th Europ. Bioenergetics Conf.,* Amsterdam.

Poindexter, J.S., 1981, Oligotrophy, fast and famine existance, *Adv. Microbial Ecol.* 5 : 63.

Raina, S., and Georgopoulos, C., 1990, A new *Escherichia coli* heat shock gene, htrC, whose product is essential for viability only at high temperatures, *J. Bacteriol.* 172 : 3417.

Robertson, L. A., and Kuenen, J.G., 1984, Anaerobic denitrification - old wine in new bottles? *Antonie van Leeuwenhoek* 50 : 525.

Robertson, L.A., and Kuenen, J.C., 1990, Mixed terminal electron acceptors (oxygen and nitrate), in: *Mixed and Multiple Substrates and Feedstocks,* (G. Hamer, T. Egli and M. Snozzi, eds.), Hartung-Gorre-Verlag, Konstanz, pp. 97-106.

Robertson, L.A., Cornelisse, R., de Vos, P., Hadioetomo, R., and Kuenen, J.G., 1989, Aerobic denitrification in various heterotrophic nitrifiers, *Antonie van Leweuwenhoek,* 56 : 289.

Rutgers, M. Teixeira de Mattos, M.J. , Postma, P.W., and van Dam, K., 1987, Establishment of the steady-state in glucose-limited chemostock cultures of *Klebsielle pneumoniae, J. Gen. Microbiol.* 133 : 445.

Schaechter, M., Maalo, O., and Kjeldgaard, N.O., 1958, dependency on medium and temperature of cell cize and chemical composition during balanced growth of *Salmonella typhimurium, J. Gen. Microbiol.* 19 : 592.

Shehata, T.E., and Marr, A.G., 1971, Effect of nutrient concentration on the growth of *Escherichia coli, J.Bacteriol.* 107 : 210.

Söhngen, N.L., 1905, Ueber Bakterien welche Methan als Kohlenstoffnahrung und Energiequelle gebrauchen, *Centralbl. f. Bakt. Parasitenk. Infekt. u. Hygiene,* Abt. 2, 15 : 513.

Spector, M.P., Aliabadi, Z., Gonzalez, T., and Foster, J.W., 1986, global control in *Salmonella typhimurium*: two-dimensional electrophoretic analusis of starvation-, anaerobiosis- and heat shock - inducible proteins, *J. Bacteriol.* 168 : 420.

Topiwala, H.H., and Hamer, G., 1971, Effect of wall growth in steady-state continuous cultures, *Biotechnol. Bioengng.* 13 : 919.

Walker, G.C., 1987, the SOS response of *Escherichia coli,* in: *Escherichia coli and Salmonella typhimurium Cellular and Molecular Biology,* Vol. 2. (F.C. Neidhardt, ed.), Am. Soc. Microbiol., Washington, pp. 1346-1357.

Wanner, O., and Gujer, W., 1986, A multispecies biofilm model, *Biotechnol. Bioengng.* 28: 314.

Wanner, U., and Egli, T., 1990, Dynamics of microbial growth and cell composition in batch culture, *FEMS Microbiol. Rev.* 75 : 19.

Westerhoff, H.V., Lolkema, J.S., Otto, R., and Hellingwerf, K.J., 1982, Thermodynamics of growth, non equilibrium thermodynamics of bacteria growth, the phenomenological and the mosaic approach, *Biochim. Biophys. Acta* 683 : 181.

Wilkinson, T.G., 1972, Interactions in a mixed bacterial population growing on methane in continuous culture, *Doctoral Thesis, Univ. of London,* pp. 1-233.

Wilkinson, T.G., Topiwala, H.H., and Hamer, G., 1974, Interactions in a mixed bacterial population growing on methane in continuous culture, *Biotechnol. Bioengng.* 16 : 41.

The Pilot Plant Testing of the Continuous Extraction of Radionuclides Using Immobilized Biomass

Marios Tsezos and Ronald G. L. McCready

INTRODUCTION

Microbial biomass has been shown to be able to sequester a variety of metal ions from aquatic solutions. The selective extraction of metal ions from dilute complex solutions by microbial biomass has been termed biosorption. The biosorption of metals is characterized by high selectivity for certain metal ions which is exhibited equally well by live and dead cells. This high selectivity is also exhibited in very complex ionic matrices as for example in acid mine drainage or biological leachate solutions (1,2).

Because of the intrinsic selectivity, biosorption has formed the basis for a new generation of metal ion sequestering technologies that make use of a novel type of adsorbents which are based on microbial biomass and have been called biosorbents (3,4,5,6,7). Biosorbents use inactive (dead) microbial biomass which results in the following advantages:

a. The solution toxicity does not affect the biomass biosorptive uptake
b. There are no biomass growth requirements to be met
c. There are no problems associated with the maintenance of culture purity

Biosorption has been demonstrated to be an excellent candidate for water pollution control applications, the recovery of metal values from dilute complex aqueous solutions, and for the extraction of radionuclides, as for example uranium, thorium, or radium, from similar solutions (1,2,6,7,8,9,10). The present work will provide an overview of the development and testing of a novel biosorptive technology that has been studied by the first author in the course of the last twelve years and was jointly tested at pilot plant scale for the extraction of uranium from industrial solutions with CANMET.

Marios Tsezos • Department of Chemical Engineering, McMaster University, Hamilton, Canada. Ronald G. L. McCready • Biotechnology Section, CANMET, Energy Mines and Resources Canada, Ottawa, Canada.

ELEMENTS OF BIOSORPTION MECHANISM

The rigorous study of the biosorption mechanism has resulted in a good understanding of the way metal ions and the inactive microbial cell interact (8,10,11,12,13,14). This understanding has subsequently been used for the development of the engineering know how necessary to transfer biosorption from the flask to the pilot plant stage.

Inactive microbial biomass, once grown and processed under the optimal conditions that maximize its biosorption uptake capacity for a specific element, has the form of a light fine powder made of individual cells or weak aggregates of cells. Previous work on the subject has demonstrated that it is necessary to immobilize the inactive microbial biomass in particles of desirable particle size and mechanical strength and favourable intraparticle kinetics (4,5,15,16). The objectives of mechanical rigidity and rapid intraparticle mass transfer are difficult to reconcile. Furthermore the immobilization process should maximize the biosorbent particle uptake capacity by protecting the intrinsic biomass biosorptive uptake capacity.

A proprietary biomass immobilization technology has been developed by the first author which produces biosorbent particles of any specified particle size, with favourable intraparticle mass transport properties and less than 15% w/w inactive additives. This technology can be used for different biomass types and has been successfully scaled up to a laboratory pilot plant scale (5,16,17). Figure 1 shows a typical photograph and an electron microscope micrograph of immobilized *Rhizopus arrhizus* biomass particles that were used for the extraction and recovery of uranium (3).

It has been shown conclusively that uranium is sequestered by the mycellial cell wall. The cell wall chitin network is primarily responsible for the biosorption of uranium (10,11,13). Uranium biosorption is fully reversible provided that the proper eluant and the appropriate elution conditions are used. Figure 2 shows typical electron micrographs of a *R. arrhizus* cell loaded with uranium before (Figure 2A) and after elution of the uranium (Figure 2B). It is important to note the complete elution of uranium as evidenced by the lack of negative electron density of the cell wall on Figure 2B, and the fact that cell wall architecture appears unaltered. The use of improper elution conditions is illustrated on Figure 3 where the use of a dilute sulfuric acid solution did not elute the uranium and at the same time substantially damaged the cell wall architecture (14).

The selectivity of biosorption has also been examined in detail. The available results suggest that alkali metals are not a problem in biosorption as they are not sequestered by the biomass. Alkaline earths are not competitors for uranium biosorption but under certain conditions may affect the biosorption of radium (9,10,11,13,18). Other elements such as aluminum, iron and zinc however require more attention (10,11,12,13,18).

A batch kinetic mass transfer model of biosorption has been developed and tested successfully on uranium biosorption (4,16). From the model we were able to determine which parameters were most important during the biosorption of metal ions by immobilized microbial biomass. The use of this mechanistic understanding from the model resulted in the substantial improvement of the particle production

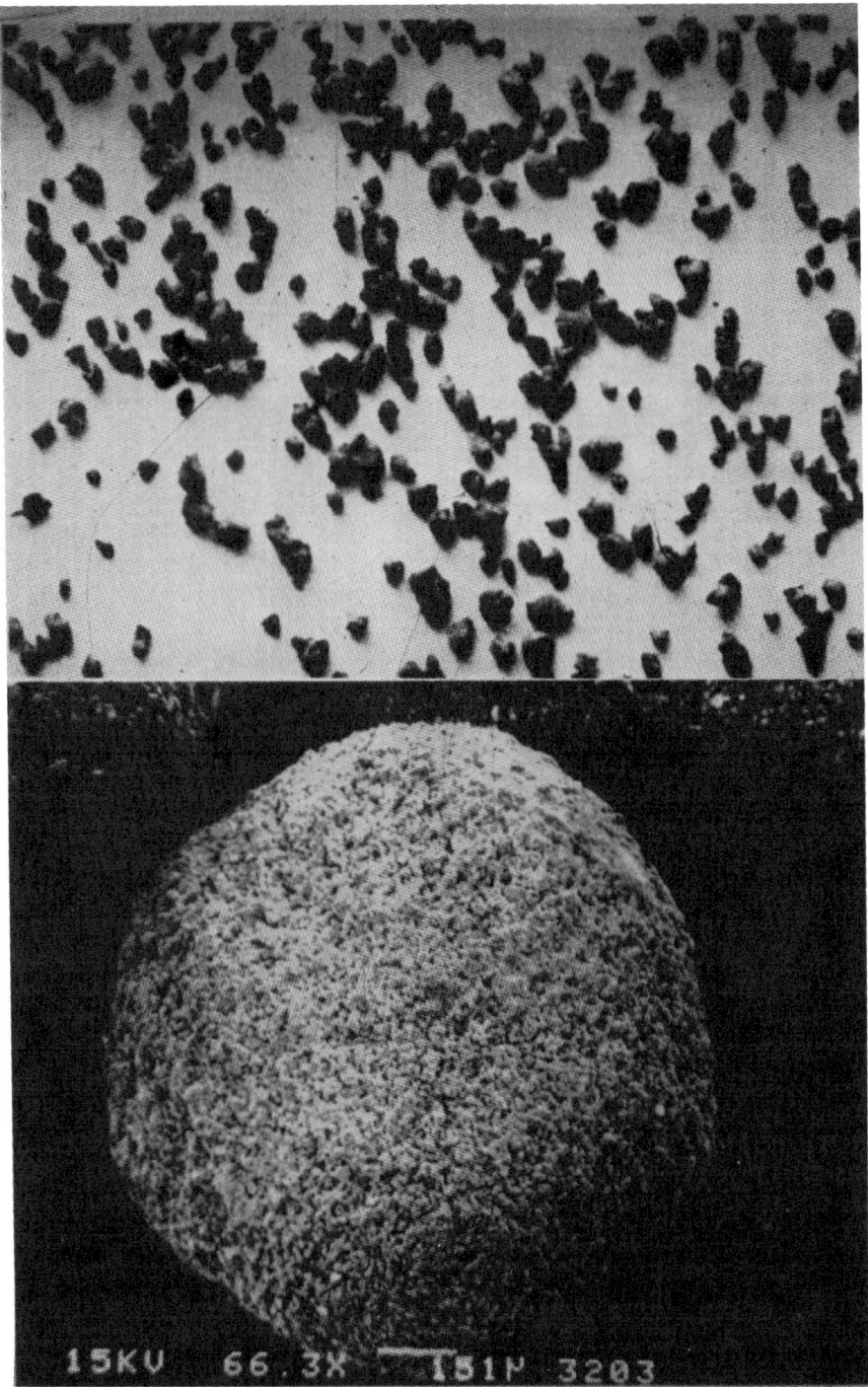

Figure 1. Typical photograph and electron micrograph of biosorbent particles.

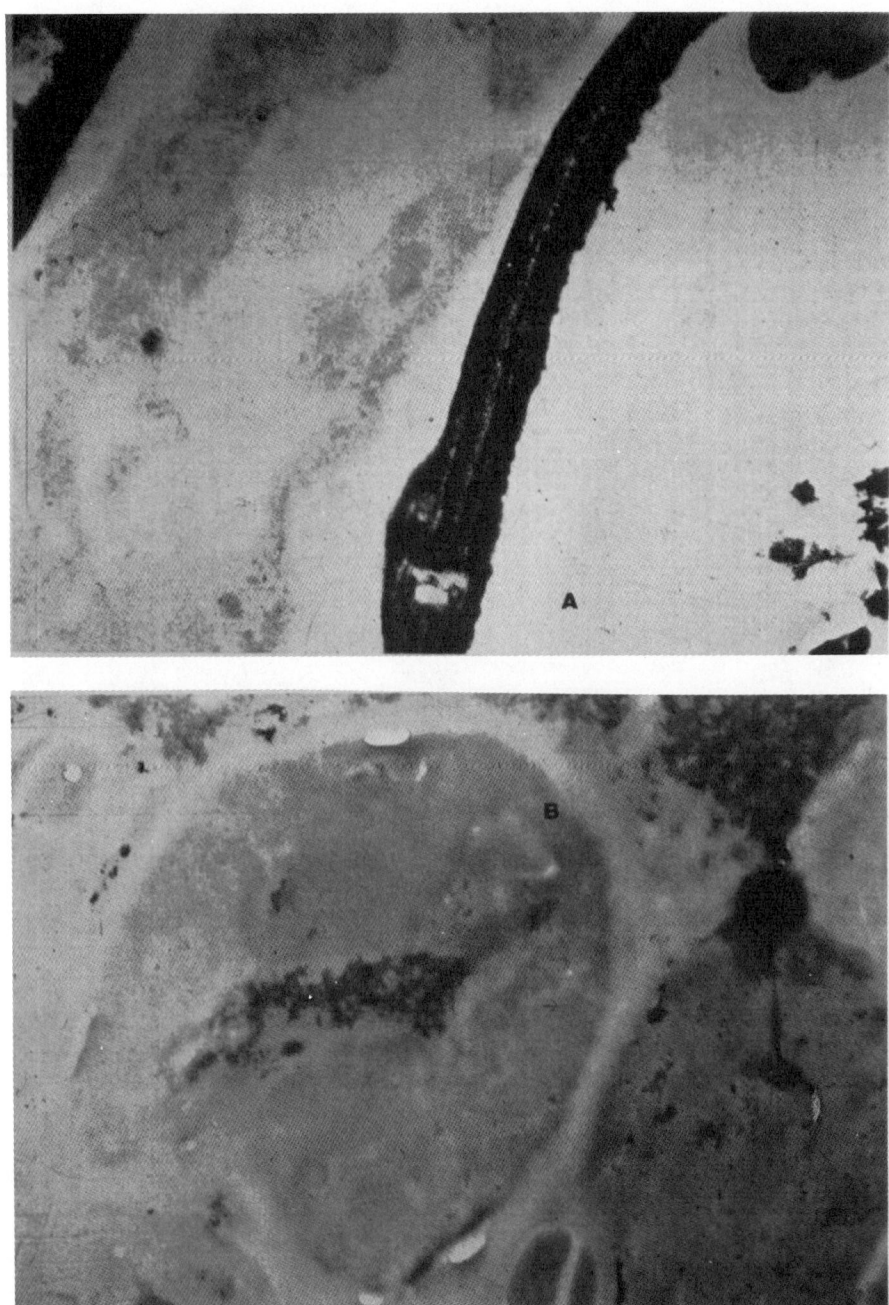

Figure 2. Typical transmission electron micrographs of *R. arrhizus* cells before (A) and after (B) successful uranium elution with $NaHCO_3$.

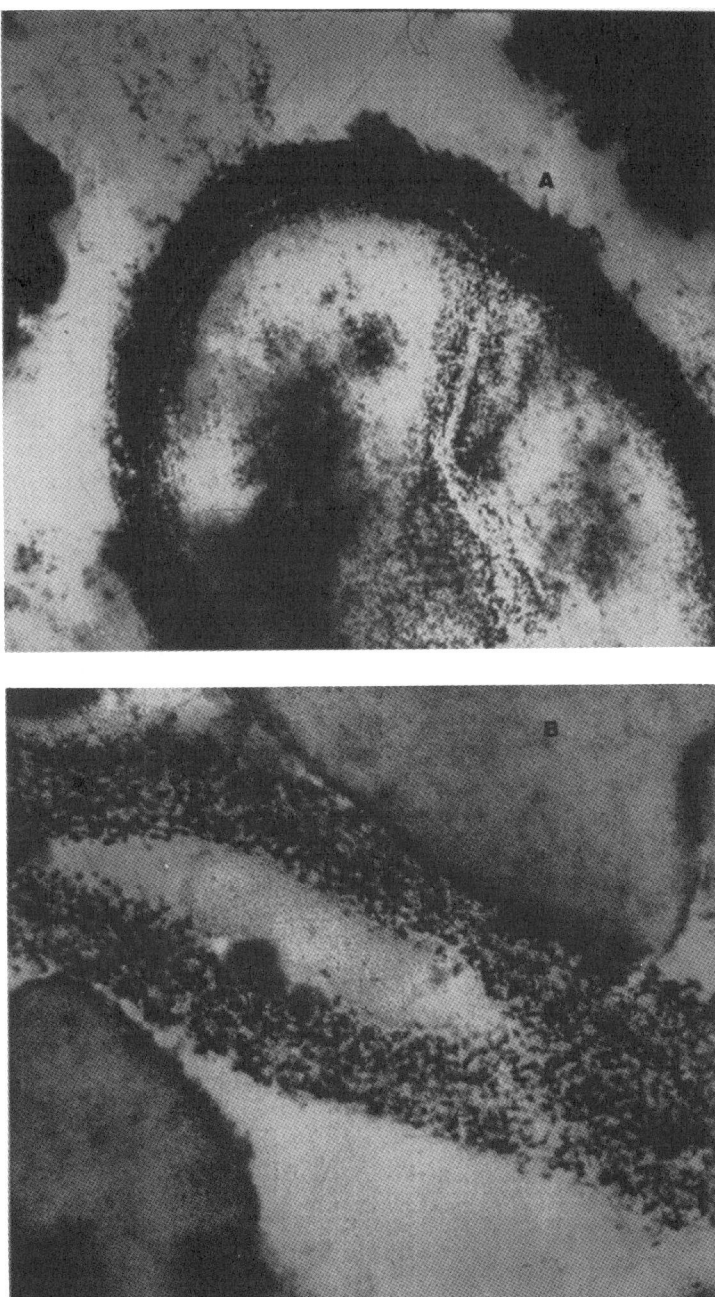

Figure 3. Typical transmission electron micrographs of *R. arrhizus* cells following unsuccessful uranium elution by H_2SO_4.

technology which produced improved intrinsic kinetic properties of the biosorbent particles. (4,5,16).

THE PILOT PLANT TESTING OF URANIUM BIOSORPTION

One of the recent success stories in the application of biotechnology in the minerals industry has been the in-situ uranium biological leaching of ores in the Elliott Lake district of Canada by Denison Mines. As a result of this technology, uranium is currently being produced at a fraction of the conventional cost. The biological pregnant leach liquors ($1.6 + 10^6$ gallons/stope) are, however, dilute and the resulting solution volumes are very large. The recovery of the uranium from these solutions by biosorption has been shown to be an excellent alternative to pumping the solutions 2500 feet up to the existing surface processing plants (3,19). In 1987 a mini bench scale continuous pilot plant was operated at McMaster University. The pilot plant was operated for 16 continuous uranium sorption/elution cycles and yielded pure uranium solutions with peak uranium concentrations 30 times the corresponding feed solution concentration. Following these encouraging results it was decided to design, construct and operate a significantly larger pilot plant at the CANMET laboratories in Ottawa. The pilot plant operation was also assisted by Denison Mines which supplied the required pregnant leach liquors from their Elliot Lake works. The pilot plant objectives were the following:

1. To examine the biosorbent mechanical stability following prolonged use under conditions that are similar to a full scale application.
2. To examine the long term stability of the biosorbent metal uptake capacity, efficiency of extraction and selectivity under the same conditions.

Table 1 shows a typical composition range for the Denison Mines pregnant leach liquor. It should be noted that the feed uranium concentration is low (100-300 ppm) and that the associated ionic matrix is very complex with other elements present in solution concentrations above that of uranium. Figure 4 is a schematic diagram of the second pilot plant. The plant was operated for about one year with very positive results. In addition to uranium, the behaviour of several other elements were monitored closely among which are thorium, aluminum, iron, silicon and yttrium.

Figure 5 illustrates the uranium biosorption breakthrough curve for the fourteenth cycle. All the uranium breakthrough curves were similar in shape and exhibited a desirable steep uranium breakthrough. The corresponding uranium elution curve is shown on Figure 6. All the uranium elution curves were characterized by a very sharp and narrow uranium concentration profile. This shape suggests a very efficient elution process whereby a small volume of eluant was capable to elute and recover the previously sequestered uranium. The very narrow and sharp elution curve also resulted in a highly concentrated (x 40) and pure uranium eluate, thus making the uranium recovery by biosorptive separation an even more attractive option. The volume of the eluant required to elute the adsorbed uranium can be optimized further so that both the uranium concentration of the eluate as well as

Table 1. Typical Composition of the Denison Mine Biologically Leached Pregnant Solution That Was the Feed to the Pilot Plant

Elements	Concentration range (pH 2.5 Bioleach solution) (mg/L)	Concentration range (pH 3.9 Bioleach solution) (mg/L)
U	95-160	95-160
Fe	259-550	0.6-7
Ca	200-320	200-315
Mg	13-24	13-23
Si	20-35	20-31
Al	30-55	27-45
Y	8-26	6-21
S	527-880	520-790
Sc	<0.5	<0.5
Th	44-181	0.5-51

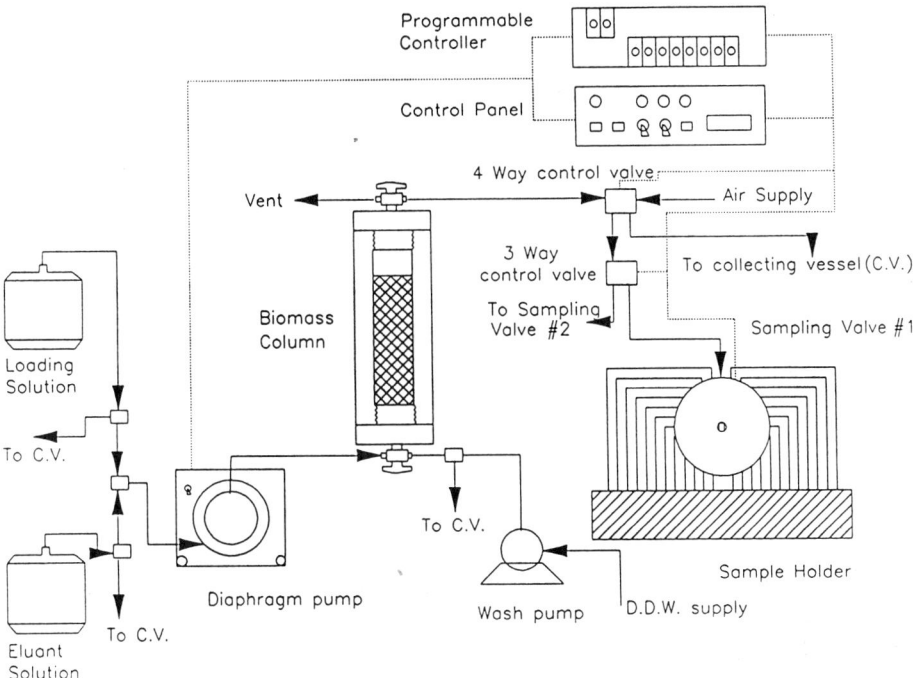

Figure 4. Schematic diagram of the continuous uranium recovery pilot plant.

its purity can be improved more. With the practice adopted during the pilot plant operation, peak uranium eluate concentrations of up to 14,000 mg/L were observed while 10,000 mg/L can be considered as an average for most cycles. Figure 7 summarizes, in a simple bar graph, the observed eluate uranium peak concentration

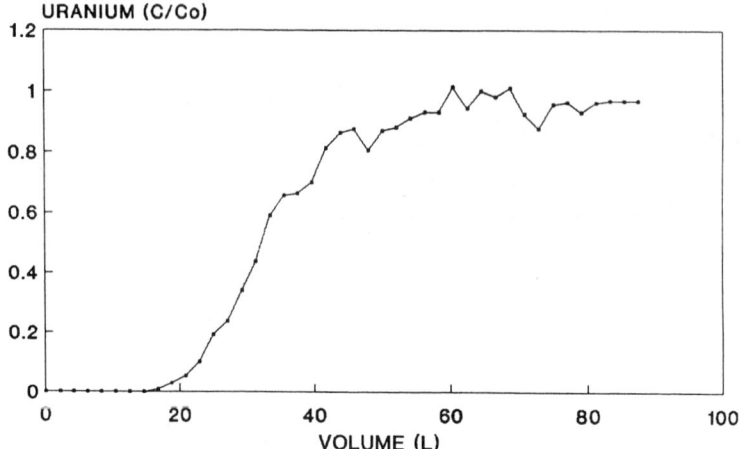

Figure 5. Typical uranium biosorption breakthrough curve (cycle 14).

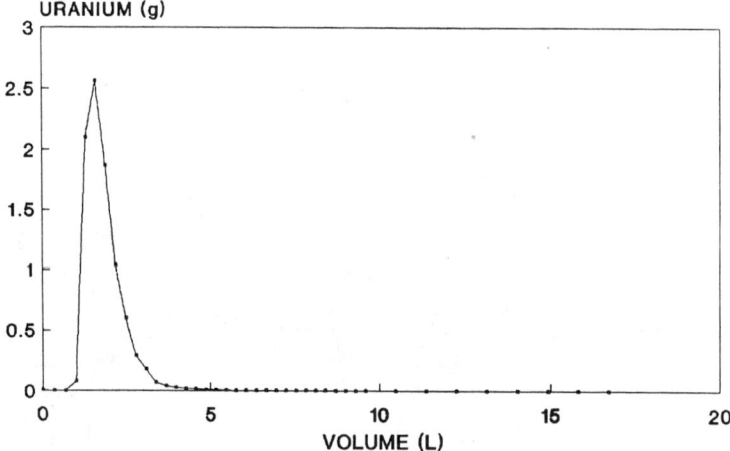

Figure 6. Typical uranium elution curve (cycle 14).

for the first 11 cycles. The efficiency of uranium elution is summarized on Figure 8, suggesting the complete elution of the adsorbed uranium in each cycle during the pilot plant operation.

In addition to uranium, thorium aluminum and yttrium were also retained by the immobilized microbial biomass. Thorium was removed completely from the feed stream and did not break through before uranium. Aluminum and yttrium both broke through the column ahead of uranium. Figure 9 is a typical example of the silicon and aluminum behaviour, while Figure 10 compares the adsorptive behaviour of yttrium to that of thorium. The self-eluting yttrium behaviour is the subject of a current project that examines the potential for the industrial separation of these two elements using biosorption.

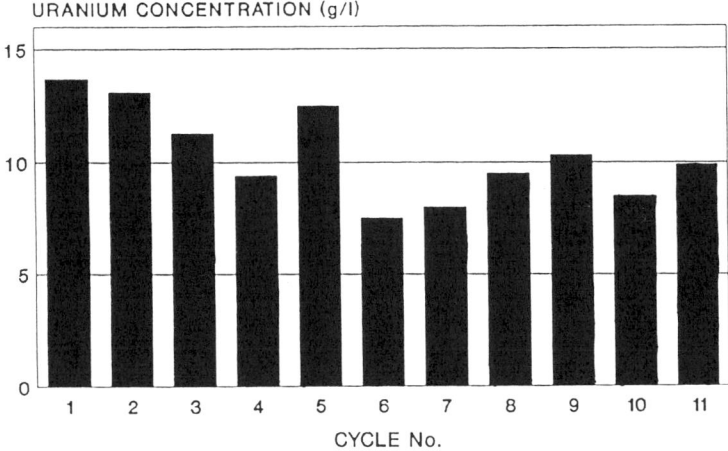

Figure 7. Summary of eluate peak uranium concentrations for the first 11 cycles of pilot plant operation.

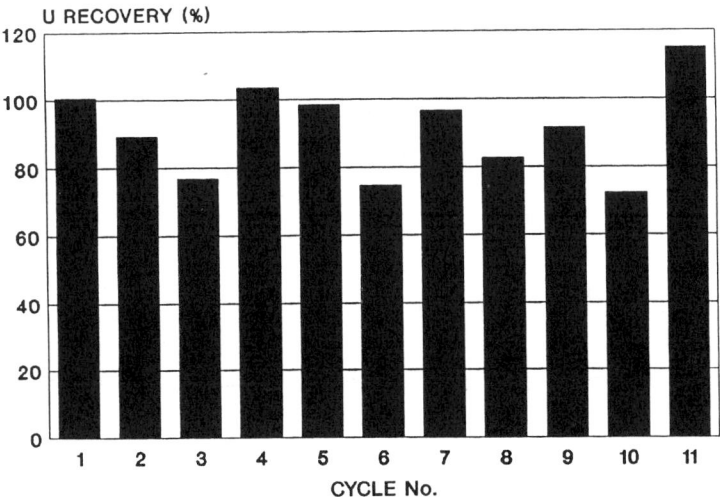

Figure 8. Summary of uranium elution efficiency observed during the first 11 cycles of the pilot plant operation.

IONIC COMPETITION EFFECTS

The experience that has accumulated thus far on the biosorptive separation and recovery of radionuclides and especially uranium has suggested that during the pilot plant testing of the biosorptive technology attention needs to be paid to possible ionic competition effects that can potentially reduce the immobilized biomass long term metal uptake capacity. A systematic study of the long term gradual re-

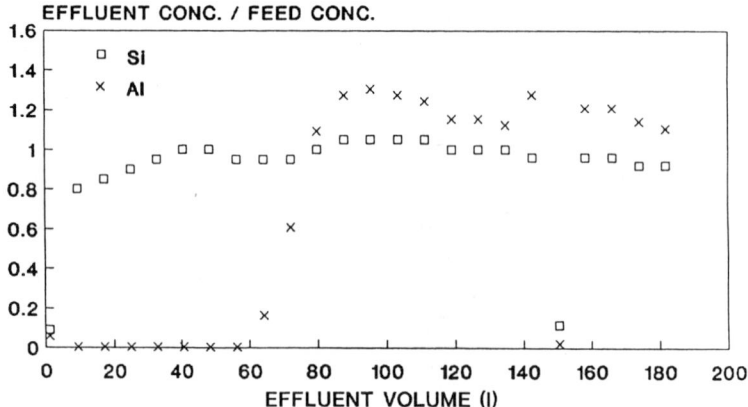

Figure 9. Typical Si and Al breakthrough curves (cycle 4).

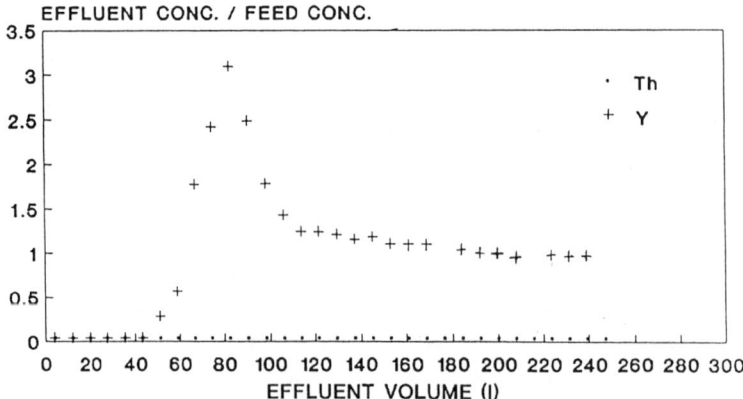

Figure 10. Typical yttrium and thorium breakthrough curve (cycle 2).

duction of the immobilized biomass uranium uptake capacity for the Elliot Lake type feed solutions has been undertaken in the last two years. This study has provided important information on the mechanistic interactions among the immobilized biomass and the metal ions in solution that are responsible for the competitive effects. As the detailed discussion of these effects is beyond the scope of the present work only a brief overview will be presented. Looking at the microbial cell wall following uranium elution, we can observe the elution of uranium and the strong residual signal of aluminum and silicon. These and other detailed equilibria, spectral, microscopic, and pilot plant experimental results have resulted in the formulation of a mechanistic hypothesis on the way aluminum interferes with uranium biosorption. Aluminum has been shown to adsorb and interfere with the cell wall active sites that are responsible for uranium adsorption. A new uranium biosorption pilot plant flow sheet has been designed by the first author, which incorporates unit operations that should control the ionic competition effects. This pilot plant is under construction at the CANMET laboratories in Ottawa and will be jointly monitored by the authors over the next year in continuous operation using biological pregnant leach liquors supplied by the Denison Mine from their Elliot Lake operation.

CONCLUSIONS

The use of immobilized microbial biomass makes possible the selective extraction and recovery of radionuclides as well as other metals from dilute, complex industrial solutions or waste waters. This technology, known as biosorption, offers selectivity and specificity higher than conventional ion exchange resins for the above types of applications.

A proprietary technology has been developed that can immobilize inactive microbial biomass in particles of specified particle size and mechanical strength with favourable mass transport properties and very low weight percentage of inactive additives. This technology also preserves the biomass biosorptive properties. The technology has been scaled up successfully to a laboratory pilot plant size.

The immobilized biomass has been tested with the help of a continuous pilot plant at the CANMET laboratories in Ottawa in collaboration with industry for the separation and extraction of uranium and other elements. Successful bench scale testing has also been carried out for the extraction of radium from mine tailings waste water. A new modified continuous pilot plant will start operation in Ottawa in the fall of 1990 in order to refine further this technology in collaboration with industry.

REFERENCES

1. Tsezos, M., "The Selective Extraction of Metals from Solution by Microorganisms. A Brief Overview". Can. Metal. Q., *24*, 2, 141-144, 1985.
2. Tsezos, M., "Engineering Aspects of Metal Binding by Biomass", Chapter 14 in *Microbial Mineral Recovery*, Eds. H.L. Ehrlich and C.L. Brierley, McGraw Hill, N.Y., 1990.
3. Tsezos, M. and McCready, R.G.L., "The Continuous Recovery of Uranium from Biologically Reached Solutions using Immobilized Biomass", Biotech. Bioeng., *34*, 10-17, 1989.
4. Tsezos, M., Noh, S.H. and Baird, M.H.I., "A batch reactor kinetic model for uranium biosorption using immobilized biomass", Biotech. Bioeng., *32*, 545-553, 1988.
5. Tsezos, M. and Deutschman, A.A., "An investigation of the engineering parameters for the use of immobilized biomass particles in biosorption", J. Chem. Tech. Biotechnol., *48*, 29-39, 1990.
6. Brierley, J.A. and Vance, D.B., "Recovery of precious metals by microbial biomass", *Biohydrometallurgy*, Proceedings of the International Symposium Warwick, 1987, Science and Technology Letters, Kew Surrey, UK, 1988.
7. Darnall, D.W., McPherson, R.A. and Gardea-Torvesdey, J., "Metal recovery from geothermal waters and ground waters using immobilized algae: *Biohydrometallurgy* Proceedings of the International Symposium, Jackson Hole, 1989, CANMET, Ottawa, Canada, 1989.
8. Tsezos, M. and Keller, D., "Adsorption of Radium-226 by Biological Origin Adsorbents", Biotechnology and Bioengineering, *25*, pp. 201-215, 1983.

9. Tsezos, M., Baird, M.H.I. and Shemilt, L.W., "The Use of Immobilized Biomass to Remove and Recover Radium from Elliot Lake Uranium Tailings Streams", Hydrometallurgy, *17*, pp. 357-368, 1987.
10. Tsezos, M. and Volesky, B., "Biosorption of Uranium and Thorium", Biotechnology and Bioengineering, *23*, pp. 583-604, 1981.
11. Tsezos, M. and Volesky, B., "The Mechanism of Uranium Biosorption by *R. arrhizus*", Biotechnology and Bioengineering, *24*, pp. 385-401, 1982.
12. Tsezos, M. and Volesky, B., "The Mechanism of Thorium Biosorption by *R. arrhizus*", Biotechnology and Bioengineering, *24*, pp. 955-969, 1982.
13. Tsezos, M., "The Role of Chitin in Uranium Adsorption by *Rhizopus Arrhizus*", Biotechnology and Bioengineering, *25*, pp. 2025-2040, 1983.
14. Tsezos, M., "Recovery of Uranium from Biological Adsorbents. Desorption Equilibrium", Biotechnology and Bioengineering, *26*, pp. 973-981, 1984.
15. Tsezos, M., "Immobilization of Ions by Naturally Occurring Materials as Alternatives to Ion Exchange Resins", *Immobilization of Ions by Biosorption*, Ellis Horwood Publishers, Editors H. Eccles, S. Hont, London, UK, 1986.
16. Tsezos, M. and Deutschman, A., "A Modelling Study of the Important Engineering Parameters in Immobilized Biomass Biosorption", Preparation, 1990.
17. Tsezos, M., "The Performance of a New Biological Adsorbent for Metal Recovery", Modelling and Experimental Results, Second International Conference on Biohydrometallurgy, Warwick, U.K., July 1987.
18. Tsezos, M., Baird, M.H.I. and Shemilt, L.W., "The Elution of Radium Adsorbed by Microbial Biomass", Chemical Engineering Journal, *34*, 3, pp. B35-B41, 1986.

Research and Development Programs for Biological Hazardous Waste Treatment in the Netherlands

Esther Soczo and Klaas Visscher

INTRODUCTION

In recent years, the development of biotechnology has made considerable progress in the Netherlands. The growing knowledge on relevant microbiological processes has also increased the possibilities for using biotechnology for the remediation and protection of the environment.

In comparison with other environmental technologies, environmental biotechnology has the advantage of using microbiological mineralization processes to convert organic compounds (e.g., xenobiotics) to inorganic end products which are part of natural recycling processes. Microorganisms are capable of very special reactions; this means that hardly any undesirable by-products and residues are produced. Other advantages are that biodegradation takes place at a moderate temperature and pressure and requires usually less energy than physical, chemical, or thermal processes.

Because of the advantages mentioned above, biotechnology can be considered as one of the promising alternatives for solving environmental problems. The development of environmental biotechnology is stimulated and financially supported by the Dutch government within the framework of different R&D programs. This paper includes a review of these R&D programs and the new developments in the field of biological hazardous waste treatment. Particular attention is given to the treatment of waste gases, wastewater, and contaminated soil. Finally, the topics of new research areas are summarized.

R&D PROGRAMS FOR ENVIRONMENTAL BIOTECHNOLOGY

As mentioned above, the Dutch government stimulate the development of environmental biotechnology within the framework of different R&D programs. The most important programs are as follows:

Esther Soczo and Klaas Visscher • National Institute of Public Health and Environmental Protection (RIVM), Laboratory for Waste Material and Emission (LAE), 3720 BA Bilthoven, The Netherlands.

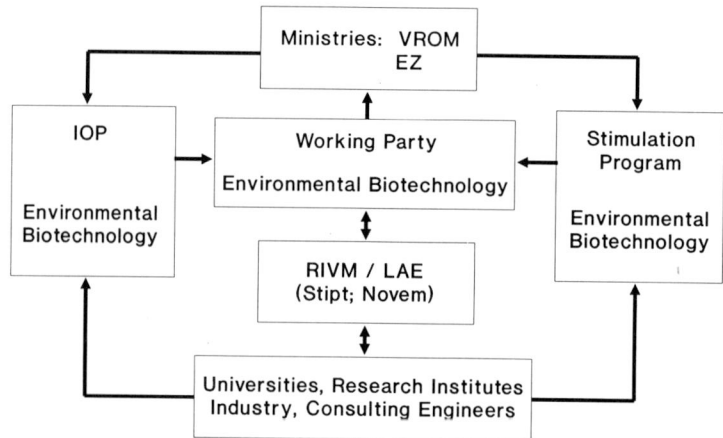

Figure 1. Management of R&D programs on environmental biotechnology.

- Innovation-Oriented Research Program (IOP) Environmental Technology
- Stimulation Program for Environmental Technology
- Programmatic Business-Oriented Technology Stimulation (PBTS)
- The Netherlands Integrated Soil Research Program

The development of environmental biotechnology in the Netherlands is strongly related to the development of environmental technology in general. The IOP-Environmental Technology Program consists of three subprograms: Environmental Biotechnology, Recycling, and Prevention. The Stimulation Program for Environmental Technology covers the following subprograms: Environmental Biotechnology, Membrane Technology, Hazardous Waste, Air, Water, and Soil. Below, special attention is paid to these two programs which cover the biggest part of fundamental and applied research in the field of environmental biotechnology. A flowchart for the management of the subprograms on Environmental Biotechnology is presented in Figure 1. These programs are financially supported by the Ministry of Housing, Physical Planning and Environment (VROM), and the Ministry of Economic Affairs (EZ). The Ministry is advised in its activities by the Working Party on Environmental Biotechnology (WMB) which was installed in 1984. The most important task of WMB, consisting of representatives from industry, research institutes and government, is the creation and execution of the environment-oriented subprograms. The Laboratory for Waste Materials and Emissions (LAE) of the RIVM coordinates the activities of the WMB and assists it in the selection of the proposed research projects and the development of the long-term research programs. RIVM/LAE also supervises the technical performance of the most important research projects. The financial management of the projects is carried out by two other institutions. The research projects are performed by existing research groups at the universities and research institutes. Due to the multidisciplinary character of the development of technologies, companies and engineering consultants participate often in these projects. This cooperation between scientists, industrial researchers and governmental representatives has considerably encouraged the development of environmental biotechnology in the Netherlands.

At an early stage, it was recognized by WMB, that in order to coordinate and stimulate environmental technology research, the development of a strategically focused, long-term R&D program is necessary. The first Dutch R&D Program on Environmental Technology was issued in 1985. In this program, the latest technology developments were summarized and the bottlenecks were identified. Based on this review and the environmental policy of the government, new research topics were established and priorities were set. The second updated edition of the program was issued in 1989 (Brinkman at al., 1989). In this program, a distinction has been made between fundamental and applied research. On the basis of the evaluation of research results, the priority was changed from applied research to fundamental research on relevant biological and technological processes (See the section on Future Research, below).

The two other programs mentioned in the beginning, also offer the possibilities for environmental biotechnology related research. In the framework of The Netherlands Integrated Soil Research Program, research is stimulated in the field of biological soil treatment. The Programmatic Business-Oriented Technology Stimulation program focuses on development and commercialization of technologies.

The development of biotechnology has been stimulated for about ten years. In this period considerable results have been achieved. More than 200 biological filters have been installed for the treatment of waste gases in the Netherlands. About 130 anaerobic wastewater treatment plants have been build inclusive, 90 of them abroad. Different soil treatment methods (e.g., landfarming and in situ biorestoration) have been developed and used for the cleanup of contaminated sites.

RECENT DEVELOPMENTS IN THE FIELD OF ENVIRONMENTAL BIOTECHNOLOGY

Treatment of Waste Gases

In recent decades, biofiltration systems have been developed in order to clean air by means of microbiological degradation processes. With biofiltration, the polluted air is led through a filter which mainly consists of a natural packing material, such as compost. In this packing material a wide variety of micro-organisms which have the potential of degrading a large number of organic compounds, can be found. In The Netherlands, the application of biofiltration started in 1978 when biofilters were put into use at sewage treatment plants to process polluted air that caused odor problems. The surface loading of these systems was relatively low (20-100 $m^3/m^2.h$)

In the last few years, the field of application was also extended to industry for the removal of smells and the biogradation of chemicals like alcohols, esters, and ketones. The development of biofiltration systems was made possible by research advances, in which a high conversion rate could be combined with high gas flows and low drops in pressure. The packing material has also been improved in recent years. The aging of the filter packing can be retarded assuring an extended period (years) of relatively high biological activity. The pressure drop of the filter bed can also be decreased considerably resulting in gas velocities of around 400 m/h. Usage of 100-200 g of organic carbon per m^3 of packing material per hour for easily biodegradable compounds can be obtained. Costs are in the range of 0.25 to 1.25 US dollars per 1000 m^3 waste gas to be treated (Ottengraf, 1987).

The treatment of xenobiotic compounds requires additional research. For example, research has been done on the treatability of polluted air from a pharmaceutical company. The air contained the compounds acetone, ethanol, isopropanol and methylene chloride. Semitechnical research showed that the component acetone could be almost completely degraded in the first step; ethanol and isopropanol in the second step. The degradation of methylene chloride, however, was not achieved. When a third step was introduced and inoculated with a culture of *Hyphomicrobium* sp., this component was also degraded with a relatively high rate. On the basis of the research results, a full-scale plant (one reactor with three steps) has been constructed.

The experiments presented above also have shown that the accumulation of degradation products in biofilters can cause problems. Accumulation of chlorides was shown to completely stop the degradation of methylene chloride. This makes a regeneration of the filter necessary. Another problem appeared to be the low rates of microbiological degradation of recalcitrant xenobiotics in general.

Because of the above mentioned difficulties in using biofilters for degradation of xenobiotics, research is now directed to the application of trickling filters in this field. In a trickling filter, the gas flow is moving con- or counter-currently in the water phase through a bed of inert packing material with a developed biolayer. First results on laboratory scale with gas flow rates up to 500 m/h show treatment capacities of up to 200 g/m^3 of reactor packing per hour for dichloromethane and 80 g/m^3 of reactor packing per hour for dichlorethane. Degrees of conversion up to 70% were achieved for both substances (Ottengraf et al., 1988).

Treatment of Wastewater

Since the enforcement in the Netherlands of the Law on the Pollution of Surface Water in 1970, the application of biological waste water treatment has increased enormously. Until 1980, effluent was treated almost exclusively in aerobic activated sludge systems. Subsequently the anaerobic technique proved a good alternative, especially in terms of saving both energy and costs. At present the anaerobic technique is used in many branches of industry.

At first, attention was almost exclusively drawn towards physical-chemical detoxification methods for xenobiotics, especially filtration stages in activated carbon columns following an aerobic biological treatment. For the last few years however, biological and combinations of biological and physical-chemical methods have come into the picture. For example, experiments have been carried out at the Institute for Inland Water Management and Waste Water Treatment (DBW/RIZA) with the addition of activated carbon to an active sludge system (PACT-system) with which waste water from a tank accommodation storage company was purified. During the experiments, shock loadings with aniline were simulated. It turned out that the addition of activated carbon to the active sludge had a stabilizing influence on the purification process (Teurlinckx and Antonijsz, 1986). To support this development, the RIVM has started a fundamental research project in which the biodegradation of cresol in chemostat cultures and gas lift reactors is investigated (De Jonge et al., 1991). When the activated sludge and the PACT-system were compared, it was found that addition of activated carbon resulted in a lower steady-state cresol effluent concentration. Results suggest that part of this extra removal is due to biodegradation.

Anaerobic waste water treatment also offers possibilities for the biodegradation of xenobiotic compounds. One of the research projects was directed to the anaerobic treatment of wastewater from the Chemical Industry Rijnmond B.V. Experiments with an UASB (Upflow Anaerobic Sludge Bed) pilot reactor showed that toxic substances such as phenol and formaldehyde could be degraded up to 99-100%. Based on this results, a full scale UASB-plant has been constructed (Gist Brocades, 1988).

A recent development concerns the biological cleaning of groundwater which derives from soil cleaning. The physical and chemical techniques which have been used so far have certain disadvantages, such as frequent high costs and environmental objections (residual material and air emissions). Research in the Netherlands concentrates on the development of bioreactors for the removal of xenobiotic compounds from groundwater. One of the research projects was directed towards the treatment of contaminated groundwater of a former pesticide production site by means of a trickling filter and a rotating biological contactor. The results were promising: 50% removal of HCH (hexachlorocyclohexane) and 99% removal of benzene and monochlorobenzene. Finally, besides the planned activated carbon filters, two rotating biological contactors were installed for the pretreatment of the groundwater. Consequently, the required activated carbon was reduced to less than 10% and biological pretreatment resulted in a cost reduction for groundwater treatment of 30-40% (Urlings et al., 1988).

Treatment of Contaminated Soil

Soil treatment is a rather new area in the application of biotechnology. The development of these technologies has been going on for about five years in the Netherlands. At present, three kinds of biological treatment methods are available: landfarming, in situ biorestoration, and bioreactors. The landfarming method and in situ biorestoration are operational. The in situ biorestoration will need to be improved and optimized at pilot- and full-scale in the future. Bioreactors are in the development stage. However, the first pilot plants are being tested this year.

During the past few years, research has mainly been focused on the biological treatment of soils contaminated with oil and PCA's (Soczo et al., 1988). Experiments on pilot- and full-scale indicated that these substances are biodegraded under aerobic conditions. One characteristic feature is that the degradation rate falls sharply at low concentration of contaminants. On the basis of these experiences, it can be concluded that the degradation rate is determined primarily by the desorption of the contaminants from soil particles and not by the microbial activity. This implies that the limiting factor in degradation must be the bioavailability of the contaminants. Consequently, the improvement of bioavailability is one of the most important research issues for the near future.

Landfarming. Landfarming has been used for the cleanup of contaminated soils for several years in the Netherlands. In the case of the most simple landfarming method, the contaminated soil is spread out over a sand layer with a drainage system to a depth of around 40 cm. Prior to this, the soil is protected by means of a plastic membrane (about 0.5 mm PVC). Landfarming is also carried out under, so-called, controlled circumstances (e.g. Cum-Bac greenhouse method). This means that during the clean-up process, a better control of parameters, such as oxygen, water content, and temperature is achieved. For the use of forced aeration, it is possible to increase the depth of the layer of the soil to be cleaned to about 100 to 150 cm and to achieve a higher biodegradation rate of the contaminants. In this case, the landfarming plots are covered and the air is cleaned by means of a compost filter.

By means of the landfarming method good results have been achieved with the treatment of sandy soils mainly contaminated with oil compounds and aromatic hydrocarbons. After one "growing season" (about 6-8 months), the final oil concentrations reached are in most cases between 400 and 1000 mg/kg dry matter. A high level of decontamination (90-99%) can be obtained in the case of lower PCA's. However, a much lower degradation level can be reached (in a longer period) in the case of higher PCA's. The clean-up costs of landfarming are in the range of 25-40 US dollars per ton of soil (Soczo and Staps, 1988).

In Situ Biorestoration. Results of research on pilot- and full-scale indicate that in situ biorestoration seems to be a satisfactory treatment for highly permeable sandy soils (Verheul et al., 1988). In the Netherlands, the groundwater is usually used as a medium for addition of oxygen, nutrients and eventually microorganisms to stimulate biodegradation of the contaminants in the soil. The water is infiltrated via drains, trenches or wells. The pumped-up water containing the dissolved degradation products and a part of the contaminants is treated above the ground and usually recycled. The addition of oxygen was found to be important for the stimulation of biodegradation. Hydrogen peroxide seems to be an excellent oxygen donor.

In situ biorestoration of petrol polluted sandy soil is being investigated at the RIVM in cooperation with the Netherlands Organization for Applied Scientific Research (Verheul et al., 1988). On the basis of promising laboratory results, a full scale experiment at a petrol station, where the subsoil has been contaminated with 30,000 liters of gasoline, started in the fall of 1989.

In the last two years, a few clean up projects have been successfully completed. The method has usually been used in case of sandy soils contaminated by oil compounds and aromatics. Field tests have shown that soil venting is a suitable method for clean up of unsaturated, sandy soils with volatile components. In the future, research will focus on the optimization of oxygen supply and the combination of water-recirculation and air venting systems to achieve more effective bioremediation.

Bioreactors. Bioreactors are in the development stage in the Netherlands. In the last three years, research has focused on the development of "dry matter" and "slurry" bioreactors. Dry matter reactors can be used for the treatment of sandy soils. The optimal soil moisture content usually is between 10 and 20 percent. In slurry reactors, all soil types may be treated. Slurries contain 50 to 60 precent solids. For rapid degradation of contaminants, intensive mixing of the soil (slurry) and sufficient oxygen supply are essential. In bioreactors, a considerably higher biodegradation rate can be achieved because the process conditions can be more accurately controlled. The average treatment time varies between 1 and 3 weeks.

Several research institutes and companies are engaged in the development of bioreactors. The Delft University of Technology is conducting research into the possibilities of microbiological degradation of contaminants in a 3-phase (air, liquid, solid) slurry reactor, the so-called Pachuca reactor (Kleijntjens et al., 1989). Laboratory experiments with a 400 liter reactor indicated that dual injection (air and liquid) for keeping the soil in suspension is successful. Kinetic studies showed a fast degradation rate for oil compounds. Based on these results a 4 m^3 pilot-plant reactor was constructed in the beginning of 1990.

The Agricultural University of Wageningen (LUW) completed laboratory research in 1987 on the degradation of hexachlorocyclohexane (α- and β-HCH) under different redox conditions. It can be concluded from the results that α-HCH is degraded most rapidly under aerobic conditions. Hardly any intermediate products accumulate. No breakdown of β-HCH was observed under any of the conditions. On the basis of this conclusion, the Research Institute for Nature Management (RIN) carried out several field experiments with "wet" and "slurry" soils. As with the laboratory experiments, the best results were obtained with a continuously aerated slurry system. The α-HCH content of the soil decreased by about 80% within four months (from 300 mg/kg d.m. to 40–60 mg/kg d.m.). Based on these results and also on other experiences, it is expected that the biodegradation of α-HCH can be optimized in a reactor (Bachmann and Zehnder, 1988). LUW has developed a bioreactor system which has already been patented in 1989. Based on this reactor concept, a demonstration project was carried out by Witteveen and Bos (Consulting Engineers) in collaboration with LUW. Initial experiments were performed with soil contaminated with oil compounds. The results will be published at the end of 1990.

Up to now, very little research has been conducted in pilot scale on the degradation of chlorinated hydrocarbons in these reactors. However, laboratory results have shown that these compounds are degradable. Inoculation with specially cultivated microorganisms was found to accelerate the degradation of chlorinated hydrocarbons considerably (Janssen et al., 1989).

FUTURE RESEARCH

The research results discussed in this paper look promising so far. Research in environmental biotechnology will be continued and partly supported by the Dutch government within the framework of the different R&D programs.

The R&D Program on Environmental Biotechnology, issued in 1989, has concluded that for balanced long-term development, more attention has to be paid to fundamental research on microbiological and technological processes (Brinkman et al., 1989). The following topics on fundamental research received a high priority in this program:

- Biodegradation of xenobiotics; identification of microorganisms and degradation pathways
- Microbial leaching of heavy metals from solid wastes; generation of know-how for process-design
- Improvement of the availability of contaminants for the microorganisms in soil treatment systems
- The use of special microorganisms in treatment systems; maintenance and immobilization of microorganisms especially in reactor systems
- Development of models for the optimization of biological treatment processes

High priority was also given to the following topics on applied research:

- The removal of toxic organic compounds from hazardous wastes
- The removal of heavy metals especially from solid wastes
- Development of compact, multi-phase (full-scale) bioreactors
- Combination of treatment systems for a more optimal processing of hazardous wastes.

A budget of 17.5 million Dutch guilders (approx. $8 million) has been allocated by the Dutch government for a four years period to conduct high priority research set by WMB. This budget will be supplemented by contributions from industry and research institutes participating in the R&D programs. Moreover, other programs also offer financial support for research in biotechnology.

Up to now, the development of environmental biotechnology has been focused on the treatment of hazardous wastes generated during the production process (so called end of pipe treatment). According to the policy of the Dutch government, in the near future, attention will also be paid to the development of process-integrated environmental biotechnology which aims at preventing environmental problems through cleaner production processes.

REFERENCES

Bachman, A., and Zehnder, A. J. B., 1988, Engineering significance of fundamental concepts in xenobiotics biodegradation in soil, in: *Contaminated Soil '88*, Ed. Wolf, K. et al., Kluwer Academic Publishers, Dordrecht, The Netherlands.

Brinkman, J., Rulkens, W. H., and Visscher, K., 1989, Environmetal biotechnology research and development program 1989-1992, RIVM report no. 738706003, Bilthoven, The Netherlands.

Gist Brocades N.V., 1988, Anaerobic wastewater treatment Chemical Industry Rijnmond B.V., Ministry of Housing, Physical Planning, and Environment, Leidschendam, The Netherlands, (in Dutch).

Janssen, D. B., Oldenhuis, R., and van den Wijngaard, A. J., 1989, Hydrolytic and oxidative degradation of chlorinated aliphatic compounds by aerobic microorganisms, in: *Biotechnology and Biodegradation*, pp. 105-125, Ed. Kamely, D. et al., Gulf Publishing Co., Houston, Texas, U.S.A.

Jonge de, R. J., Breure, A. M., and van Andel, J. G., 1991, Enchanced biodegradation of o-cresol by activated sludge in the presence of powdered activated carbon, *Applied Microbiology and Biotechnology* (in press).

Kleijntjens, R. H., and Luyben, K. Ch. A. M., 1989, Technological and kinetical aspects of microbial soil decontamination in a slurry reactor, Delft University of Technology, Department of Biochemical Engineering, Delft, The Netherlands.

Ottengraf, S. P. P., 1987, Biological system for waste gas elimination, *Trends in Biotechnology* 5:132-136.

Ottengraf, S. P. P., Diks, R., and van Lith, C., 1988, Technological aspects of the removal of xenobiotics from waste gases, in: Proceedings 2nd Dutch Biotechnology Congress, Dutch Biotechnological Society, Zeist, The Netherlands.

Soczo, E. R., and Staps, J. J. M., 1988, Review of biological soil treatment techniques in the Netherlands, in: *Contaminated Soil '88*, Ed. Wolf, K. et al., Kluwer Academic Publisher, Dordrecht, The Netherlands.

Soczo, E. R., Staps, J. J. M., and Visscher, K., 1988, Research results regarding biological soil treatment techniques in the Netherlands, *Resourse, Conservation, and Recycling*, 1:65-76, Elservier Science Publisher B.V./Pergamon Press plc, Printed in the Netherlands.

Teurlinckx, L.V.M., and Anthonijsz, R., 1986, Biological treatment of industrial wastewater with activated carbon addition, Inland Water Management and Wastewater Treatment (DBW/RIZA), report no. 86-022, Lelystad, The Netherlands, (in Dutch).

Urlings, L. G. C. M., Spuij, F., and van der Hoek, J. P., 1988, Biological treatment of contaminated groundwater from a former pesticide production site in Bunschoten, TAUW Infra Consult B.V., Deventer, The Netherlands.

Verheul, J. H. M., van de Berg, R., and Eikelboom, D. H., 1988, In situ biorestoration of a subsoil contaminated with gasoline, in: *Contaminated Soil '88*, pp. 705-715, Ed. Wolf, K. et al., Kluwer Academic Publisher, Dordrecht, The Netherlands.

Environmental Biotechnology—from Flask to Field: A Review

Robert A. Goldstein, Al W. Bourquin, Thomas W. Federle, C.P. Leslie Grady, and William D. Mahaffey

INTRODUCTION

The conference organizers assigned the authors of this paper the task of preparing a critical review and summary of the Conference, that not only included major themes discussed by the participants but in addition identified important points that were not covered. During the conference, each author took responsibility for covering a different session; however, since many themes came up in multiple sessions, this paper has been organized according to theme and not session.

PARTICIPANTS AND CONTENTS

The community of people involved with environmental biotechnology is widely spread geographically and institutionally. Participants at the Conference represented a broad geographical distribution from North America and Europe, as well as a diversity of organizations that included universities, government regulatory agencies, research institutes, government and industrial consumers of biotechnology, vendors of biotechnology, and biotechnology consultants.

The Conference covered highly varied subject matter ranging from technical posters on new molecular methods to measure gene expression in the environment to oral presentations on the sociology of communication. This diversity of subject matter is consistent with the diversity of disciplines that compose environmental biotechnology.

The organizers never precisely defined the scope of activities that they conceived as being covered by the adjective "environmental" in the Conference's title.

Robert A. Goldstein • Electric Power Research Institute, Palo Alto, California 94303. Al W. Bourquin • Ecova Italia, 35129 Padova, Italy. Thomas W. Federle • Proctor and Gamble, Ivorydale Technical Center, Cincinnati, Ohio 45217. C.P. Leslie Grady • Clemson University, Clemson, South Carolina 29634. William D. Mahaffey • Ecova Corporation, Redmond, Washington 98052.

Almost all of the papers focused on bioremediation; however, there was some discussion of waste minimization. There was also discussion of agricultural biotechnologies, but only in the context of giving examples of communication strategies being used to gain public acceptance.

WHAT IS ENVIRONMENTAL BIOTECHNOLOGY?

There were disagreements among the speakers as to whether bioremediation was a new or old technology; high tech or low tech. Discussions of lux genes that lit up every time a degradative operon was turned on created an impression of a novel high tech activity, while discussions of excavating, fertilizing and irrigating soil sounded more low tech. Some speakers referred to bioremediation as innovative. Others claimed that it was as old as sewage and waste water treatment. If composting and land farming are biotechnologies, then is not soybean farming? Is any technology that has a biological component a biotechnology? What exactly is an environmental biotechnology? This is not simply an academic question. It has important real world ramifications. Should environmental biotechnology regulations be applicable to all bioremediation technologies? Should procedures that consist of nothing more that plowing, irrigating and fertilizing land be governed by the same laws that manage risk associated with the release of genetically engineered organisms in the environment?

Grady (1990) has suggested that the use of any one of three characteristics would qualify an activity as an environmental biotechnology: innovative engineering systems; nontraditional microorganisms; and molecular biology tools. Fluidized bed bioreactors are a good example of innovative systems that have moved from the research bench to the field, increasing the throughput of biological treatment systems and allowing their application for the biodegradation of complex waste streams that would not be treatable by conventional technology. Fungal systems are a good example of the application of nontraditional microorganisms. Because environmental engineers have traditionally treated contaminated wastewaters, they have focused on the use of bacteria. It is apparent from recent research, however, that fungi have unique capabilities that can be exploited for biotransformation and biodegradation in both aquatic and terrestrial systems. Finally, the new tools of molecular biology have opened up a whole new vista of opportunities for understanding microbial ecology. Such understanding will ultimately result in an ability to design bioreactors to foster the growth of microorganisms containing desired genetic traits.

However, let us not forget how many speakers reminded us of the important role of communication in winning societal and regulatory acceptance of new technologies. It is highly pertinent to ask what the term biotechnology connotes to the general public. Does it imply some aspect of genetic engineering? If yes, then is it wise to label technologies, regardless of how new or innovative they might be, biotechnologies if they do not involve genetic engineering?

FROM THE FLASK TO THE FIELD

While some presentations dealt with exciting new developments in the laboratory and others discussed applications of more established techniques in the field, the process by which research results in the laboratory are transferred to application in the field was not explicitly addressed. Is there a smooth transition? Is there a lag time? If yes, what is it and what factors affect it? How can lag time be minimized?

At present, our scientific understanding of biotreatment far exceeds our ability to apply this knowledge. During the meeting, information was presented on the bioremediation of alkanes, benzene-toluene-xylene (BTX), and even some chlorinated hydrocarbons. Was this new? Probably not, but the explanations of what was happening in the field were much better than we would have heard even two years ago. Why? Because we are witnessing better integration of science into the engineering applications arena. But, for the most part what was presented in the field application papers was science that was in the literature 10-20 years ago. Our scientific understanding exceeds the implementation or engineering applications of the science.

What can accelerate technology development and market expansion in the future is a better integration of scientific understanding with engineering applications. A number of participants addressed this subject and the beginnings of such integration was indicated in some of the papers from the USA. However, the paper from Canada and especially the one from the Netherlands exemplify the potential efficiency and expediency of environmental technology development when such integration occurs. Dr. Tsezos's paper (Canada) on metals removal and the chemical understanding behind that process should help move that technology from the "flask to the field" much more rapidly than the normal trial and error method of gross process control. The Dutch, probably because of the very small size of their country and their commitment to developing technologies to continuously upgrade their industrial base, have a well developed system for moving scientific research findings into the field. The applications are usually considered at the beginning of the project in close cooperation with engineers. This results in much less lag time when significant results are obtained. An example is the recent work at the University of Groningen on vapor-phase bioreactors. A pilot plant was operational prior to the basic research reaching the literature.

POTENTIAL

There was general agreement among the meeting's participants that environmental biotechnology has high potential value for waste management. The ability of microorganisms to reduce the potential toxicity of substances to higher organisms through processes of biodegradation, biotransformation and bioaccumulation can easily be demonstrated in the laboratory. In theory, there is no reason why biotechnologies based on such capabilities cannot be successfully developed and applied. However, in the field, these processes may be limited by environmental conditions, hence, the engineering challenge of realizing in the field the potential observed in the flask.

It is important to remember that there is unlikely to be a "silver bullet," a single biotechnology that will work in all situations or that will by itself totally remediate a complex waste. Most likely, waste site cleanups will require "treatment trains," a sequence of applications of different technologies, some biotic, some abiotic.

FIELD APPLICATIONS

To successfully implement environmental biotechnology many technical details must simultaneously fall into place. These critical details exist at the molecular, cellular, population, habitat and engineering levels. For biotechnology to work in the case of an onsite cleanup, the following conditions must occur:

1. The chemicals to be cleaned up must be inherently biotransformable or biodegradable. Furthermore, they must be at low enough levels as to be nontoxic to the detoxifiers, but high enough to select and/or support microbial populations capable of detoxification. In addition, the chemicals must be in an environmental form, such that they are bioavailable to the detoxifiers. For example, chemicals intercalated between the bilayers of smectite clay would not be available to microbial attack and would not be amenable to microbial cleanup.
2. Organisms must exist with the ability to detoxify the target chemicals. For this to occur, several things must be present at the cellular level. Genes must exist that code for degradative or transformation enzymes that attack the chemical. These genes must be expressed. There must be a mechanism for transporting the chemical to the enzymes or vice versa and there must be no buildup of toxic intermediates or by-products.
3. These organisms must be present at the cleanup site and there needs to be a sufficient supply of electron acceptors and other nutrients (e.g., nitrogen and phosphorus) to support their activity. Other physical and chemical conditions (e.g., moisture, pH, and temperature) also must be within a set range for the organisms to be active. The organisms must grow at a sufficient rate to escape predation and wash-out and effectively compete with other populations for key nutrients.

Typically not all these conditions are met, which makes engineering intervention necessary. Engineering solutions can include: constructing new metabolic pathways through the use of genetic manipulation, inoculating sites with detoxifier populations, and increasing the supply of electron acceptors and nutrients at a given site or otherwise modifying the environment (e.g., raising pH). Although simple in concept, these engineering solutions can be complicated in their own right and are fraught with tiny details that can hamper their successful implementation.

FIELD VERIFICATION

Many claims are being made of successful bioremediation, but data proving the importance of biological activity are often weak or nonexistent. Inaccurate or exaggerated claims can hinder the development and adoption of bioremediation as a viable cleanup strategy because they will foster suspicion on the part of decision makers and the public. Consequently speakers throughout the Conference stressed the need to rigorously document the contribution of biological activity when asserting that it is the major mechanism of pollutant loss at a remediation site. That this is difficult to do in full-scale facilities was evident from the reported case studies.

The major difficulty in documenting biological activity is associated with the fact that its existence often must be deduced from failure of mass balances of the targeted chemical to close. In other words, since detoxification is a destructive technique, whereas the other primary mechanisms of removal (sorption and volatilization) are conservative, it is often assumed that any chemical not trapped in offgases or remaining attached to solids in the system was indeed biodegraded. In order to use such reasoning correctly, several things must be done. First, offgases must be trapped in a way that mass fluxes away from the site can be calculated accurately. Second, extraction techniques for removing contaminant sorbed to solids must be quantitative and reproducible. Third, it must be shown that the contaminant is not subject to photo and chemical degradation under the conditions at the site. Unfortunately, all of these criteria are seldom met during field studies, giving rise to questions about claims of biodetoxification.

Several speakers suggested ways to better document the magnitude of biodegradation occurring in the field. We will not recapitulate those suggestions, but instead point out two additional approaches worthy of consideration. It is common during laboratory studies to maintain a large continuously operated reactor from which aliquots are removed and placed into small batch reactors for detailed analyses of rates, mass balances, etc. The same concept can be applied to field-scale bioremediation projects to prove the existence of biodegradation. For example, in the French Limited study it would have been possible to remove aliquots of the pond, including the biomass, and place them into small-scale reactors in which tight mass balances could be made. In addition, proof of mineralization of key pollutants could have been obtained by spiking the small reactors with radiolabeled pollutants and measuring the release of radioactive CO_2. Similar procedures could be also employed with soil composting or land farming systems while they are in operation.

When only a few organic pollutants are involved, biodegradation can also be proven by performing an electron or energy balance as is commonly employed by environmental engineers during wastewater treatment studies. The chemical oxygen demand (COD) test is a measure of the number of electrons available in an organic substrate. When that substrate undergoes biodegradation, all of those electrons must end up either in the biomass formed or in the terminal electron acceptor. In other words, in an aerobic environment the amount of COD removed from solution must equal the COD of the biomass formed plus the amount of oxygen consumed. If there is little loss of COD, then carbon mineralization has not occurred. Use of this simple technique in combination with compound specific analysis can greatly strengthen the verification of biodegradation.

The U.S. Environmental Protection Agency (USEPA) has stated that it plans to become active in promoting and funding the collection of scientifically rigorous data in connection with field bioremediation demonstrations to document the biological component of the remediation. We applaud such a position. A major impediment to the acceptance of environmental biotechnologies is their lack of predictability. Who wants to purchase a product of unknown performance specifications? It is essential to be able to predict prior to application cleanup effectiveness as a function of time. Environmental technology should not be a "black box"; put the wastes in at one end and innocuous substances somehow come out at the other. The processes and mechanisms that make it work must be understood.

European governments have developed a system to produce good demonstration data while at the same time promoting science, technology and industry. The European Community (EC) is funding projects to combined university (100% costs) and industry (50% costs) teams. Individual governments, UK, Germany, the Netherlands, and Italy, are funding similar programs. A major difference from the US is the emphasis on implementation and industry development. Although the US is a leader in basic research, it lags in application. USEPA has the opportunity to aid not only the cleanup of waste sites but also promote American industry through demonstration projects cooperatively funded with industry.

WASTE MINIMIZATION

In the past, industry and society utilized what were then "cost minimization" waste management practices. Years ago, waste minimization was considered uneconomical because it increased manufacturing costs. Likewise, resource recovery was not practiced unless the recovered resource was cheaper than other sources. The most cost effective waste management strategies were to release wastes to the atmosphere or a nearby water body, or bury them. Today, there are economic, societal and regulatory incentives to minimize waste production and recover resources. It is very expensive to treat wastes at the end of the pipe and extremely expensive to remediate waste sites. This experience provides society with some valuable lessons:

1. When chemicals are discharged to the environment, they rarely disappear, and given sufficient time they often are widely dispersed from the point of original release.
2. "Cost effective" today can be extremely costly tomorrow, just as what was a remote isolated location years ago, can be a suburban school playground today.
3. Long term viability of a technology depends upon understanding the little details. A great concept goes only so far. Many great concepts have never succeeded because of executional details.

These lessons of the past need to be considered when embracing environmental technology as a solution for today and the future. We need to recognize that implementing environmental biotechnology is a very complex and complicated process, whose small details must not be overlooked or trivialized.

TECHNOLOGY TRANSFER

The importance of technology transfer was emphasized throughout the meeting. It is important to communicate with technical peers, regulators and the general public. The purpose of this communication is to accelerate transfer from the flask to the field, and to gain acceptance and support from potential consumers, regulators and the general public.

The appropriate communication medium of course depends on the intended audience, but one should remember that in today's busy society, senior scientists often do not have the time to study technical journal manuscripts and hence even for technical peers brief, clear, well illustrated documents should be used in addition to technical manuscripts. But do not forget, it is imperative to produce detailed scientific publications that rigorously support claimed results.

All speakers who discussed communication with the general public agreed on the need to be open and understandable. One should avoid being perceived as arrogant or as talking down to the audience, and be wary of overselling. Credibility is a priceless commodity that once lost is almost impossible to regain.

EDUCATION

There is a need for both science and engineering in environmental biotechnology. This suggests that environmental biotechnology should be a graduate program in which students are educated in a multidisciplinary environment. The creation of such an environment requires either a new university department containing faculty from relevant scientific and engineering disciplines or a multidisciplinary research center such as the Center for Environmental Biotechnology at the University of Tennessee. The faculty needs to shed its disciplinary chauvinism and dedicate itself to teaching students to work as team members.

There are certain things that can be done at the undergraduate level to prepare students to study environmental biotechnology and work effectively in teams. First engineering students should be exposed to the fundamental concepts of biology, particularly molecular biology and microbial physiology. Engineering will become more and more involved with biology in the decades ahead and an understanding of such subjects will be just as important as knowledge of physics and chemistry has been. Second, scientists should be knowledgeable in mathematics through differential equations. Environmental problems are extremely complex and hypotheses can often be tested only by a combination of experimentation and computer simulation. Scientists are usually skilled in the first, but weak in the second. Finally, engineers and scientists should both be trained in the concepts and applications of transport phenomena. Many systems are limited by the fluxes of components through them rather than by the potential activity of the microbial organisms present. A better mechanistic understanding of transport processes in combination with an ability to develop and use mathematical models of complex environmental systems will enable better development of concepts as well as better communication between scientists and engineers as they study to become environmental technologists.

CONCLUSION

For environmental biotechnology to become commonly applied and accepted, its end points must be predictable. The achievement of predictability depends on mechanistic understanding, integration of science and engineering, rigorous scientific analysis in both laboratory and field, enhanced transfer of laboratory results to field application, thorough documentation of field results, and standardization of field treatment protocols.

While the above criteria are necessary for acceptance, they are not sufficient. To obtain public acceptance, information must be transferred to the public and government in a clear and totally open manner. In addition public needs and concerns must be understood and responded to.

Environmental biotechnology is a great concept on the brink. This conference in communicating an appreciation of its technical and nontechnical promises and problems, provides valuable information for developers, consumers, consultants, educators, regulators and the general public.

REFERENCES

Grady, C.P.L., Jr., 1990. Applications of biotechnology in waste treatment, Proceedings of the 4th WPCF/JSWA Joint Technical Seminar on Sewage Treatment Technology, pp. 4-19.

Contributors

Edgar Berkey
National Environmental Technology
　Applications Corporation
University of Pittsburgh Applied
　Research Center
615 William Pitt Way
Pittsburgh, PA 15238

James W. Blackburn
Energy, Environment, and Resource
　Center
University of Tennessee
Knoxville, TN 37996-0710

Al W. Bourquin
Ecova Italia
Via G. Savelli, 13/B
35129 Padova, Italy

Jessica M. Cogan
National Environmental Technology
　Applications Corporation
University of Pittsburgh Applied
　Research Center
615 William Pitt Way
Pittsburgh, PA 15238

Geoffrey C. Compeau
ECOVA Corporation
3820 159th Avenue, N.E.
Redmond, WA 98052

John C. Corey
Westinghouse Savannah River
　Company
P.O. Box 616
Aiken, SC 29802

Sue Markland Day
The Center for Environmental
　Biotechnology
and The Waste Management
　Research and Education Institute
University of Tennessee
327 South Stadium Hall
Knoxville, TN 37996-0710

Thomas W. Federle
Proctor and Gamble
Ivorydale Technical Center
Cincinnati, OH 45217

Michael R. Fitzgerald
Energy, Environment, and Resources
　Center
and Department of Political Science
University of Tennessee
Knoxville, TN 37996-0710

Robert Fox
International Technology Corporation
312 Directors Drive
Knoxville, TN 37923

Robert A. Goldstein
Electric Power Research Institute
P.O. Box 10412
Palo Alto, CA 94303

C.P. Leslie Grady
Clemson University
Clemson, SC 29634-0919

Duane A. Graves
International Technology Corporation
312 Directors Drive
Knoxville, TN 37923

B.E. Haigler
U.S. Air Force Engineering and
    Services Center
Tyndall Air Force Base, FL 32403

Geoffrey Hamer
Institute of Aquatic Sciences and
    Water Pollution Control
Swiss Federal Institute of Technology
    - Zürich
8600 Dübendorf, Switzerland

Karolyn L. Hardaway
Texas Eastman Company
Utilities Laboratories
P.O. Box 7444
Longview, TX 75607

Armin Heitzer
Institute of Aquatic Sciences and
    Water Pollution Control
Swiss Federal Institute of Technology
    - Zürich
8600 Dübendorf, Switzerland

Gerald J. Hooker
Westinghouse Savannah River
    Company
P.O. Box 616
Aiken, SC 29802

Mark S. Katterjohn
International Technology Corporation
Austin, TX 78746

A. Keith Kaufman
Applied BioTreatment Association
4887 East LaPalma Avenue, Suite 701
Anaheim, CA 92807

Val J. Kelmeckis
National Environmental Technology
    Applications Corporation
University of Pittsburgh Applied
    Research Center
615 William Pitt Way
Pittsburgh, PA 15238

Walter W. Kovalick, Jr.
Technology Innovation Office
Office of Solid Waste and Emergency
    Response
United States Environmental
    Protection Agency
410 M Street, S.W.
Washington, DC 20460

Fran V. Kremer
United States Environmental
    Protection Agency
26 West Martin Luther King Drive
Cincinnati, OH 45268

William J. Lacy
Lacy & Company
9114 Cherrytree Drive
Alexandria, VA 22309

Craig A. Lang
International Technology Corporation
312 Directors Drive
Knoxville, TN 37923

Maureen E. Leavitt
International Technology Corporation
312 Directors Drive
Knoxville, TN 37923

Contributors

David G. Linz
Gas Research Institute
8800 Bryn Mawr Avenue
Chicago, IL 60631

Carol D. Litchfield
Environment America, Inc.
P.O. Box 9056
Newark, DE 19714

William D. Mahaffey
ECOVA Corporation
3820 159th Avenue, N.E.
Redmond, WA 98052

Amy S. McCabe
Energy, Environment, and Resources Center
University of Tennessee
327 South Stadium Hall
Knoxville, TN 37996-0710

Ronald G.L. McCready
Biotechnology Section
CANMET
Energy, Mines, and Resources Canada
Ottawa, Ontario, Canada

Michael J. McFarland
Utah Water Research Laboratory
Utah State University
Logan, UT 84322-8200

Lawrence T. McGeehan
National Environmental Technology Applications Corporation
University of Pittsburgh Applied Research Center
615 William Pitt Way
Pittsburgh, PA 15238

A. Thomas Merski
National Environmental Technology Applications Corporation
University of Pittsburgh Applied Research Center
615 William Pitt Way
Pittsburgh, PA 15238

Andrew C. Middleton
Remediation Technologies, Inc.
Pittsburgh, PA 15219

Craig A. Myler
U.S. Army Toxic and Hazardous Materials Agency
Aberdeen Proving Ground, MD 21010-5401

Edward F. Neuhauser
Niagra Mohawk Power Corporation
Syracuse, NY 13202

Lori Patras
Unocal Corporation
Brea, CA 92621

Frank R. Peduto
Bureau of Spill Prevention and Response
New York State Department of Environmental Conservation
500 Wolfe Road, Room 209
Albany, NY 12233

C.A. Pettigrew
U.S. Air Force Engineering and Services Center
Tyndall Air Force Base, FL 32403

Gary S. Sayler
Center for Environmental Biotechnology
University of Tennessee
10515 Research Drive, Suite 200
Knoxville, TN 37932

Ronald C. Sims
Utah Water Research Laboratory
Utah State University
Logan, UT 84322-8200

Wayne Sisk
U.S. Army Toxic and Hazardous Materials Agency
Aberdeen Proving Ground, MD 21010-5401

John H. Skinner
Office of Research and Development
United States Environmental
  Protection Agency
Washington, DC 20460

Esther Soczo
National Institute of Public Health
  and Environmental Protection
Laboratory for Waste Materials and
  Emissions
(RIVM/LAE)
P.O. Box 1
3720 BA Bilthoven, The Netherlands

J.C. Spain
U.S. Air Force Engineering and
  Services Center
Tyndall Air Force Base, FL 32403

Cindy K. Tew
Westinghouse Savannah River
  Company
P.O. Box 616
Aiken, SC 29802

Godfred E. Tong
Monsanto Corporate Research
Monsanto Company
800 North Lindbergh Boulevard
St. Louis, MO 63167

Marios Tsezos
Department of Chemical Engineering
McMaster University
Hamilton, Ontario, Canada L8S 4L7

Ronald Unterman
Envirogen, Inc.
Princeton Research Center
4100 Quakerbridge Road
Lawrenceville, NJ 08648

Klaas Visscher
National Institute of Public Health
  and Environmental Protection
Laboratory for Waste Materials and
  Emissions
(RIVM/LAE)
P.O. Box 1
3750 BA Bilthoven, The Netherlands

Patricia Taylor Woodyard
CH2M HILL, Inc.
8425 Christie Avenue, Suite 500
Emeryville, CA 94506

Thomas G. Zitrides
BioScience Management, Inc.
1530 Valley Center Parkway, Suite 120
Bethlehem, PA 18017

Index

A

Abiotic, 10–11, 77, 167, 169, 240, 274
Absorption, 3, 134
Acclimation periods, 175, 184
Acetonitrile, 154, 157
Acid mine drainage, 5, 69, 249
Activated sludge, 2, 12, 149–150, 173, 187, 238, 264–265, 269
Adsorption, 2–4, 31, 114, 118, 135, 137, 144, 171, 247, 259–260
Aerated static pile, 140
Aerobic microbes, 4
Agricultural chemicals, 132, 135
Alaska Bioremediation Research Project, 64, 66
Alcaligenes, 8, 180, 238
Aliphatic hydrocarbons, 85, 157
Alkali addition, 99
Alkane degradation, 88
Alternative Treatment Technology Information Center (ATTIC), 66, 70, 189
Ames assay, 170
Ammonia, 25, 79, 93, 95, 114–115, 117–118, 134, 159, 237, 239, 243
Applied BioTreatment Association, 47, 49, 67, 195
Aroclors, 160-162
Aromatic hydrocarbons, 68, 85–86, 266
Atlantic Research, 154
ATTIC, *see* Alternative Technologies Information Center
Aviation fuel, 64, 152, 157

B

Bench scale, 11, 32, 35, 40, 71, 101–102, 106, 109, 118, 143, 163-166, 203, 254, 259

Benzenes, 64, 86, 117, 161, 175–176, 179, 182, 207, 221–222, 265, 273
Benzoates, 8, 150, 176
Best Demonstrated Available Technologies (BDAT), 225
Bio-scrubbers, 5
Bioanalytical method, 9
Bioassay, 69, 89, 131, 169–170, 174
Bioavailability, 6, 9, 11, 114, 224, 228, 231, 266
Biocatalysts, 130, 132–133, 135
Biochemical pathway, 3, 12, 169, 230
Biocides, 131, 235
Bioconversion, 61
Biocorrosion, 131
Biodegradation, 1, 5–6, 9–13, 29–33, 61–65, 68–72, 78, 86, 92, 101, 104–105, 108, 113, 115, 118–120, 125, 140, 145, 148, 153, 156–157, 160–162, 165, 168–169, 175, 179–180, 182, 184, 186–187, 192, 195, 197, 201, 203, 205–206, 226, 235, 237, 239, 244, 246, 261, 265–269, 272–273, 275–276
Biofilm, 30–33, 148–149, 244, 248
Biofilter, 134–135, 263–264
Biofouling, 131
Biological process, 10–11
Biological processing, 201
Bioluminescence (*lux*) genes, 9
Bioluminescent reporter technology, 9, 13
Biomining, 201
Bioreactor, 4–5, 11, 89, 101-102, 106-107, 132, 147-150, 153–154, 155, 161-163, 186, 221-223, 225, 265, 267-268, 272-273
Bioreclamation, 112–113, 121, 125, 156, 201
Bioremediation, 3, 9, 12, 20–21, 25–26, 29, 31–33, 35–36, 39–40, 45, 47–51, 53–54, 56, 59–71, 78–79, 83, 85–92, 99, 101, 104–105, 109, 111–113, 115, 117–118, 121, 123, 125, 129, 145, 147–149, 151–153, 155, 157, 159–162, 174–175, 185–186, 189, 191–195, 197, 201,

Bioremediation *(cont.)*
    204, 206–208, 211–215, 217–218, 221–222, 224–227, 230–231, 233–236, 242, 245–246, 266, 272–273, 275–276
Bioremediation Action Committee, 66
Bioremediation Field Initiative, 59–60
Bioremediation Field Program, 185
Biosorption, 135–136, 157, 249–250, 254, 257, 259–260
Biotransformation, 2, 132, 135, 145, 159, 167, 169, 171–172, 184, 236, 272–273
Biotreatability study, 104
Biotreatment, 5, 26, 47–51, 54–57, 59–60, 91, 129, 132–135, 139, 148, 153, 155–157, 175–176, 185, 203–204, 218, 221–224, 226, 228, 230–231, 235, 273
BioTrol, 149
BOD, 2
BTX, 7, 273
Buffering capacity, 99
By-product, 4, 25, 40, 71, 128–130, 132–133, 135, 137, 140, 144, 148-149, 153, 187, 238–239, 251, 274

C

Cadmium, 69, 207, 242
Carbon cycle, 159
Carbon dioxide, 62, 68, 101, 112, 129, 139, 160, 166, 195, 207, 218, 237
CERCLA, *see* Comprehensive Environmental Reclamation and Liability Act
Certification program, 49
Chem Waste Management, 150
Chemical oxygen demand (COD), 2, 150, 275
Chemostat, 181–182, 234, 241, 244, 247, 265
Chlorinated hydrocarbons, 68, 134, 267, 273
Chlorobenzenes (CLBZ), 7, 117, 152, 156, 176, 179, 181–182, 184
Chlorobenzoates, 8, 12, 176, 181, 184
Chlorobenzoic acid, 150
Chlorocatechols, 152, 179, 181-182
Chloromethylcatechol, 177, 179
p-Chlorotoluene, 176–179, 182, 184
Chlorotoluenes, 149
Clay, 26, 27, 31, 122, 151, 165, 188, 235–236, 274
Clean Air Act, 221
Clean Water Act (CWA), 38, 204, 218, 221-222
Cloning, 9
Co-mingling, 5
Coal carbonization, 25
Coal tar, 25
Cobble beaches, 89
COD, *see* Chemical Oxygen Demand

Coefficient of variation, 172
Coking waste lagoon, 71, 83
Cometabolism, 179, 184
Compost, 138, 140–142, 145, 163, 263, 266
Composting, 5, 138–140, 142, 144–147, 151, 155, 157, 163, 272, 275
Comprehensive Environmental Response, Compensation, and Liability Act (CERCLA), 38, 163, 166, 174, 189, 205, 222-223, 226, 228
Congeners, 7, 160, 162
Consortium, 101, 104–108, 144, 148–149, 152, 185, 198
Cooperative Research and Development Agreements (CRASDAS), 198, 199-200
Copper, 69, 207
Creosote, 98, 186, 188–189, 221
Cross-training, 12
CWA, *see* Clean Water Act

D

DDT, 7
Dechlorination activity, 102
Degraders, 2, 32
Degreasing operation, 187
Dehydrogenase activity, 170
Delisted, 221
Denitrifying bacteria, 118
Desorption testing, 32
Desulfurizing, 201
Detection limit, 26, 82, 114, 117, 172, 182
Detoxification, 166, 169–170, 174, 201, 206, 224, 264, 274–275
Dichlorobenzene, 176, 184
p-Dichlorobenzene, 177, 182
Diesel fuel, 2, 4, 71, 152
Dioxins, 7, 207
Dioxygenase, 9, 176–177, 179, 181–182, 184
DNA hybridization, 10, 12
DNA probe technology, 9

E

E. coli, *see* *Escherichia coli*
Ecotoxicology, 87
ECOVA, 98, 106, 149
Efficacy, 47–48, 87, 96, 101, 154, 161, 233
Electron acceptor, 118, 239, 243, 248, 274, 275
Electron donor, 240
Emerging Technologies Program, 69
Environmental fate, 6, 166

Index

Environmental Protection Agency, *see* United
    States Environmental Protection Agency
Environmental Remediation, Inc., 88–89
Enzyme activity, 181, 242
EPA, *see* United States Environmental
    Protection Agency
Escherichia coli, 8, 179, 242, 247–248
Ethylbenzene, 117, 175
Ethylene glycol, 149, 187
Evaporative loss, 161
Explosives, 137–140, 142–146, 151, 157
Exxon Company, USA, 64

F

Feasibility study, 71, 112, 118, 162-163, 213–214
Federal Insecticide, Fungicide, and Rodenticide
    Act, 205, 228, 229
Fertilizer, 64, 65–66, 86, 88–89, 107, 215
Field data, 37, 97, 99
Financial barriers, 59
Fixed film bioreactor, 186
Flavobacterium, 149, 239
Food-processing industry, 61, 132
French Limited site, 64, 70, 207
French Limited study, 275

G

Gas industry, 25, 29
Gas Research Institute (GRI), 25, 26, 29, 32, 35
Gasoline, 2, 4, 71, 151–152, 159, 175, 188, 214,
    226, 266
Gene, 3, 8–11, 13, 176, 177–179, 274
Gene expression, 10, 242, 271
Gene probe, 2, 9, 13
Genetic engineering, 7, 9, 19, 159, 201, 272
Genetic engineers, 193
Genetically engineered, 205, 272
Gravel, 26, 27, 28, 85-86, 188
GRI, *see* Gas Research Institute
Groundwater, 4, 45, 47, 50, 53–54, 61–62, 64, 69–
    70, 91, 98, 111–115, 117–118, 121–123, 125,
    139, 148–149, 151–152, 154, 156, 175, 184,
    186–187, 205, 214–215, 225–226, 265–266,
    269

H

Halogenated compounds, 48
Hazardous waste, 1, 5–6, 8-9, 12, 15, 37, 45, 50,
    53, 56, 57, 60, 61-63, 66, 68, 70, 109,
Hazardous waste *(cont.)*
    143–144, 147, 156–157, 159–160, 173–174,
    178–179, 185, 189, 197, 203, 204–209, 217–
    218, 221–222, 224–227, 231–232, 261–262,
    268
Heap leaching, 5
Heavy metals, 135–136, 154, 156, 207, 226, 268
Henry's Law constant, 3, 165
Heterotrophs, 80
Hexahydro-1,3,5-trinitro-1,3,5-triazine (RDX),
    137-138, 151
High salt content waste streams, 134
High-energy beaches, 85
HMX, *see* 1,3,5,7-tetranitro-1,3,5,7-tetraazocine
Hydrocarbon contamination, 4, 123, 150, 157
Hydrocarbon degraders, 80, 113
Hydrogen peroxide, 64, 78, 80, 114–116, 118–
    120, 121, 152, 188, 242, 246, 266
Hydrogeology, 3, 5, 112–113, 121, 125, 148, 214–
    215

I

Idaho National Engineering Laboratory, 201
in situ, 3–5, 9, 11, 35, 37, 39, 50, 62–64, 67, 69–
    70, 91, 111–114, 117, 121–123, 125, 147–
    148, 151, 155–157, 163, 184, 186–187, 201,
    203, 206, 213–214, 222–223, 225, 239, 246,
    263, 265–266, 269
Incineration, 3, 26, 38–40, 51, 53, 61–62, 139,
    144–145, 160, 207, 224–225, 227
Indigenous bacteria, 26, 112, 143, 149, 152
Induced cells, 179
Inducer, 153, 179, 183
Inoculum, 96, 98, 101, 104, 107–108, 149, 161,
    186
Interdisciplinary, 6
Intermediates, 37, 152, 169, 175, 237, 267, 274
International Technology Corporation, 111–112
Iron, 69, 114–115, 117, 154, 215, 250, 254
Isomers, 7, 132, 135–136
Isopropanol, 149, 264

K

Kerosene-like solvent residues, 92

L

Lagoon, 64, 71–72, 78, 89, 138, 150–151
Lampblack, 25, 27

Land Disposal Restrictions (LDRs), 40, 50, 223–226, 231
Land farming, 89, 91, 140, 147, 151, 225, 263, 272, 275
Land treatment, 5, 26, 37, 92, 98, 135, 174, 186–187, 225
LDRs, see Land Disposal Restrictions
Leachate, 4, 140, 150–151, 156, 171–172, 249
Liability, 59, 62, 209, 215, 226
lux genes, 9, 272

M

Manganese, 69, 215
Manufactured gas plant (MGP), 25-26, 27, 29, 36
Manufactured Gas Plant soil, 9, 10–11
Markets, 16, 47–48, 54, 58, 128, 136, 208–209, 217, 223, 224, 227–228, 231, 233, 273
Mass balance, 160–161, 165–167, 169, 239, 275
Mass media, 16–17, 19, 22, 192
Mega-Borg spill, 194
Mercury, 13, 73–77
Mesophilic (35 °C), 142
Metal resistances genes, 9
Metals, 2, 4–5, 9, 37, 62, 64, 69, 187, 249–250, 254, 258–260, 273
Methane, 4, 132–134, 152, 159, 187, 237, 239, 243, 246–248
Methane monooxygenase, 152, 239
Methyl benzoate, 8, 180, 181, 184
Methyl ethyl ketone, 149
Methyl isobutyl ketone, 149
Methylotrophs, 179, 239
MGP, see Manufactured Gas Plant
Microbe Masters, Inc., 96
Microbial community, 2, 13, 94, 170
Microbial ecology, 13, 193, 195, 245, 247, 272
Microcosm, 10, 96, 118–119, 121, 164, 167, 169
Microtox^{TM}, 170–172, 174
Mineralization, 2–3, 101, 155, 166, 218, 224, 227, 237, 261, 275
Molecular ecology, 3
Mutant strain, 177-179, 182

N

Naphthalene, 9–11, 13, 86, 117, 150, 176, 182, 207
NAPL, see Nonaqueous phase liquid
National Bioremediation Spill Response Subcommittee, 67
National Contingency Plan (NCP), 228–229

National Contingency Plan, product list, 229
National Environmental Technology Applications Corporation (NETAC), 65, 66, 85, 86-90
National Governors Association, 67
National Pollutant Discharge Elimination System (NPDES), 187, 223, 231
National Priorities List (NPL), 54, 98, 139, 204
Naturally-occurring microorganisms, 1, 49, 66, 206, 218, 228, 230
NCP, see National Contingency Plan
NETAC, see National Environmental Technologies Applications Corporation
Nitrification, 170, 240
Nitrogen fixation, 170, 239, 246
p-Nitrophenol (PNP), 153, 154, 156
Nonaqueous phase liquid (NAPL), 26
NPDES, see National Pollutant Discharge Elimination System
NPL, see National Priorities List
Nucleic acid analysis, 9
Nutrient, 1, 3-6, 26, 29, 33, 37, 61, 63, 64, 65, 66-67, 69, 73, 78-79, 83, 92, 93-95, 96, 99, 112-115, 117-121, 123-125, 131, 151-152, 156, 164-165, 175, 186-188, 213-214, 215, 238, 240, 242-248, 266, 274
Nutrient-augmented slurries, 114
Nutrient-augmented water, 78
Nutrient-enhanced bioremediation, 65

O

Occidental Chemical Company, 150
Odor control, 134
Office of Research and Development (ORD), 59, 86, 173-174, 185, 203
Oil spill, 19, 20, 21, 64–67, 70, 85–86, 87, 88–89, 90, 208, 211–212, 215
Oily sludge bottoms, 148
Olefin feedstocks, 111
Optimization, 5–6, 9, 12, 97, 148, 266, 268
Organic pollutants, 1, 69, 275
Ortho pathway, 176–182
Osmiophilic biocatalysts, 134
Oxidation, 2, 8, 120, 134, 176
Oxidation-reduction potential, 80, 118
Oxygenated, 4, 114

P

PAHs, see Polynuclear aromatic hydrocarbons
Pan studies, 26, 29, 33, 34, 35
Partition coefficient, 3, 165, 171

Index

PCBs, *see* Polychlorinated biphenyls
PCP, *see* Pentachlorophenol
Pentachlorophenol (PCP), 69, 98-102, 104-109, 149, 156, 184
Perturbation techniques, 10
Pesticides, 67, 132, 150, 154, 159, 174–175, 205, 206, 229, 265, 269
Phenols, 4, 8, 70, 150, 153, 159, 161, 176, 179, 206, 207, 221, 223, 265
Phthalates, 7, 149
Pilot scale, 69, 71, 78, 92, 138, 140, 163-166, 267
Planktonic algae, 65
Plant Pest Act, 230
Plasmid, 9, 11
PNP, *see* p-Nitrophenol
Polychlorinated biphenyls (PCBs), 5, 7, 23, 37, 39, 159, 160-162, 203, 206, 207, 223, 227
Polynuclear aromatic hydrocarbons (PAHs), 10, 25, 26-29, 30-33, 35, 36, 71-72, 77, 98, 101, 151, 160, 172, 186
Predictive capabilities, 37, 44, 171
Prince William Sound, 48, 64, 85-86, 88-89
Priority pollutants, 2, 221
Proof of completion, 214
Propane, 159
Pseudomonas fluorescens, 11
Pseudomonas putida, 8, 149, 179, 184
Public perception, 16, 20, 23, 45, 112, 125, 191, 193
Public policy makers, 15
Publicly Owned Treatment Works (POTWs), 222
Pulping process, 132
Pump and treat, 4, 62, 69, 111–112, 125, 187

R

RCRA, *see* Resource Conservation and Recovery Act
RDX, *see* hexahydro-1,3,5-trinitro-1,3,5-triazine
Reaction kinetics, 3
Recalcitrant, 1, 80, 112, 159, 162, 184, 218, 230, 264
Redox potential, 165, 247
Regulatory and Investigative Treatment Zone (RITZ) Model, 171, 174
Regulatory barriers, 57
Reilly, William K., 50, 86, 185, 203
Release of microorganisms, 20, 159, 231, 272
Removal efficiency, 3, 150, 154, 157, 182, 265
Research and Development (R&D), 57
Research and development permit, 58
Resource Conservation and Recovery Act (RCRA), 38, 45, 50, 51, 54, 56-58,

Resource Conservation and Recovery Act *(cont.)* 111–112, 125, 166, 185, 189, 204, 206, 217–218, 221–227
Responsible party (RP), 211–213, 215
Rifkin, Jeremy, 20
Risk, 9, 20, 23, 38, 40, 41, 44, 51, 53, 59, 62, 65, 67, 70, 88, 112, 125, 138–139, 156-157, 203, 204-205, 215, 223, 225, 227, 231, 272
Risk Reduction Engineering Laboratory's Treatability Database Program, 189
RITZ, *see* Regulatory and Investigative Treatment Zone Model
RP, *see* Responsible party

S

Salicylate, 9, 176
Sand, 26-28, 30-31, 121-123, 149, 172, 187-188, 235, 266–267
Saturated zone, 5, 78, 80, 82, 113–114, 117
Savannah River Laboratory, 201
Science, 13, 17–18, 184
Scientific American, 17–18
Sediment, 5, 65, 121–122, 146, 157, 160, 202, 235, 237, 247
Selenium, 201
Sequencing batch reactor, 150, 156, 187
Shock loads, 175
Shoreline, 64
Shoreline cleanup, 64–65, 85
Silt, 26–27, 31, 122, 235
Silty, 122, 188
Site characterization, 70, 186, 188, 224, 226, 228
Slag, 71
Sludge, 2, 5, 64, 111, 134, 145, 192, 206, 207, 221, 234, 264–265
Slurry reactor, 5, 10, 32–33, 35, 109, 143, 163, 222, 267, 269
Soil, 2, 4–5, 25–33, 35-37, 39-40, 45, 47–48, 50, 53–56, 58, 61, 64, 67, 71, 78–79, 91–101, 104-106, 108-109, 111–112, 113–118, 122–123, 125, 134–135, 137–145, 148-150, 151–152, 155, 157, 160-161, 163-175, 180, 184, 186, 187-188, 202, 203, 205-206, 207, 213-215, 221-222, 225-226, 230-231, 233, 235-238, 240, 243, 246, 247, 261-263, 265-269, 272, 275
Soil washing, 101, 106–108
Soil water partitioning coefficient (K_p), 32
Solid-phase biological treatment, 91, 98, 140, 147
Source reduction, 127–129, 135
Starvation conditions, 4
Stripping, 2–4, 50, 187
Subsurface contamination, 3

Sulfur, 5, 154, 157, 221
Superfund Innovative Technology Evaluation (SITE), 57, 69–70, 145
Sybron Chemicals, Inc., 88, 89
System analysis, 10, 11, 173

T

Tar refinery wastes, 150, 156
TCE, see Trichloroethylene
Technology Innovation Office, 53, 58, 185
Technology transfer, 21, 23, 67, 70, 197–200, 277
Technology Transfer Acts, 202
Television News Archive, 15, 17, 24
Temperature inducible, 8
1,3,5,7-tetranitro-1,3,5,7-tetraazocine (HMX), 137-138, 144-145, 151
Texas Eastman Company, 111–112
The New York Times, 17–18
Thermal cracking, 25
Thermophilic, 140, 142, 145, 151
Thiobacillus, 154, 238
TNT, see 2,4,6-Trinitrotoluene
Toluene, 4, 8, 64, 86, 117, 134, 153, 159, 161, 175–176, 179, 181–182, 184, 207, 221, 273
Total Petroleum Hydrocarbons (TPH), 2-3, 72, 78, 82-82, 92-97, 117-119, 152
Toxic Substances Control Act (TSCA), 38, 49, 50, 152, 205, 223, 227-228, 230
Toxicity, 6, 67, 69, 88–89, 138–140, 145, 169–174, 218, 221–222, 224, 231, 249, 273
TPH, see Total Petroleum Hydrocarbons
Trade press, 193
Transport, chemical, 3
Transposon introduction, 9
Traverse City, 157
Traverse City Coast Guard Station, 64, 152
Treatability database, 189
Treatability studies, 6, 40, 45, 58–60, 70, 91, 98, 101, 105, 106, 118-119, 123, 163–167, 169–172, 174, 188–189, 207, 224, 227
Treatment train, 4, 151, 155, 193, 207, 274
Trichloroethylene (TCE), 4, 7-8, 13, 152, 153, 157, 159, 161-162, 179, 184, 187, 203, 206, 221
2,4,6-Trinitrotoluene (TNT), 7, 137-138, 144-145, 151
TSCA, see Toxic Substances Control Act

U

United States Environmental Protection Agency, 44, 50, 53-54, 57-61, 63-70, 86-90, 92, 98, 118, 125, 145, 152, 163, 171, 173-174, 184-185, 188-189, 191, 192, 194, 203-208, 217-218, 221, 223-228, 230, 232
Ubiquity principle, 192
Uncertainty, 20, 38–39, 41, 44–45, 59, 228
Underground Storage Tank, 38, 54, 70, 152, 175, 185, 187, 208, 217, 226, 232
USEPA, see United States Environmental Protection Agency

V

Vadose zone, 78, 79–80, 171, 174
Vadose Zone Interactive Processes (VIP) Model, 171, 174
Vendor Identification Database, 59
Vibrio fischeri, 9
Vinyl chloride (VC), 64, 153, 187
VIP, see Valdose Zone Interactive Processes Model
Volatilized components, 85

W

Waste destruction technologies, 129, 132
Waste minimization, 127–130, 132–133, 135, 153, 218, 222, 272, 276
Weathered crude oil, 66, 88
White rot fungus (*Phenerocheate chrysosporium*), 4, 69-70, 203
Windrowing, 140

X

Xylenes, 64, 86, 117, 149, 161, 175, 273

Z

Zinc, 69, 207, 250